W9-CNF-515

An Introduction to Nonlinear Boundary Value Problems

This is Volume 109 in
MATHEMATICS IN SCIENCE AND ENGINEERING
A series of monographs and textbooks
Edited by RICHARD BELLMAN, *University of Southern California*

The complete listing of books in this series is available from the Publisher upon request.

An Introduction to Nonlinear Boundary Value Problems

Stephen R. Bernfeld

Department of Mathematics
Memphis State University
Memphis, Tennessee

V. Lakshmikantham

Department of Mathematics
University of Texas
Arlington, Texas

Academic Press, Inc.

New York and London 1974

A Subsidiary of Harcourt Brace Jovanovich, Publishers

ACADEMIC PRESS, INC.
111 Fifth Avenue, New York, New York 10003

United Kingdom Edition published by
ACADEMIC PRESS, INC. (LONDON) LTD.
24/28 Oval Road, London NW1

Library of Congress Cataloging in Publication Data

Bernfeld, Stephen R
An introduction to nonlinear boundary value problems.

(Mathematics in science and engineering, v.)
Bibliography: p.
1. Boundary value problems. 2. Nonlinear theories.
I. Lakshmikantham, V., joint author. II. Title.
III. Series.
QA379.B47 515'.35 73-21996
ISBN 0-12-093150-8

PRINTED IN THE UNITED STATES OF AMERICA

Contents

Preface

The theory of nonlinear boundary value problems is an extremely important and interesting area of research in differential equations. Due to the entirely different nature of the underlying physical processes, its study is substantially more difficult than that of initial value problems and consequently belongs to a third course in differential equations. Although this sophisticated branch of research has, in recent years, developed significantly, the available books are either more elementary in nature, for example the book by Baily, Shampine, and Waltman, or directed to a particular method of importance, such as that by Bellman and Kalaba. Hence it is felt that a book on an advanced level that exposes the reader to this fascinating field of differential equations and provides a ready access to an up-to-date state of this art is of immense value. With this as motivation, we present in our book a variety of techniques that are employed in the theory of nonlinear boundary value problems. For example, we discuss the following:

(i) methods that involve differential inequalities;
(ii) shooting and angular function techniques;
(iii) functional analytic approaches;
(iv) topological methods.

We have also included a chapter on nonlinear boundary value problems for functional differential equations and a chapter covering special topics of interest.

The main features of the book are

(i) a coverage of a portion of the material from the contribution of Russian mathematics of which the English speaking world is not well aware;

(ii) the use of several Lyapunov-like functions and differential inequalities in a fruitful way;

(iii) the inclusion of many examples and problems to help the reader develop an expertise in the field.

This book is an outgrowth of a seminar course given by the authors. We have assumed the reader is familiar with the fundamental theory of ordinary differential equations, including the theory of differential inequalities, as well as the basic theory of real and functional analysis. It is designed to serve as a textbook for an advanced course and as a research monograph. It is therefore useful to the specialist and the nonspecialist alike. The reader who is familiar with the contents of the book, it is hoped, is fully equipped to contribute to the area.

Acknowledgments

We wish to express our warmest thanks to Professor Richard Bellman whose interest and enthusiastic support made this work possible. We are immensely pleased that our book appears in his series. The staff of Academic Press has been most helpful.

We thank our colleagues who participated in the seminar on boundary value problems at the University of Rhode Island in 1971-1972. In particular, we appreciate the comments and criticism of Professors E. Roxin, R. Driver, and M. Berman. Moreover, we gratefully acknowledge several helpful suggestions offered by Professor L. Jackson. We are very much indebted to Professors G. S. Ladde and S. Leela for their enthusiastic support in many stages of the development of this monograph and to Mr. T. K. Teng for his careful proofreading. Moreover, we wish to thank Mrs. Rosalind Shumate and Mr. Sreekantham for their excellent typing of the manuscript, and we wish to express our appreciation to Ms. Elaine Barth for her superb typing of the final copy.

The first-mentioned author would like to acknowledge some interesting helpful discussions on boundary value problems with the differential equation's group at the University of Missouri at Columbia.

Finally, the final preparation of this book was facilitated by a National Science Foundation Grant GP-37838.

Chapter 1
METHODS INVOLVING DIFFERENTIAL INEQUALITIES

1.0 INTRODUCTION

A variety of techniques are employed in the theory of nonlinear boundary value problems. This chapter is primarily concerned with the methods involving differential inequalities. The basic idea is to modify the given boundary value problem suitably, and then to use the theory of differential inequalities and the existence theorems in the small to establish the desired existence results in the large.

After presenting needed existence theorems in the small, we first concentrate on scalar second-order differential equations and associated boundary value problems. We then introduce upper and lower solutions, discuss the modification technique, and utilize Nagumo's condition to obtain a priori bounds on solutions and their derivatives. Once we have these bounds at our disposal, to prove existence theorems on finite or infinite intervals is relatively simple and straightforward. Boundary value problems subjected to nonlinear boundary conditions as well are treated in this framework. We then develop Lyapunov-like theory for boundary value problems employing several Lyapunov-like functions and the theory of differential inequalities in a fruitful way. We also treat in detail Perron's method of proving existence in the large by utilizing the properties of sub- and superfunctions and the existence results in the small. This technique works well for scalar equations.

We next extend the results considered for scalar equations to a finite system of second-order differential equations. Here there are two directions to follow, that is, either try to obtain the required bounds componentwise or in terms of a convenient norm. We offer results from both points of view indicating their relative merits and using Lyapunov-like theory, whenever possible, to derive general results.

1.1 EXISTENCE IN THE SMALL

Let R^n denote the real n-dimensional, Euclidean space and for $x \in R^n$, let $\|x\|$ denote any convenient norm of x. Let J be the interval $[a,b]$. We shall mean by $C^{(n)}[A,B]$ the class of n-times continuously differentiable functions from a set A into a set B.

We will be concerned, in this section, with the existence of solutions of the second-order differential equations of the form

$$(1.1.1) \qquad x'' = f(t,x,x'),$$

satisfying the boundary conditions

$$(1.1.2) \qquad x(t_1) = x_1, \quad x(t_2) = x_2, \quad t_1,t_2 \in J,$$

where $f \in C[J \times R^n \times R^n, R^n]$. For the purposes of this chapter, we also need an existence result under more general boundary conditions. This we do consider for the scalar case, leaving a thorough discussion of the general theory to a later chapter.

First of all, we observe that the only solution of

$$(1.1.3) \qquad x'' = 0,$$

subject to the boundary conditions

$$x(t_1) = 0, \qquad x(t_2) = 0, \tag{1.1.4}$$

is the trivial solution. This implies, from the theory of linear differential equations, that there exists a unique solution of

$$x'' = h(t), \tag{1.1.5}$$

satisfying (1.1.4) for each $h \in C[J,R^n]$. Moreover, since the problem (1.1.3), (1.1.4) possesses the two linearly independent solutions $u(t) = (t - t_1)$, $v(t) = (t_2 - t)$, the method of variation of parameters readily gives the integral equation

$$x(t) = \frac{1}{t_1 - t_2}\left[\int_{t_1}^{t} (t_2 - t)(s - t_1)h(s)\,ds + \int_{t}^{t_2} (t - t_1)(t_2 - s)h(s)\,ds\right] \tag{1.1.6}$$

for the solution $x(t)$ of (1.1.5) subject to (1.1.4). Relation (1.1.6) can be written in the familiar form

$$x(t) = \int_{t_1}^{t_2} G(t,s)h(s)\,ds, \tag{1.1.7}$$

where

$$G(t,s) = \begin{cases} (t_2 - t)(s - t_1)/(t_1 - t_2), & t_1 \le s \le t \le t_2, \\ (t_2 - s)(t - t_1)/(t_1 - t_2), & t_1 \le t \le s \le t_2. \end{cases}$$

This function $G(t,s)$ is usually referred to as the Green's function for the boundary value problem in question. Hence the solution of (1.1.5) verifying conditions (1.1.2) takes the form

$$x(t) = \int_{t_1}^{t_2} G(t,s)h(s)\,ds + w(t), \tag{1.1.8}$$

where $w''(t) = 0$ and $w(t_1) = x_1$, $w(t_2) = x_2$. It therefore follows that if $x(t)$ is a solution of (1.1.1), (1.1.2), then

$$x(t) = \int_{t_1}^{t_2} G(t,s)f\big(s, x(s), x'(s)\big)\,ds + w(t). \tag{1.1.9}$$

Conversely, if $x(t)$ is a solution of (1.1.9), we can verify by differentiation of (1.1.6) that $x(t)$ satisfies (1.1.1), (1.1.2).

Let us next recall some properties of the function $G(t,s)$ for later use. For a fixed t, the maximum of $|G(t,s)|$ is attained at $s = t$ and $|G(t,t)|$ has its maximum value at $t = (t_1 + t_2)/2$, that is,

$$|G(t,s)| \le (t_2 - t_1)/4\,. \tag{1.1.10}$$

Furthermore,

$$\int_{t_1}^{t_2} |G(t,s)|\,ds = (t_2 - t)(t - t_1)/2$$

and consequently

$$\int_{t_1}^{t_2} |G(t,s)|\,ds \le (t_2 - t_1)^2/8\,. \tag{1.1.11}$$

Moreover,

$$\int_{t_1}^{t_2} |G_t(t,s)|\,ds = \big((t - t_1)^2 + (t_2 - t)^2\big)/2(t_2 - t_1)\,,$$

the maximum of which is attained at $t = t_1$ and $t = t_2$. Hence, we obtain

$$\int_{t_1}^{t_2} |G_t(t,s)|\, ds \le (t_2 - t_1)/2 . \tag{1.1.12}$$

We are now ready to prove an existence and uniqueness result by using the contraction mapping theorem.

THEOREM 1.1.1. Let $f \in C[J \times R^n \times R^n, R^n]$ and for (t,x_1,y_1), $(t,x_2,y_2) \in J \times R^n \times R^n$,

$$\|f(t,x_1,y_1) - f(t,x_2,y_2)\| \le K\|x_1 - x_2\| + L\|y_1 - y_2\|, \tag{1.1.13}$$

where $K, L > 0$ are constants such that

$$K\left((t_2 - t_1)^2/8\right) + L\left((t_2 - t_1)/2\right) < 1 . \tag{1.1.14}$$

Then the boundary value problem (1.1.1), (1.1.2) has a unique solution.

Proof: Let B be the Banach space of functions $u \in C^{(1)}\left[[t_1,t_2], R^n\right]$ with the norm

$$\|u\|_B = \max_{t_1 \le t \le t_2} \left[K\|u(t)\| + L\|u'(t)\|\right] .$$

Define the operator $T: B \to B$ by

$$Tu(t) = \int_{t_1}^{t_2} G(t,s) f\big(s,u(s),u'(s)\big)\, ds + w(t).$$

We then have, by (1.1.11) and (1.1.13),

$$\begin{aligned}\|Tu_2(t) - Tu_1(t)\| &\le \frac{(t_2 - t_1)^2}{8} \left[K\|u_2(t) - u_1(t)\| + L\|u_2'(t) - u_1'(t)\|\right] \\ &\le \frac{(t_2 - t_1)^2}{8} \|u_2 - u_1\|_B .\end{aligned}$$

Also, because of (1.1.12),

$$\|Tu_2'(t) - Tu_1'(t)\| \leq \frac{(t_2 - t_1)}{2} \left[K\|u_2(t) - u_1(t)\| + L\|u_2'(t) - u_1'(t)\|\right]$$

$$\leq \frac{(t_2 - t_1)}{2} \|u_2 - u_1\|_B .$$

It then follows that

$$\|Tu_2 - Tu_1\|_B \leq \left[K \frac{(t_2 - t_1)^2}{8} + L \frac{(t_2 - t_1)}{2} \|u_2 - u_1\|_B\right].$$

This, in view of assumption (1.1.14), shows that T is a contraction mapping and thus has a unique fixed point which is the solution of the problem (1.1.1), (1.1.2). The proof is complete.

An interesting problem is to find the largest possible interval in which the preceding theorem is valid. In the case when $f \in C[J \times R \times R, R]$, one can offer such a best possible result. We have intentionally given such a result in the following exercise with generous hints.

EXERCISE 1.1.1. Assume that $f \in C[J \times R \times R, R]$ and satisfies (1.1.13). Let $u(t)$ be any solution of

$$u'' + Lu' + Ku = 0 \tag{1.1.15}$$

which vanishes at $t = t_1$ and let $\alpha(L,K)$ be the first unique number such that $u'(t) = 0$ for $t = t_1 + \alpha(L,K)$. Show that the boundary value problem (1.1.1), (1.1.2) has a unique solution if $t_2 - t_1 < 2\alpha(L,K)$ and that this result is best possible.

<u>Hints</u>: Step 1. First show that there is a unique solution to the boundary value problem $x'' = f(t,x,x')$, $x(t_1) = x_1$, $x'(t_3) = x_3$ if $(t_3 - t_1) < \alpha(L,K)$. This can be shown by applying the contraction mapping theorem relative to the Banach space $E = C^{(1)}\left[[t_1, t_3], R\right]$ with the norm

$$\|v\|_E = \max\left[\max_{t_1 \le t \le t_3} |v(t)|/u_0(t), \max_{t_1 \le t \le t_3} |v'(t)|/u_0'(t)\right],$$

where $u_0(t) > 0$ is a solution of

$$u'' + \frac{Lu' + Ku}{\alpha} = 0$$

for α sufficiently close to 1.

Step 2. Show that the existence of unique solutions of (1.1.1), (1.1.2) and of (1.1.1) with either $x(t_1) = x_1$, $x'(t_3) = x_3$ or $x'(t_1) = x_3$, $x(t_2) = x_2$ on any interval of length less than d implies the existence of a unique solution of (1.1.1), (1.1.2) on any interval of length less than 2d.

Step 3. To prove that the result is the best possible show that $u'' + L|u'| + Ku = 0$ has a nontrivial solution verifying $u(t_1) = u(t_2) = 0$, where $t_2 - t_1 = 2\alpha(L,K)$. Since $u(t) \equiv 0$ also satisfies the problem, argue that the result is best possible. Observe that $\alpha(L,K)$ can be explicitly computed.

EXERCISE 1.1.2. Show that it is sufficient to define f in Theorem 1.1.1 for $t \in [t_1, t_2]$, $\|x\| \le N$, $\|x'\| \le 4N/(t_2 - t_1)$, where N satisfies either

$$m\,\frac{(t_2 - t_1)^2}{8} \le N\left[1 - \left(K\,\frac{(t_2 - t_1)^2}{8} + L\,\frac{(t_2 - t_1)}{2}\right)\right],$$

if $m = \max_{t_1 \le t \le t_2} \|f(t,0,0)\|$, or $M\,(t_2 - t_1)^2/8 \le N$,

if $M = \max \|f(t,x,x')\|$ for $t \in [t_1, t_2]$, $\|x\| \le N$, $\|x'\| \le 4N/(t_2 - t_1)$.

<u>Hint</u>: Apply the contraction mapping theorem on the ball $\|u\|_0 \le N$ where

$$\|u\|_0 = \max\left[\max_{t_1 \le t \le t_2} \|u(t)\|, \frac{(t_2 - t_1)}{4} \max_{t_1 \le t \le t_2} \|u'(t)\|\right].$$

To obtain merely an existence result we employ Schauder's fixed point theorem as is usual.

THEOREM 1.1.2. Let $M > 0$, $N > 0$ be given numbers and let, for $t \in J$,

$$\|x\| \le 2M, \qquad \|y\| \le 2N, \qquad \|f(t,x,y)\| \le q,$$

where $f \in C[J \times R^n \times R^n, R^n]$. Suppose that $\delta = \min[(8M/q)^{\frac{1}{2}}, 2N/q]$. Then any problem (1.1.1), (1.1.2) such that $[t_1,t_2] \subset J$, $t_2 - t_1 \le \delta$, $\|x_1\| \le M$, $\|x_2\| \le M$, and $\|x_1 - x_2\|/(t_2 - t_1) \le N$, has a solution. Furthermore, given any $\varepsilon > 0$, there is a solution $x(t)$ such that $\|x(t) - w(t)\| < \varepsilon$, $\|x'(t) - w'(t)\| < \varepsilon$ on $[t_1,t_2]$, provided $t_2 - t_1$ is sufficiently small.

Proof: Consider the Banach space $B = C^{(1)}\left[[t_1,t_2],R^n\right]$ with the norm $\|x\|_B = \max_{t_1 \le t \le t_2} \|x(t)\| + \max_{t_1 \le t \le t_2} \|x'(t)\|$. Notice that the set

$$B_0 = [x \in B: \|x\| \le 2M, \|x'\| \le 2N]$$

is a closed, convex subset of B. Define the mapping $T: B \to B$ by

$$Tx(t) = \int_{t_1}^{t_2} G(t,s)\, f\big(s,x(s),x'(s)\big)\, ds + w(t).$$

Using now estimates (1.1.11), (1.1.12), we obtain

$$\|Tx(t)\| \le \big((t_2 - t_1)^2/8\big)\, q + M$$

and

$$\|(Tx)'(t)\| \leq ((t_2 - t_1)/2)\, q + N.$$

Hence, for $t_2 - t_1 \leq \delta$, T maps B_0 into itself. Also, since

$$\|(Tx)''(t)\| \leq \|f(t, x(t), x'(t))\| \leq q,$$

by Ascoli's theorem it follows that T is completely continuous. Schauder's fixed point theorem therefore assures that T has a fixed point in B_0, which is a solution of (1.1.1), (1.1.2).

If $x(t)$ is a solution of the boundary value problem (1.1.1), (1.1.2), with $x \in B_0$, then we have

$$\|x(t) - w(t)\| \leq ((t_2 - t_1)^2/8)\, q,$$

and

$$\|x'(t) - w'(t)\| \leq ((t_2 - t_1)/2)\, q,$$

for $t \in [t_1, t_2]$. Consequently the last assertion follows and the proof of Theorem 1.1.2 is complete.

COROLLARY 1.1.1. Assume that $f \in C\left[[t_1, t_2] \times R^n \times R^n, R^n\right]$ and is bounded on $[t_1, t_2] \times R^n \times R^n$. Then every boundary value problem (1.1.1), (1.1.2) has a solution.

Proof: Pick $M > 0$ sufficiently large so that

$$\|x_1\| \leq M, \qquad \|x_2\| \leq M, \qquad \|x_1 - x_2\|/(t_2 - t_1) \leq M$$

and such that

$$(t_2 - t_1) \leq 8M^{\frac{1}{2}}/q, \qquad (t_2 - t_1) \leq 2M/q,$$

where q is an upper bound of f. The conclusion then follows from Theorem 1.1.2.

As was indicated earlier, we shall next discuss the

situation where $f \in C[J \times R \times R, R]$. Then a natural question is under what general boundary conditions can one construct a Green's function for the scalar equation (1.1.3). Let us look at the boundary conditions

$$a_1 x(t_1) - a_2 x'(t_1) = 0, \tag{1.1.16}$$

$$b_1 x(t_2) + b_2 x'(t_2) = 0. \tag{1.1.17}$$

As long as there exist two linearly independent solutions of (1.1.3) satisfying (1.1.16), (1.1.17), it is possible to construct the desired Green's function. One sufficient condition is to find two nonparallel lines satisfying (1.1.16) and (1.1.17), that is, demanding that

$$a_1 b_2 + a_2 b_1 \neq 0. \tag{1.1.18}$$

Another sufficient condition, for example, is to require that $a_2 b_2 = 0$. In view of these remarks, it is sufficient to assume that

$$a_1, a_2, b_1, b_2 \geq 0, \qquad a_1 + b_1 > 0, \qquad \text{and} \qquad a_2 + b_2 > 0. \tag{1.1.19}$$

Let us now consider the nonhomogeneous boundary conditions

$$a_1 x(t_1) - a_2 x'(t_1) = A, \tag{1.1.20}$$

$$b_1 x(t_2) + b_2 x'(t_2) = B. \tag{1.1.21}$$

It is not difficult to conclude the existence of a solution $\varphi(t)$ satisfying (1.1.3) and (1.1.20), (1.1.21). Thus any solution $x(t)$ of (1.1.1) obeying the boundary conditions (1.1.20) and (1.1.21), takes the form

$$x(t) = \int_{t_1}^{t_2} G_0(t,s) f\big(s, x(s), x(s)\big)\, ds + \varphi(t), \tag{1.1.22}$$

where

$$G_0(t,s) = \begin{cases} (1/c)\, u(t)v(s), & t_1 \le s \le t \le t_2, \\ (1/c)\, u(s)v(t), & t_1 \le t \le s \le t_2, \end{cases}$$

and $c = u(t)v'(t) - v(t)u'(t)$. Here we assume $u(t)$, $v(t)$ are two linearly independent solutions of (1.1.3) which satisfy (1.1.16), (1.1.17). This discussion leads to the following result.

THEOREM 1.1.3. Let $f \in C\big[[t_1,t_2] \times R \times R, R\big]$, bounded on $[t_1,t_2] \times R \times R$. Then the boundary value problem (1.1.1), (1.1.20), (1.1.21) has a solution, whenever (1.1.18) or (1.1.19) holds.

Proof: Let M be the bound of f on $[t_1,t_2] \times R \times R$. Define a mapping $T\colon E \to E$ by

$$Tx(t) = \int_{t_1}^{t_2} G_0(t,s) f\big(s,x(s),x'(s)\big)\, ds + \varphi(t),$$

where the Banach space $E = C^{(1)}\big[[t_1,t_2],R\big]$ with the norm

$$\|x\|_E = \max_{t_1 \le t \le t_2} |x(t)| + \max_{t_1 \le t \le t_2} |x'(t)|.$$

Letting

$$N = \max_{s,t \in [t_1,t_2]} |G_0(t,s)(t_2 - t_1)|,$$

$$N_1 = \max_{s,t \in [t_1,t_2]} |G_{0t}(t,s)|(t_2 - t_1)|,$$

$$L = \max_{t_1 \le t \le t_2} |\varphi(t)|, \text{ and } L_1 = \max_{t_1 \le t \le t_2} |\varphi'(t)|,$$

it follows that

$$|Tx(t)| \le NM + L, \qquad |(Tx)'(t)| \le N_1M + L_1.$$

Hence, T maps the closed, bounded, and convex set

$$B_0 = [x \in E: \ |x(t)| \le NM + L, \ |x'(t)| \le N_1M + L_1]$$

into itself. Moreover, since $|(Tx)''| \le M$, T is completely continuous by Ascoli's theorem. The Schauder's fixed point theorem then yields the fixed point of T which is a solution of (1.1.1), (1.1.20), (1.1.21), thus completing the proof of the theorem.

EXERCISE 1.1.3. Find the solution of the boundary value problem $x'' = t$, $x(0) - x'(0) = 1$ and $x(1) = 0$.

1.2 UPPER AND LOWER SOLUTIONS

Let us consider the second-order equation

$$x'' = f(t,x,x'), \tag{1.2.1}$$

where $f \in C[J \times R \times R, R]$, J being the interval $[a,b]$ as before. The interior of J will be denoted by J^0. Let us define certain types of solutions of differential inequalities that will play a prominent part in the subsequent work.

DEFINITION 1.2.1. A function $\alpha \in C[J,R] \cap C^{(1)}[J^0,R]$ is said to be

(i) a <u>lower solution</u> of (1.2.1) on J if

$$D_-\alpha'(t) \equiv \liminf_{h \to 0} \frac{\alpha'(t+h) - \alpha'(t-h)}{2h} \ge f(t,\alpha(t),\alpha'(t)), \qquad t \in J^0;$$

(ii) an <u>upper solution</u> of (1.2.1) on J if

$$D^-\alpha'(t) \equiv \limsup_{h \to 0} \frac{\alpha'(t+h) - \alpha'(t-h)}{2h} \le f(t,\alpha(t),\alpha'(t)), \qquad t \in J^0.$$

A fundamental result concerning the upper and lower

solutions is the following.

THEOREM 1.2.1. Assume that

(i) $f \in C[J \times R \times R, R]$ and $f(t,x,y)$ is nondecreasing in x for each $(t,y) \in J \times R$;

(ii) α is a lower solution and β is an upper solution of (1.2.1) on J;

(iii) $\alpha(a) \leq \beta(a)$ and $\alpha(b) \leq \beta(b)$.

If one of the differential inequalities involved is strict, then we have $\alpha(t) < \beta(t)$ on J^0.

Proof: Suppose, on the contrary, that $\alpha(t) \geq \beta(t)$ for some $t \in J^0$. Then, setting $m(t) = \alpha(t) - \beta(t)$, we notice that $m(t)$ has a nonnegative maximum at some $t_0 \in J^0$. Consequently, $m(t_0) \geq 0$, $m'(t_0) = 0$, and $D_-m'(t_0) \leq 0$. Hence

$$\alpha(t_0) \geq \beta(t_0), \qquad \alpha'(t_0) = \beta'(t_0), \qquad \text{and} \quad D_-\alpha'(t_0) \leq D^-\beta'(t_0).$$

Thus

$$\begin{aligned} f\big(t_0,\alpha(t_0),\alpha'(t_0)\big) &\leq D_-\alpha'(t_0) \leq D^-\beta'(t_0) \\ &\leq f\big(t_0,\beta(t_0),\beta'(t_0)\big) \\ &\leq f\big(t_0,\alpha(t_0),\beta'(t_0)\big) \end{aligned}$$

in view of the monotonic character of f. This is a contradiction since one of the differential inequalities is assumed to be strict. The proof is therefore complete.

We observe that the proof of Theorem 1.2.1 breaks down if one of the differential inequalities is not assumed strict. Nonetheless, the following assertion is true.

THEOREM 1.2.2. Assume hypotheses (i)-(iii) of Theorem 1.2.1 hold. Suppose further that $f(t,x,y)$ obeys a one-sided

Lipschitz condition in y on each compact subset of $J \times R \times R$, that is,

$$f(t,x,y_1) - f(t,x,y_2) \le L(y_1 - y_2), \qquad y_1 \ge y_2.$$

Then the inequality $\alpha(t) \le \beta(t)$ on J, is true.

Proof: Suppose that $\alpha(t) > \beta(t)$ for some $t \in J^0$. Let $2\varepsilon = \max[\alpha(t) - \beta(t)]$ on J. Let $[c,d] \subset J$ be such that

$$\alpha(c) - \beta(c) = \varepsilon, \quad \alpha(d) - \beta(d) = \varepsilon, \quad \alpha(t) - \beta(t) \ge \varepsilon, \quad t \in [c,d]$$

and there exists a $t_0 \in (c,d)$ such that $\alpha(t_0) - \beta(t_0) = 2\varepsilon$.

Consider the compact subset given by

$$S = \Big[(t,x,y)\colon t \in [c,d],\ |\beta(t) - x| \le 1,\ |\beta'(t) - y| \le 1\Big].$$

For any $\delta > 0$ sufficiently small, let $\rho(t)$ satisfy the conditions

$$\rho'' = (L+1)\rho', \qquad -\min(\delta,\varepsilon) \le \rho(t) \le 0, \qquad t \in [c,d],$$
$$0 < \rho'(t) \le 1, \qquad t \in (c,d).$$

Set $m(t) = \beta(t) - \rho(t)$. Note that $m(t) \ge \beta(t)$, $t \in [c,d]$, and $\beta'(t) \ge m'(t)$, $t \in (c,d)$. Moreover, for $t \in (c,d)$,

$$\begin{aligned} D^-m'(t) &= D^-\beta'(t) - \rho''(t) \le f\big(t,\beta(t),\beta'(t)\big) - (L+1)\rho'(t) \\ &\le f\big(t,m(t),\beta'(t)\big) - (L+1)\rho'(t) \\ &\le f\big(t,m(t),m'(t)\big) + L\rho'(t) - (L+1)\rho'(t) \\ &< f\big(t,m(t),m'(t)\big), \end{aligned}$$

using successively the assumptions on f.

If we now let $v(t) = m(t) + \varepsilon$, we see, because of the monotonicity of f,

$$D^-v'(t) = D^-m'(t) < f\big(t,m(t),m'(t)\big) \le f\big(t,v(t),v'(t)\big), \qquad t \in (c,d).$$

Also,

$$\alpha(c) = \beta(c) + \varepsilon \leq m(c) + \varepsilon = v(c),$$

$$\alpha(d) = \beta(d) + \varepsilon \leq m(d) + \varepsilon = v(d).$$

An application of Theorem 1.2.1 yields that

$$\alpha(t) < v(t), \qquad t \in (c,d).$$

From the definition of $m(t)$, we obtain

$$\beta(t) \leq m(t) \leq \beta(t) + \varepsilon$$

and, as a result, we obtain at $t = t_0$,

$$\alpha(t_0) < v(t_0) = m(t_0) + \varepsilon \leq \beta(t_0) + 2\varepsilon.$$

This contradiction proves the theorem.

We can deduce a uniqueness result for a certain boundary value problem from Theorem 1.2.2. This we state as a corollary.

COROLLARY 1.2.1. Let the assumptions of Theorem 1.2.2 hold except (ii) and (iii). Suppose that $x,y \in C^{(2)}[[t_1,t_2],R]$ are solutions on $[t_1,t_2] \subset J$ such that $x(t_1) = y(t_1)$, $x(t_2) = y(t_2)$. Then $x(t) \equiv y(t)$ on $[t_1,t_2]$.

REMARK 1.2.1. If the Lipschitz condition on $f(t,x,y)$ with respect to y on compact sets is omitted, Corollary 1.2.1 is no longer valid. For example, the boundary value problem

$$x'' = \left(\frac{3}{2}\right)\left(\frac{5}{2}\right)^{2/3}|x'|^{1/3}, \qquad x(-1) = x(1) = 1,$$

has solutions $x(t) \equiv 1$ and $x(t) = |t|^{5/2}$ on $[-1,1]$.

Instead of the one-sided Lipschitz condition on f, one could assume that solutions of initial value problems $x'' = f(t,x,x')$ are unique, to conclude the validity of Theorem 1.2.2. This is precisely what the next theorem does.

THEOREM 1.2.3. Suppose that

(i) $f \in C[J \times R \times R, R]$ and $f(t,x,y)$ is nondecreasing in x for each $(t,y) \in J \times R$;

(ii) α is a lower solution and β is an upper solution of (1.2.1) on $[c,d] \subset J$;

(iii) solutions of initial value problems for (1.2.1) are unique;

(iv) $\alpha(c) \leq \beta(c)$ and $\alpha(d) \leq \beta(d)$.

Then we have $\alpha(t) \leq \beta(t)$ on $[c,d]$.

The proof of this theorem rests on a result that we are going to consider in the following section and therefore will be given there.

COROLLARY 1.2.2. If $f(t,x,x')$ satisfies hypotheses (i) and (iii) of Theorem 1.2.3, and if $x,y \in C^{(2)}\left[[t_1,t_2],R\right]$ are solutions on $[t_1,t_2] \subset J$ such that $x(t_1) = y(t_1)$, $x(t_2) = y(t_2)$, then $x(t) \equiv y(t)$ on $[t_1,t_2]$.

Another interesting result concerning lower and upper solutions which would yield uniqueness of solutions with general linear boundary conditions is the following.

THEOREM 1.2.4. Assume that

(i) $f \in C[J \times R \times R, R]$ and $f(t,x,y)$ is strictly increasing in x for each $(t,y) \in J \times R$;

(ii) α is a lower solution and β is an upper solution of (1.2.1) on J;

(iii) for each (t,x,y_1), $(t,x,y_2) \in J \times R \times R$,

$$|f(t,x,y_1) - f(t,x,y_2)| \leq L|y_1 - y_2|, \qquad L > 0;$$

(iv) $a_1\alpha(a) - a_2\alpha'(a) \leq a_1\beta(a) - a_2\beta'(a)$, $b_1\alpha(b) + b_2\alpha'(b) \leq b_1\beta(b) + b_2\beta'(b)$, where $a_1 + a_2 > 0$, $b_1 + b_2 > 0$, $a_1, a_2, b_1, b_2 \geq 0$, and $a_1 + b_1 > 0$.

Then $\alpha(t) \leq \beta(t)$ on J.

Proof: We shall first show that at the end points of J the desired inequality holds.

Let $\alpha(a) > \beta(a)$ which implies from (iv) that $a_2 \neq 0$ and hence $\alpha'(a) \geq \beta'(a)$. Then there exists a $\delta > 0$ such that $\alpha(t) > \beta(t)$ on $[a,\delta)$. Consequently, using the strict monotony of f and condition (iii), we obtain

$$
\begin{aligned}
(1.2.2) \qquad D_-\alpha'(t) - D^-\beta(t) &\geq f\big(t,\alpha(t),\alpha'(t)\big) - f\big(t,\beta(t),\beta'(t)\big) \\
&> f\big(t,\beta(t),\alpha'(t)\big) - f\big(t,\beta(t),\beta'(t)\big) \\
&\geq -L|\alpha'(t) - \beta'(t)|.
\end{aligned}
$$

We claim that there exists a $t_0 \in [a,\delta)$ such that $\alpha'(t_0) - \beta'(t_0) > 0$. If not, suppose that $\alpha'(t) - \beta'(t) \leq 0$ on $[a,\delta]$. Then, by (1.2.2), we would have

$$D_-\alpha'(t) - D^-\beta'(t) \geq L[\alpha'(t) - \beta'(t)],$$

which, by the theory of differential inequalities, yields

$$\alpha'(t) - \beta'(t) \geq [\alpha'(a) - \beta'(a)]e^{L(t-a)}.$$

This, together with $\alpha'(t) - \beta'(t) \leq 0$, assures us that $\alpha'(t) \equiv \beta'(t)$ on $[a,\delta]$. In view of (1.2.2), we then arrive at the contradiction

$$D_-\alpha'(t) - D^-\beta'(t) > 0$$

which establishes the claim $\alpha'(t_0) - \beta'(t_0) > 0$ for some $t_0 \in [a,\delta)$. Again, we appeal to the differential inequality (1.2.2) to conclude $\alpha'(t) - \beta'(t) > 0$ on $[t_0,\delta]$. This argument may be extended to show that $\delta = b$ which implies that $\alpha(b) > \beta(b)$. The last conclusion, in its turn, leads to the contradiction $\alpha'(b) \leq \beta'(b)$ because of the second inequality in (iv). Thus $\alpha(a) > \beta(a)$ is impossible.

If we assume that $\alpha(b) > \beta(b)$, we are lead to the relation $\alpha(a) > \beta(a)$ arguing as before, which we have shown is impossible. It therefore remains to be shown that $\alpha(t) \leq \beta(t)$ on (a,b), which is exactly similar to the proof of Theorem 1.2.1. The proof of the theorem is therefore complete.

1.3 THE MODIFIED FUNCTION

Employing the notion of upper and lower solutions and the existence theorem in the small, it is possible to establish the existence of solutions in the large of some boundary value problems for a modified form of the differential equation

$$x'' = f(t,x,x'), \tag{1.3.1}$$

where $f \in C[J \times R \times R, R]$. Let us first define the modified function.

DEFINITION 1.3.1. Let $\alpha, \beta \in C^{(1)}[J,R]$ with $\alpha(t) \leq \beta(t)$ on J and let $c > 0$ be such that $|\alpha'(t)|, |\beta'(t)| < c$ on J. Then define

$$F^*(t,x,x') = \begin{cases} f(t,x,c) & \text{for} \quad x' \geq c, \\ f(t,x,x') & \text{for} \quad |x'| \leq c, \\ f(t,x,-c) & \text{for} \quad x' \leq -c, \end{cases}$$

and

$$F(t,x,x') = \begin{cases} F^*\big(t,\beta(t),x'\big) + [x-\beta(t)]/1+x^2 & \text{for} \quad x > \beta(t), \\ F^*(t,x,x') & \text{for} \quad \alpha(t) \leq x \leq \beta(t), \\ F^*\big(t,\alpha(t),x'\big) + [x-\alpha(t)]/1+x^2 & \text{for} \quad x < \alpha(t). \end{cases}$$

The function $F(t,x,x')$ will be called the <u>modification</u> of $f(t,x,x')$ associated with the triple $\alpha(t)$, $\beta(t)$, c. It follows from the definition that $F(t,x,x')$ is continuous on $J \times R \times R$ and that

$$|F(t,x,x')| \leq M \quad \text{on} \quad J \times R \times R,$$

where $M = M_0 + 1$ and

$$M_0 = \max\Big[|f(t,x,x')|: t \in J,\ \alpha(t) \leq x \leq \beta(t),\ |x'(t)| \leq c\Big]$$
$$+ \max_{t \in J} |\alpha(t)| + \max_{t \in J} |\beta(t)|.$$

Let us consider now the modified boundary value problem

$$(1.3.2) \qquad x'' = F(t,x,x'), \qquad x(a) = \gamma, \quad x(b) = \delta,$$

relative to which we have the following.

THEOREM 1.3.1. Let $\alpha, \beta \in C^{(1)}[J,R]$ be, respectively, lower and upper solutions of (1.3.1) on J such that $\alpha(t) \leq \beta(t)$ on J. Then the boundary value problem (1.3.2), where F is the modification of f associated with the triple $\alpha(t)$, $\beta(t)$, c, has a solution $x \in C^{(2)}[J,R]$ satisfying

$$(1.3.3) \qquad \alpha(t) \leq x(t) \leq \beta(t) \quad \text{on} \quad J,$$

provided that $\alpha(a) \leq \gamma \leq \beta(a)$, $\alpha(b) \leq \delta \leq \beta(b)$.

Proof: By Corollary 1.1.1, the boundary value problem (1.3.2) has a solution $x \in C^{(2)}[J,R]$. Thus we only need to show (1.3.3). We shall only prove that $x(t) \leq \beta(t)$ on J. The arguments are essentially the same for the case $\alpha(t) \leq x(t)$. Assume, if possible, that $x(t) > \beta(t)$ for some $t \in J$. Then $x(t) - \beta(t)$ has a positive maximum at a point $t_0 \in J^0$. Hence it follows that $x'(t_0) = \beta'(t_0)$, $|x'(t_0)| < c$ and

$$x''(t_0) = F\big(t_0, x(t_0), x'(t_0)\big)$$
$$= f\big(t_0, \beta(t_0), \beta'(t_0)\big) + \frac{[x(t_0) - \beta(t_0)]}{1 + x^2(t_0)}.$$

Since β is an upper function on J,

$$D^{-}\beta'(t_0) \le f(t_0,\beta(t_0),\beta'(t_0))$$

and therefore, we arrive at

$$D_{-}[x'(t_0)-\beta'(t_0)] = x''(t_0) - D^{-}\beta'(t_0) \ge \frac{[x(t_0)-\beta(t_0)]}{((1+x^2(t_0))} > 0$$

which is impossible at a maximum of $x(t)-\beta(t)$. We conclude that $x(t) \le \beta(t)$ on J. The proof is complete.

Let us next consider the general linear boundary conditions

(1.3.4) $a_1x(a) - a_2x'(a) = A, \qquad b_1x(b) + b_2x'(b) = B,$

where $a_1 + a_2 > 0$, $b_1 + b_2 > 0$, $a_1, a_2, b_1, b_2 \ge 0$, and $a_1 + b_1 > 0$. We wish to prove a result analogous to Theorem 1.3.1 with respect to conditions (1.3.4). First of all, we need the following lemma.

LEMMA 1.3.1. Assume that

(i) $\alpha,\beta \in C^{(1)}[J,R]$ are, respectively, lower and upper solutions of (1.3.1) on J such that $\alpha(t) \le \beta(t)$ on J;

(ii) $f \in C[J\times R\times R,R]$ and for $t \in J$, $\alpha(t) \le x \le \beta(t)$, $f(t,x,x')$ satisfies a Lipschitz condition in x' for a constant $L > 0$.

Then, there exists a function $f^* \in C[J\times R\times R,R]$ which is bounded on $J\times R\times R$ and is Lipschitzian for the same constant $L > 0$ whenever $t \in J$ and $\alpha(t) \le x \le \beta(t)$.

<u>Proof</u>: As before, let $c > 0$ be such that $|\alpha'(t)|, |\beta'(t)| < c$ on J. Then define $f^*(t,x,x')$ for $t \in J$ and $\alpha(t) \le x \le \beta(t)$ by setting

$$f^*(t,x,x') = \begin{cases} f(t,x,c) & \text{for} \quad x' \geq c, \\ f(t,x,x') & \text{for} \quad |x'| \leq c, \\ f(t,x,-c) & \text{for} \quad x' \leq -c. \end{cases}$$

We then extend the domain of definition of $f^*(t,x,x')$ to $J \times R \times R$ by letting

$$f^*(t,x,x') = \begin{cases} f^*(t,\beta(t),x') & \text{for} \quad x > \beta(t), \\ f^*(t,\alpha(t),x') & \text{for} \quad x < \alpha(t). \end{cases}$$

It is easy to see that the function f^* so defined possesses all the stated properties. Hence the proof is complete.

THEOREM 1.3.2. Let hypotheses (i) and (ii) of Lemma 1.3.1 hold. Then there exists a solution $x \in C^{(2)}[J,R]$ of

$$x'' = f^*(t,x,x') \tag{1.3.5}$$

which satisfies the boundary conditions (1.3.4), provided that

$$a_1\alpha(a) - a_2\alpha'(a) \leq A \leq a_1\beta(a) - a_2\beta'(a), \tag{1.3.6}$$

$$b_1\alpha(b) + b_2\alpha'(b) \leq B \leq b_1\beta(b) + b_2\beta'(b). \tag{1.3.7}$$

Proof: Define the function $F(t,x,x')$ on $J \times R \times R$ as

$$F(t,x,x') = \begin{cases} f^*(t,\beta(t),x') + ([x-\beta(t)]/1+x^2) & \text{for} \quad x > \beta(t), \\ f^*(t,x,x') & \text{for} \quad \alpha(t) \leq x \leq \beta(t), \\ f^*(t,\alpha(t),x') + ([x-\alpha(t)]/1+x^2) & \text{for} \quad x < \alpha(t), \end{cases}$$

where $f^*(t,x,x')$ is the function obtained in Lemma 1.3.1. Since f^* is bounded, F is also bounded. Hence, by Corollary 1.1.1, there exists a solution $x \in C^{(2)}[J,R]$ of

$$x'' = F(t,x,x') \tag{1.3.8}$$

satisfying the boundary conditions (1.3.4). We now show that for $t \in J$, $\alpha(t) \leq x(t) \leq \beta(t)$, which implies that $x(t)$ is a solution of the problem (1.3.4), (1.3.5).

Assume that $x(a) > \beta(a)$ in which case a_2 cannot be zero. Then there exists a $\delta > a$ such that $x(t) > \beta(t)$ on $[a,\delta)$. By definition of f^*, F, we have for $a \leq t \leq \delta$

$$(1.3.9) \qquad \begin{aligned} x''(t) - D^-\beta'(t) &\geq f^*\big(t,\beta(t),x'(t)\big) + \frac{[x(t)-\beta(t)]}{1+x^2(t)} \\ &\quad - f^*\big(t,\beta(t),\beta'(t)\big) \\ &\geq -L|x'(t)-\beta'(t)|, \end{aligned}$$

since $\beta(t)$ is also an upper solution of (1.3.8). Also, relation (1.3.6) implies, because of the assumption $x(a) \geq \beta(a)$, that $x'(a) \geq \beta'(a)$. We claim that there is a $t_0 \in [a,\delta)$ such that $x'(t_0) - \beta'(t_0) > 0$. If not, suppose that $x'(t) - \beta'(t) \leq 0$ on $a \leq t \leq \delta$. Then, by (1.3.9), we would obtain

$$x''(t) - D^-\beta'(t) \geq L[x'(t) - \beta'(t)]$$

which, by the theory of differential inequalities, yields

$$x'(t) - \beta'(t) \geq [x'(a) - \beta'(a)]e^{L(t-a)}.$$

This, together with $x'(t) - \beta'(t) \leq 0$, implies that $x'(t) \equiv \beta'(t)$ on $a \leq t \leq \delta$. On the basis of (1.3.9), we then arrive at the contradiction

$$x''(t) - D^-\beta'(t) \geq \frac{[x(t)-\beta(t)]}{1+x^2(t)} > 0$$

which establishes the claim, $x'(t_0) - \beta'(t_0) > 0$ for some $t_0 \in [a,\delta)$. Again appealing to the differential inequality (1.3.9), we conclude that $x'(t) - \beta'(t) > 0$ on $[t_0,\delta]$. In fact, this argument leads to $\delta = b$ and, consequently, it follows that $x(b) > \beta(b)$. This, in turn, leads to the contradiction $x'(b) \leq \beta'(b)$, in view of the relations (1.3.4) and (1.3.7).

If, on the other hand, we assume that $x(b) > \beta(b)$, we are lead to the relation $x(a) > \beta(a)$ arguing as before, which has just been shown to be impossible. It therefore remains to show that $x(t) \leq \beta(t)$ for $t \in J^0$.

Assume then there exist $t \in J^0$ for which $x(t) > \beta(t)$. This implies the existence of $t_1, t_2 \in J$ such that

$$x(t_1) = \beta(t_1), \quad x(t_2) = \beta(t_2) \quad \text{and} \quad x(t) > \beta(t), \quad t_1 < t < t_2.$$

The difference $x(t) - \beta(t)$, therefore, assumes a positive maximum at some $t_0 \in (t_1, t_2)$. Clearly, $x'(t_0) = \beta'(t_0)$ and $x''(t_0) - D^-\beta'(t_0) \leq 0$. A computation, however, leads to the contradiction

$$x''(t_0) - D^-\beta''(t_0) \geq f^*(t_0, \beta(t_0), \beta'(t_0)) + \frac{[x(t_0) - \beta(t_0)]}{1 + x^2(t_0)}$$
$$- f^*(t_0, \beta(t_0), \beta'(t_0)) > 0,$$

noting $|x'(t_0| = |\beta'(t_0)| < c$ and the definition of f^*. We have thus established $x(t) \leq \beta(t)$ on J. One can similarly verify that $\alpha(t) \leq x(t)$ on J completing the proof.

Let us now proceed to supply the proof of Theorm 1.2.3. For this purpose we need the following result.

THEOREM 1.3.3. Assume that $f \in C[J \times R \times R, R]$ and that solutions of initial value problems for (1.3.1) are unique. Let $\alpha, \beta \in C^{(1)}[J, R]$ be lower and upper solutions of (1.3.1) on J, respectively, such that $\alpha(t) \leq \beta(t)$ on J. Then, if $\alpha(t_0) = \beta(t_0)$ and $\alpha'(t_0) = \beta'(t_0)$ for some $t_0 \in J$, it follows that $\alpha(t) \equiv \beta(t)$ on J.

<u>Proof</u>: Suppose that $\alpha(t) \not\equiv \beta(t)$ on J. We will consider the case where there is an interval $(t_0, t_1] \subset J$ such that $\alpha(t_0) = \beta(t_0)$, $\alpha'(t_0) = \beta'(t_0)$, and $\alpha(t) < \beta(t)$ on $[t_0, t_1]$.

Let $F_1(t,x,x')$ be the modification function of $f(t,x,x')$ according to Definition 1.3.1 for the interval $[t_0,t_1]$ and the triple α,β,c_1. Then, if $\alpha(t_1) < \delta_1 < \beta(t_1)$, we have by Theorem 1.3.1 that the boundary value problem (BVP)

$$x'' = F_1(t,x,x'), \qquad x(t_0) = \alpha(t_0) \qquad x(t_1) = \delta_1$$

has a solution $x_1 \in C^{(2)}[[t_0,t_1],R]$ such that $\alpha(t) \leq x_1(t) \leq \beta(t)$ on $[t_0,t_1]$. Therefore, $x_1(t_0) = \alpha(t_0)$, $x'(t_0) = \alpha'(t_0)$ and since $|x_1'(t)| < c_1$ on some neighborhood of t_0, it follows by the definition of F_1, there is a maximal interval $[t_0,t_2] \subset [t_0,t_1]$ on which $x_1(t)$ is a solution of (1.3.1). If $t_2 = t_1$,

$$\alpha(t_2) = \alpha(t_1) < \delta_1 = x_1(t_2) = x_1(t_1) < \beta(t_1).$$

If $t_0 < t_2 < t_1$, it is still true that $\alpha(t_2) < x_1(t_2) < \beta(t_2)$ for, if either inequality were an equality, we would have $|x_1'(t)| < c_1$ and the interval $[t_0,t_2]$ would not be maximal.

We can therefore construct another modification function $F_2(t,x,x')$ of $f(t,x,x')$ on $[t_0,t_2]$ relative to the triple α,β,c_2. An application of Theorem 1.3.1 again with $\alpha(t_2) < \delta_2 < \beta(t_2)$, shows that there is a solution $x_2 \in C^{(2)}[[t_0,t_2],R]$ of the BVP

$$x'' = F_2(t,x,x'), \qquad x(t_0) = \alpha(t_0), \qquad x(t_2) = \delta_2$$

satisfying $\alpha(t) \leq x_2(t) \leq \beta(t)$ on $[t_0,t_2]$. As before, it follows that there is a maximal interval $[t_0,t_3] \subset [t_0,t_2]$ on which $x_2(t)$ is a solution of (1.3.1) and that $\alpha(t_3) < x_2(t_3) < \beta(t_3)$. This contradicts the assumption that solutions of initial value problems for (1.3.1) are unique, since $x_1(t_0) = x_2(t_0)$ and $x_1'(t_0) = x_2'(t_0)$. We therefore conclude that $\alpha(t) \equiv \beta(t)$ on J.

We are now ready to give the proof of Theorem 1.2.3.

Proof of Theorem 1.2.3: Assume that $\alpha(t) > \beta(t)$ at some points of $[c,d]$ and let $N = \max[\alpha(t) - \beta(t)]$ on $[c,d]$. Clearly there exists a $t_0 \in [c,d]$ where $\alpha(t_0) - \beta(t_0) = N$ and $\alpha'(t_0) = \beta'(t_0)$. The functions α and $\beta_1 = \beta + N$ satisfy the conditions of Theorem 1.3.3 in view of the nondecreasing nature of $f(t,x,x')$ in x. Consequently it follows that $\alpha(t) \equiv \beta(t) + N$ on $[c,d]$, a contradiction which proves the theorem.

1.4 NAGUMO'S CONDITION

As we have seen, the proofs for the existence theorems for solutions of boundary value problems depend on finding a priori bounds for the solution and its derivative. Here we present some sufficient conditions for obtaining such bounds.

DEFINITION 1.4.1. Let $f \in C[J \times R \times R; R]$ and $\alpha, \beta \in C[J,R]$ with $\alpha(t) \leq \beta(t)$ on J. Suppose that for $t \in J$, $\alpha(t) \leq x \leq \beta(t)$ and $x' \in R$,

$$|f(t,x,x')| \leq h(|x'|) \tag{1.4.1}$$

where $h \in C[R_+, (0,\infty)]$. If

$$\int_\lambda^\infty \frac{s\,ds}{h(s)} > \max_{t \in J} \beta(t) - \min_{t \in J} \alpha(t), \tag{1.4.2}$$

where

$$\lambda(b - a) = \max\big[|\alpha(a) - \beta(b)|, |\alpha(b) - \beta(a)|\big], \tag{1.4.3}$$

we say that f satisfies Nagumo's condition on J relative to α, β.

THEOREM 1.4.1. Assume that f satisfies Nagumo's condition on J with respect to the pair α, β. Then for any solution $x \in C^{(2)}[J,R]$ of (1.3.1) with $\alpha(t) \leq x(t) \leq \beta(t)$ on J, there exists an $N > 0$ depending only on α, β, h such that

$$|x'(t)| \leq N \quad \text{on} \quad J. \tag{1.4.4}$$

<u>Proof</u>: Because of (1.4.2), we can choose an $N > \lambda$ such that

$$\int_{\lambda}^{N} \frac{s\,ds}{h(s)} > \max_{t \in J} \beta(t) - \min_{t \in J} \alpha(t)$$

If $t_0 \in J_0$ is such that $(b-a)x'(t_0) = x(b) - x(a)$, then by (1.4.3), we have $|x'(t_0)| \leq \lambda$. Assume that (1.4.4) is not true. Then there exists an interval $[t_1, t_2] \subset J$ such that the following cases hold:

(i) $x'(t_1) = \lambda$, $x'(t_2) = N$, and $\lambda < x'(t) < N$, $t \in (t_1, t_2)$,

(ii) $x'(t_1) = N$, $x'(t_2) = \lambda$, and $\lambda < x'(t) < N$, $t \in (t_1, t_2)$,

(iii) $x'(t_1) = -\lambda$, $x'(t_2) = -N$, and $-N < x'(t) < -\lambda$, $t \in (t_1, t_2)$,

(iv) $x'(t_1) = -N$, $x'(t_2) = -\lambda$, and $-N < x'(t) < -\lambda$, $t \in (t_1, t_2)$.

Let us consider case (i). On $[t_1, t_2]$, we obtain, by (1.4.1),

$$|x''(t)|x'(t) = |f(t, x(t), x'(t))|x'(t) \leq h(|x'(t)|)x'(t)$$

and as a result

$$\left| \int_{t_1}^{t_2} \frac{x''(s)x'(s)\,ds}{h(|x'(s)|)} \right| \leq \int_{t_1}^{t_2} \frac{|x''(s)|x'(s)\,ds}{h(|x'(s)|)}$$

$$\leq \int_{t_1}^{t_2} x'(s)\,ds = x(t_2) - x(t_1).$$

This leads to the contradiction

$$\int_{\lambda}^{N} \frac{s\,ds}{h(s)} \leq \max_{t\in J} \beta(t) - \min_{t\in J} \alpha(t).$$

We can deal with the remaining possibilities in a similar way and therefore we conclude that (1.4.4) is valid.

COROLLARY 1.4.1. Let $h \in C[R_+,(0,\infty)]$, $f \in C[J\times R\times R,R]$, and for $t \in J$, $|x| \leq M$, and $x' \in R$, $|f(t,x,x')| \leq h(|x'|)$. Assume that

$$\int^{\infty} \frac{s\,ds}{h(s)} = \infty.$$

Then, for any solution $x \in C^{(2)}[J,R]$ with $|x(t)| \leq M$, there exists an $N > 0$ depending only on $M,h,(b-a)$ such that $|x'(t)| \leq N$ on J. Also, $N \to 0$ as $M \to 0$.

REMARK 1.4.1. The conclusion of Corollary 1.4.1 is false, if $f \in C[J\times R^n\times R^n,R^n]$ and absolute values are replaced by norms, as the following example shows. Let $x(t) = (\cos nt, \sin nt)$ so that $\|x(t)\| = 1$, $\|x'(t)\| = n$, $\|x''(t)\| = n^2 = \|x'(t)\|^2$. Thus the assumptions of Corollary 1.4.1 hold with $M = 1$, $h(s) = s^2 + 1$. However, there does not exist an $N > 0$ such that $\|x'(t)\| \leq N$ for all choices of n. Thus it is clear that Nagumo's condition is not sufficient to obtain the desired conclusion in the case of systems. We shall discuss this further when we treat systems of equations later in this chapter.

EXERCISE 1.4.1. Let $\alpha,\beta \in C^{(1)}[J,R]$ with $\alpha(t) \leq \beta(t)$ on J. Suppose that $f \in C[J \times R \times R, R]$ and for $t \in J$, $\alpha(t) \leq x \leq \beta(t)$, $y_1, y_2 \in R$,

$$|f(t,x,y_1) - f(t,x,y_2)| \leq L|y_1 - y_2|,$$

where $L > 0$ is a constant. Show that f satisfies Nagumo's condition.

THEOREM 1.4.2. Assume that

(i) $\alpha,\beta \in C^{(1)}[J,R]$ such that $\alpha(a) = \beta(a)$, $\alpha(b) = \beta(b)$, and $\alpha(t) \leq \beta(t)$ on J;

(ii) $\varphi,\psi \in C^{(1)}[\omega,R]$ with $\varphi(t,x) \leq \psi(t,x)$ on ω, where $\omega = [(t,x)\colon \alpha(t) \leq x \leq \omega(t),\ t \in J]$ and on ω,

$$\varphi_t(t,x) + \varphi_x(t,x)\varphi(t,x) < f\big(t,x,\varphi(t,x)\big), \tag{1.4.5}$$

$$\psi_t(t,x) + \psi_x(t,x)\psi(t,x) > f\big(t,x,\psi(t,x)\big), \tag{1.4.6}$$

where $f \in C[J \times R \times R, R]$;

(iii) $\varphi\big(a,\alpha(a)\big) \leq \alpha'(a)$, $\beta'(a) \leq \psi\big(a,\beta(a)\big)$.

Then for any solution $x \in C^{(2)}[J,R]$ of (1.3.1) such that $\big(t,x(t)\big) \in \omega$, $t \in J$, we have $\varphi\big(t,x(t)\big) \leq x'(t) \leq \psi\big(t,x(t)\big)$, $t \in J$.

<u>Proof</u>: Let $x(t)$ be any solution of (1.3.1) such that $\alpha(t) \leq x(t) \leq \beta(t)$ on J. Then, by (i) and (iii) it follows that

$$\varphi\big(a,x(a)\big) \leq x'(a) \leq \psi\big(a,x(a)\big).$$

Assume now that there exists a $t_0 \in (a,b)$ such that $x'(t_0) < \varphi\big(t_0,x(t_0)\big)$. Then there must exist an interval $[t_1,t_2] \subset J$ with $t_1 < t_0 < t_2$ such that $x'(t_i) = \varphi\big(t_i,x(t_i)\big)$, $i = 1,2$ and $x'(t) < \varphi\big(t,x(t)\big)$, $t_1 < t < t_2$, and further that

$$x''(t_1) - [\varphi_t(t_1,x(t_1)) + \varphi_x(t_1,x(t_1))x'(t_1)] \leq 0.$$

This is a contradiction because of assumption (ii). We therefore conclude that $x'(t) \geq \varphi(t,x(t))$ on J. A similar argument shows that $x'(t) \leq \psi(t,x(t))$ on J and the proof is complete.

We remark that the inequality sign in (1.4.5) could be reversed in which case the contradiction is arrived at because of

$$x''(t_2) - [\varphi_t(t_2,x(t_2)) + \varphi_x(t_2,x(t_2))x'(t_2)] \geq 0.$$

A similar comment holds relative to inequality (1.4.6). Of course, in this case, we have to replace (iii) by

$$\varphi(b,\alpha(b)) \leq \alpha'(b), \qquad \beta'(b) \leq \psi(b,\beta(b)).$$

COROLLARY 1.4.2. Assume (i) of Theorem 1.4.2. Let there exist constants N_1,N_2 such that $\alpha'(a),\ \alpha'(b) \geq N_1$, $\beta'(a),\beta'(b) \leq N_2$, and $f(t,x,N_i) \neq 0$, $i = 1,2$. Then the conclusion of Theorem 1.4.2 is true with $\varphi(t,x) = N_1$ and $\psi(t,x) = N_2$.

EXERCISE 1.4.2. Let hypothesis (i) of Theorem 1.4.2 hold. Assume that $f \in C[J \times R \times R, R]$ and $|f(t,x,x')| \to \infty$ as $|x'| \to \infty$ uniformly on compact (t,x) sets. Show that the conclusion of Corollary 1.4.2 is true.

EXERCISE 1.4.3. Show that if the assumptions of Corollary 1.4.1 hold and $h(s)$ is nondecreasing, there always exist functions $\varphi(t,x)$, $\psi(t,x)$ (independent of t) satisfying the hypotheses of Theorem 1.4.2.

Consider the following example on $J = [-1,1]$,

$$x'' = e^{-2(x+1)} - (x')^{2n}, \qquad x(-1) = 0 = x(1),$$

where n is a positive integer. Take $\alpha(t) = t^2 - 1$ and $\beta(t) = 0$. Choose $N_1 = -2$ and $N_2 = 2$. Then we see all the hypotheses of Corollary 1.4.2 are satisfied. Hence we conclude that any solution $x(t)$ such that $t^2 - 1 \leq x(t) \leq 0$ satisfies $|x'(t)| \leq 2$ on J. However, we notice that for $n > \frac{1}{2}$, Corollary 1.4.1 is not applicable.

A variant of Theorem 1.4.2 which is more useful is the following result.

THEOREM 1.4.3. Suppose that

(i) $\alpha, \beta \in C^{(1)}[J,R]$ such that $\alpha(t) \leq \beta(t)$ on J;

(ii) $\varphi, \psi \in C^{(1)}[\omega,R]$ with $\varphi(t,x) \leq \psi(t,x)$ on ω, where

$$\omega = [(t,x)\colon t \in J,\ \alpha(t) \leq x \leq \beta(t)]$$

and on ω

$$\varphi_t(t,x) + \varphi_x(t,x)\varphi(t,x) \leq f\big(t,x,\varphi(t,x)\big),$$

$$\psi_t(t,x) + \psi_x(t,x)\psi(t,x) \geq f\big(t,x,\psi(t,x)\big),$$

where $f \in C[J \times R \times R, R]$.

Then, for any solution $x \in C^{(2)}[J,R]$ of (1.3.1) such that $\alpha(t) \leq x(t) \leq \beta(t)$ on J and $\varphi\big(a,x(a)\big) \leq x'(a) \leq \psi\big(a,x(a)\big)$, we have

$$\varphi\big(t,x(t)\big) \leq x'(t) \leq \psi\big(t,x(t)\big), \qquad t \in J.$$

<u>Proof</u>: Let $x \in C^{(2)}[J,R]$ be any solution of $x'' = f(t,x,x')$ such that $\varphi\big(a,x(a)\big) \leq x'(a) \leq \psi\big(a,x(a)\big)$ and $\alpha(t) \leq x(t) \leq \beta(t)$, $t \in J$. Suppose that there exists a $t_0 \in (a,b]$ such that $x'(t_0) > \psi\big(t_0,x(t_0)\big)$. Define

$$V(t) = \left[x'(t) - \psi\big(t,x(t)\big)\right] \exp\left[-\int_t^{t_0} \psi_x\big(s,x(s)\big)\,ds\right].$$

Then,

$$\begin{aligned}
V'(t) \exp\left[\int_t^{t_0} \psi_x\big(s,x(s)\big)\,ds\right] \\
&= x''(t) - \left[\psi_t\big(t,x(t)\big) + \psi_x\big(t,x(t)\big)x'(t)\right] \\
&\quad + \left[x'(t) - \psi\big(t,x(t)\big)\right]\psi_x\; t,x(t) \\
&= x''(t) - \left[\psi_t\big(t,x(t)\big) + \psi_x\big(t,x(t)\big)\psi\big(t,x(t)\big)\right] \\
&\le f\big((t,x(t),x'(t)\big) - f\big(t,x(t),\psi(t,x(t))\big) \\
&= 0,
\end{aligned}$$

in some interval to the left of t_0. This implies that $V(t)$ is nonincreasing as t increases and positive near t_0 and hence on $[a,t_0]$. Thus we have $V(a) > 0$ or equivalently $x'(a) > \psi\big(a,x(a)\big)$ which is a contradiction. Hence $x'(t) \le \psi\big(t,x(t)\big)$ on J. Similarly we can verify $\varphi\big(t,x(t)\big) \le x'(t)$ on J.

1.5 EXISTENCE IN THE LARGE

We are now ready to prove theorems on existence in the large. We begin with one of the basic results in this direction.

THEOREM 1.5.1. Let $\alpha,\beta \in C^{(1)}[J,R]$ be, respectively, lower and upper solutions of (1.3.1) on J such that $\alpha(t) \le \beta(t)$ on J. Suppose further that $f(t,x,x')$ satisfies Nagumo's condition on J relative to the pair α,β. Then, for any $\alpha(a) \le c \le \beta(a)$, $\alpha(b) \le d \le \beta(b)$, the BVP

$$(1.5.1) \qquad x'' = f(t,x,x'), \qquad x(a) = c, \quad x(b) = d,$$

has a solution $x \in C^{(2)}[J,R]$ with $\alpha(t) \le x(t) \le \beta(t)$ and

$|x'(t)| \leq N$ on J, where N depends only on α,β and the Nagumo's function h.

Proof: By Theorem 1.4.1, there is an $N > 0$ depending only on α,β,h such that $|x'(t)| \leq N$ on J for any solution with $\alpha(t) \leq x(t) \leq \beta(t)$ on J. Choose a $c_1 > N$ so that $|\alpha'(t)| < c_1$, $|\beta'(t)| < c_1$ on J. Then, by Theorem 1.3.1, the BVP

$$x'' = F(t,x,x'), \qquad x(a) = c, \quad x(b) = d,$$

has a solution $x \in C^{(2)}[J,R]$ such that $\alpha(t) \leq x(t) \leq \beta(t)$ on J, where $F(t,x,x')$ is the modification function of $f(t,x,x')$ with respect to α,β, and c_1. By the mean value theorem there is a $t_0 \in J_0$ such that

$$(b - a)x'(t_0) = x(b) - x(a),$$

and as a result using (1.4.3) it follows that $|x'(t_0)| \leq \lambda < N < c_1$. This implies that there is an interval containing t_0, where $x(t)$ is a solution of $x'' = f(t,x,x')$. By Theorem 1.4.1, we have $|x'(t)| \leq N < c_1$ on this interval. However, $x(t)$ is a solution of $x'' = f(t,x,x')$ as long as $|x'(t)| < c_1$. We conclude that $x(t)$ is a solution of (1.5.1) on J. The proof is complete.

COROLLARY 1.5.1. Under the assumptions of Theorem 1.5.1, any infinite sequence of solutions of $x'' = f(t,x,x')$ obeying the relation $\alpha(t) \leq x(t) \leq \beta(t)$ on J, has a uniformly convergent subsequence converging to a solution of $x'' = f(t,x,x')$ on J.

A conclusion similar to Theorem 1.5.1 may be drawn relative to the boundary value problem (1.3.1), (1.3.4), on the basis of Theorem 1.3.2, Exercise 1.4.1, and the proof of Theorem 1.5.1.

We state it in the following exercise.

EXERCISE 1.5.1. Let hypotheses (i) and (ii) of Lemma 1.3.1 hold. Show that for any A,B for which relations (1.3.6) and (1.3.7) are valid, the BVP (1.3.1), (1.3.4) has a solution $x \in C^{(2)}[J,R]$ such that $\alpha(t) \leq x(t) \leq \beta(t)$ on J and $|x'(t)| \leq N$ on J, where $N > 0$ depends only on α,β,L.

EXERCISE 1.5.2. Let $f \in C[J \times R \times R, R]$, $f(t,x,x')$ be nondecreasing in x for each (t,x') and satisfy $|f(t,x,y_1) - f(t,x,y_2)| \leq L\,|y_1 - y_2|$ for $(t,x) \in J \times R$ and $y_1,y_2 \in R$. Show that the BVP (1.3.1), (1.3.4) has a solution. If f is strictly increasing in x, then show that the solution is unique.

<u>Hint</u>: Set $M = \max_J f(t,0,0)$, $m = \min_J f(t,0,0)$ and let $\beta(t) \geq 0$, $\alpha(t) \leq 0$ be solutions of $x'' + L|x'| - m = 0$, $x'' - L|x'| - M = 0$, respectively, subject to (1.3.6), (1.3.7) which may be computed explicitly. Then, using monotony of f, show α,β are lower, upper solutions. For uniqueness, use Theorem 1.2.4.

EXERCISE 1.5.3. Verify that $\beta(t) = t$, $\alpha(t) = t^2 - 3t + 2$ are upper, lower solutions for the problem $x'' = -|x|^{\frac{1}{2}} + t$, $x(1) = 0$, $x'(2) = 1$, on $J = [1,2]$. Draw conclusions on the applicability of Exercises 1.5.1, 1.5.2 to this problem. Discuss the merits.

The foregoing discussion concerning the problem (1.3.1), (1.3.4) imposes more restrictions on the function f compared to the BVP (1.5.1). This is apparently due to the techniques involved in the process rather than the need, because the result that we present next involves nonlinear boundary conditions more general than (1.3.4) and yet the assumptions on f are just those required for the BVP (1.5.1).

We shall first prove the following.

THEOREM 1.5.2. Assume that

(a) $\alpha,\beta \in C^{(1)}[J,R]$, respectively, are lower and upper solutions of (1.3.1) on J such that $\alpha(t) \leq \beta(t)$ on J and $\alpha(a) < \beta(a)$;

(b) $f(t,x,x')$ satisfies Nagumo's condition on J with respect to the pair α,β;

(c) $g \in C\big[[\alpha(a),\beta(a)] \times R,R\big]$, $g(x,y)$ is nondecreasing in y for each x, and

$$g\big(\alpha(a),\alpha'(a)\big) \geq 0, \qquad g\big(\beta(a),\beta'(a)\big) \leq 0. \tag{1.5.2}$$

Then, for any $\alpha(b) \leq d \leq \beta(b)$, the BVP

$$x'' = f(t,x,x'), \qquad g\big(x(a),\, x'(a)\big) = 0, \quad x(b) = d$$

has a solution $x_d \in C^{(2)}[J,R]$ with $\alpha(t) \leq x_d(t) \leq \beta(t)$ on J.

Proof: It is enough to show that given $\varepsilon > 0$ and $\alpha(b) \leq d \leq \beta(b)$, there is a solution $x(t,\varepsilon)$ of (1.3.1) on J that fulfills

$$x(b,\varepsilon) = d, \quad \big|g\big(x(a,\varepsilon),\, x'(a,\varepsilon)\big)\big| < \varepsilon$$

and $\alpha(t) \leq x(t,\varepsilon) \leq \beta(t)$ on J.

For a fixed $\alpha(b) \leq d \leq \beta(b)$, let $\pi(c)$ denote the set of solutions $x(t)$ to the BVP (1.5.1) such that $\alpha(a) \leq c \leq \beta(a)$ and $\alpha(t) \leq x(t) \leq \beta(t)$ on J. By Theorem 1.5.1, it is clear that $\pi(c)$ is nonempty for all $\alpha(a) \leq c \leq \beta(a)$. Suppose now that the theorem is not true. Then there exists an $\varepsilon_0 > 0$, such that for every $\alpha(a) \leq c \leq \beta(a)$ and $x(t) \in \pi(c)$, we have

$$\big|g\big(x(a),x'(a)\big)\big| \geq \varepsilon_0.$$

Define the set S by

$$S = [x(t) \in \pi(c): \alpha(a) \leq c \leq \beta(a) \text{ and } g(x(a),x'(a)) \geq \varepsilon_0]$$

and let $c_0 = \sup[x(a) = c: x(t) \in S]$. We notice that $x(t) \in \pi(\alpha(a))$ implies $\alpha'(a) \leq x'(a)$ which by (1.5.2) yields $g(x(a),x'(a)) \geq \varepsilon_0$. Similarly, $x(t) \in \pi(\beta(a))$ implies $x'(a) \leq \beta'(a)$ so that $g(x(a),x'(a)) \leq -\varepsilon_0$. This observation proves that $c_0 < \beta(a)$. Let $y_0(t)$ be a solution of (1.3.1) which is obtained as a uniform limit of members of S so that

$$(1.5.3) \quad y_0(a) = c_0, \quad y_0(b) = d, \quad \text{and} \quad g(y_0(a), y_0'(a)) \geq \varepsilon_0.$$

Now let $N \geq 1$ be such that $c_0 + (1/N) \leq \beta(a)$. For $n \geq N$, let $y_n(t) \in \pi(c_0 + (1/n))$ satisfying, in addition, the inequality $y_n(t) \geq y_0(t)$ on J. This is clearly possible since $y_0(t)$ may be treated as a lower solution of (1.3.1). Then a subsequence of $y_n(t)$ converges to a solution $x_0(t) \in \pi(c_0)$ obeying $x_0(t) \geq y_0(t)$ on J. By definition of c_0, we have $g(y_n(a), y_n'(a)) \leq -\varepsilon_0$ and consequently $g(x_0(a), x_0'(a)) \leq -\varepsilon_0$. Since $x_0'(a) \geq y_0'(a)$ and $x_0(a) = y_0(a)$, this leads to $g(y_0(a), y_0'(a)) \leq -\varepsilon_0$ which is a contradiction in view of (1.5.3). The proof is complete.

Employing similar arguments as in the proof of Theorem 1.5.2, which were based on a consequence of Theorem 1.5.1, namely Corollary 1.5.1, we may prove the following theorem.

THEOREM 1.5.3. Assume that

(a) $\alpha,\beta \in C^{(1)}[J,R]$, respectively, are lower and upper solutions of (1.3.1) on J such that

$$\alpha(t) \leq \beta(t) \text{ on } J \text{ and } \alpha(b) < \beta(b);$$

(b) $f(t,x,x')$ satisfies Nagumo's condition on J relative to the pair α,β;

(c) $h \in C[[\alpha(b),\beta(b)] \times R, R]$, $h(x,y)$ is nondecreasing in y for each x and $h(\alpha(b), \alpha'(b)) \leq 0$, $h(\beta(b),\beta'(b)) \geq 0$.

Then, for any $\alpha(a) \leq c \leq \beta(a)$ there is a solution $x_c \in C^{(2)}[J,R]$ of the BVP

$$x'' = f(t,x,x'), \qquad x(a) = c, \qquad h(x(b),x'(b)) = 0$$

which satisfies $\alpha(t) \leq x_c(t) \leq \beta(t)$ on J.

Combining the proofs of Theorems 1.5.2 and 1.5.3, we may obtain our main result relative to nonlinear boundary conditions.

THEOREM 1.5.4. Assume that

(a) $\alpha,\beta \in C^{(1)}[J,R]$, respectively, are lower and upper solutions of (1.3.1) on J such that $\alpha(t) \leq \beta(t)$ on J and $\alpha(a) < \beta(a)$, $\alpha(b) < \beta(b)$;

(b) $f(t,x,x')$ satisfies Nagumo's condition on J with respect to the pair α,β;

(c) $g \in C[[\alpha(a),\beta(a)] \times R,R]$, $h \in C[[\alpha(b),\beta(b)] \times R,R]$, $g(x,y)$, $h(x,y)$ are nondecreasing in y for each x and

$$g(\alpha(a),\alpha'(a)) \geq 0, \qquad g(\beta(a),\beta'(a)) \leq 0,$$
$$h(\alpha(b),\alpha'(b)) \leq 0, \qquad h(\beta(b),\beta'(b)) \geq 0.$$

Then there is a solution $x \in C^{(2)}[J,R]$ of the BVP

$$x'' = f(t,x,x'), \qquad g(x(a),x'(a)) = 0, \quad h(x(b), x'(b)) = 0,$$

which satisfies $\alpha(t) \leq x(t) \leq \beta(t)$ on J.

<u>Proof</u>: For each $\alpha(b) \leq d \leq \beta(b)$, let $\pi(d)$ denote the set of solutions of the BVP

$$x'' = f(t,x,x'), \qquad g(x(a),x'(a)) = 0, \quad x(b) = d,$$

which satisfy $\alpha(t) \leq x(t) \leq \beta(t)$ on J. By Theorem 1.5.2, it is clear that the set $\pi(d)$ is nonempty for all

$d \in [\alpha(b),\beta(b)]$. Suppose that the conclusion of the theorem is not true. Then $h(x(b),x'(b)) \neq 0$ for $x(t) \in \pi(d)$, $\alpha(b) \leq d \leq \beta(b)$. Thus, if $x(t) \in \pi(\beta(b))$, then $h(x(b), x'(b)) > 0$ and if $x(t) \in \pi(\alpha(b))$, $h(x(b), x'(b)) < 0$. Let $d_0 = \sup[x(b) = d\colon x(t) \in \pi(d), \alpha(b) \leq d \leq \beta(b)$ and $h(x(b),x'(b)) < 0]$. It follows that $d_0 < \beta(b)$. Let $y(t)$ be a solution of (1.3.1) such that $\alpha(t) \leq y(t) \leq \beta(t)$ on J, $g(y(a),y'(a)) = 0$, $y(b) = d_0$, and $h(y(b),y'(b)) < 0$. If $y(a) = \beta(a)$, then an application of Theorem 1.5.3 to the pair $y(t),\beta(t)$ yields a solution $y_1(t)$ of (1.3.1) satisfying $y(t) \leq y_1(t) \leq \beta(t)$ on J and $h(y_1(b),y_1'(b)) = 0$. However, $y'(a) \leq y_1'(a) \leq \beta'(a)$ implies that

$$0 = g(y(a),y'(a)) \leq g(y_1(a),y_1'(a)) \leq g(\beta(a),\beta'(a)) \leq 0$$

which contradicts the supposition that the theorem is not true. Hence $y(a) < \beta(a)$. Now use $y(t)$ in place of $\alpha(t)$ in Theorem 1.5.2 to conclude that the set

$$D = [x(t)\colon x(t) \in \pi(d),\ d_0 < d \leq \beta(b),\ y(t) \leq x(t) \leq \beta(t) \text{ on } J]$$

in nonempty. Moreover, $x(t) \in \pi(d)$ implies that $h(x(b), x'(b)) > 0$. However, $\inf[x(b) = d\colon x(t) \in D] = d_0$, which leads to a contradiction by a convergence argument and the fact that $h(x,y)$ is nondecreasing in y. Hence the proof is complete.

On the basis of Theorem 1.4.3, it is possible to relax the condition of the basic Theorem 1.5.1 which is the content of the next result.

THEOREM 1.5.5. Suppose that assumptions (i) and (ii) of Theorem 1.4.3 hold. Let α,β be lower and upper solutions of (1.3.1), respectively, on J, such that $\alpha(a) = \beta(a)$ and $\varphi(t,\gamma(t)) \leq \gamma'(t) \leq \psi(t,\gamma(t))$, for $\gamma(t) = \alpha(t),\beta(t)$, $t \in J$.

Then, for any $\alpha(b) \leq d \leq \beta(b)$ and $\alpha(a) = c = \beta(a)$, the BVP (1.5.1) has a solution $x \in C^{(2)}[J,R]$ such that

$$\bigl(t,x(t),x'(t)\bigr) \in \Omega, \qquad t \in J,$$

where Ω is the set given by

$$\Omega = [(t,x,x'): (t,x) \in \omega \text{ and } \varphi(t,x) \leq x' \leq \psi(t,x)].$$

Proof: Define a modified function $F(t,x,x')$ on $\omega \times R$ by

$$F(t,x,x') = \begin{cases} f\bigl(t,x,\varphi(t,x)\bigr) & \text{if } x' < \varphi(t,x), \\ f(t,x,x') & \text{if } (t,x,x') \in \Omega, \\ f\bigl(t,x,\psi(t,x)\bigr) & \text{if } x' > \psi(t,x), \end{cases}$$

and extend $F(t,x,x')$ to $J \times R \times R$ by setting

$$F(t,x,x') = \begin{cases} F\bigl(t,\alpha(t),x'\bigr) & \text{if } x < \alpha(t), \\ \\ F\bigl(t,\beta(t),x'\bigr) & \text{if } x > \beta(t). \end{cases}$$

Observe that $F(t,x,x')$ is bounded on $J \times R \times R$ and therefore satisfies Nagumo's condition. Furthermore (because of the assumptions on α',β'), α,β, are also lower and upper solutions relative to

$$x'' = F(t,x,x'). \tag{1.5.4}$$

Consequently, by Theorem 1.5.1 there exists a solution $x \in C^{(2)}[J,R]$ to the modified BVP

$$x'' = F(t,x,x'), \qquad x(a) = c, \quad x(b) = d,$$

such that $\alpha(t) \leq x(t) \leq \beta(t)$, $t \in J$ for $\alpha(a) = c = \beta(a)$ and $\alpha(b) \leq d \leq \beta(b)$. We now apply Theorem 1.4.3 to conclude that this $x(t)$ is actually a solution of the BVP (1.5.1) satisfying $\bigl(t,x(t),x'(t)\bigr) \in \Omega$, $t \in J$. Hence the proof is complete.

1.6. LYAPUNOV-LIKE FUNCTIONS

We wish to employ, in this section, Lyapunov-like functions and the theory of differential inequalities to establish existence in the large for the BVP (1.5.1). Let $V \in C[J \times R \times R, R]$. We define the generalized derivative $D^+V_f(t,x,x')$ relative to the differential equation $x'' = f(t,x,x')$ by

$$D^+V_f(t,x,x') \equiv \limsup_{h \to 0+} \frac{1}{h} \left[V\big(t+h, x+hx', x'+hf(t,x,x')\big) - V(t,x,x')\right].$$

We shall often use several functions V_i, $i = 1,2,\ldots,n$, which we call Lyapunov-like functions and for notational consistency we write $D^+V_{if}(t,x,x')$ to denote the generalized derivative of V_i with respect to the differential equation

$$x'' = f(t,x,x'). \tag{1.6.1}$$

THEOREM 1.6.1. Assume that

(i) $\alpha, \beta \in C^{(1)}[J,R]$ such that $\alpha(t) \le \beta(t)$ on J and $F(t,x,x')$ is the modified function of $f(t,x,x')$ relative to the triple α, β, c, as given in the Definition 1.3.1;

(ii) there exist two Lyapunov functions $W_i \in C[D_i, R]$, $i = 1,2$,. where $D_1 = J \times [x\colon x \le \alpha(t)] \times R$, $D_2 = J \times [x\colon x \ge \beta(t)] \times R$, such that $W_1(t,x,x') = 0$ if $x = \alpha(t)$, $W_1(t,x,x') > 0$ if $x < \alpha(t)$, $W_2(t,x,x') = 0$ if $x = \beta(t)$, $W_2(t,x,x') > 0$ if $x > \beta(t)$;

(iii) for $i = 1,2$, $g_i \in C[J \times R^+, R]$, $W_i(t,x,x')$ is locally Lipschitzian in (x,x') and

$$D^+W_{1F}(t,x,x') \ge g_1\big(t, W_1(t,x,x')\big) \quad \text{in the interior of } D_1,$$

$$D^+W_{2F}(t,x,x') \ge g_2\big(t, W_2(t,x,x')\big) \quad \text{in the interior of } D_2;$$

(iv) the only solution of $r' = g_i(t,r)$, $r(t_0) = 0$, $t_0 \in J$, is $r(t) \equiv 0$ on J.

Then, if $\alpha(a) \leq c \leq \beta(a)$, $\alpha(b) \leq d \leq \beta(b)$, the BVP

$$x'' = F(t,x,x'), \qquad x(a) = c, \quad x(b) = d, \tag{1.6.2}$$

has a solution $x \in C^{(2)}[J,R]$ satisfying $\alpha(t) \leq x(t) \leq \beta(t)$ on J.

Proof: By Corollary 1.1.1, the BVP (1.6.2) has a solution $x \in C^{(2)}[J,R]$. It suffices to show that $\alpha(t) \leq x(t) \leq \beta(t)$ on J. We will only prove that $\alpha(t) \leq x(t)$ on J since the proof for $x(t) \leq \beta(t)$ is essentially the same. Assume that $x(t) < \alpha(t)$ for some $t \in (a,b)$. Then there exists an interval $[t_1,t_2] \subset (a,b)$ such that $x(t_1) = \alpha(t_1)$, $x(t_2) = \alpha(t_2)$, and $x(t) < \alpha(t)$ for $t \in (t_1,t_2)$. We thus have by (ii)

$$W_1\big(t_2,x(t_2),x'(t_2)\big) = 0. \tag{1.6.3}$$

From (iii) and the theory of differential inequalities, we obtain

$$W_1\big(t,x(t),x'(t)\big) \leq r(t,t_2,r_0), \qquad a \leq t \leq t_2, \tag{1.6.4}$$

where $r(t,t_2,r_0)$ is the left maximal solution of $r' = g_1(t,r)$, $r(t_2) = r_0 = W_1\big(t_2,x(t_2),x'(t_2)\big)$. By (iv) and (1.6.3), it is clear that $r(t,t_2,0) \equiv 0$. However, for any $t \in (t_1,t_2)$, we obtain, because of (ii) and (1.6.4), the contradiction

$$0 < W_1\big(t,x(t),x'(t)\big) \leq 0.$$

The proof is therefore complete.

We notice that this theorem generalizes Theorem 1.3.1. In case $\alpha(t)$, $\beta(t)$ are lower and upper solutions of (1.6.2), respectively, on J, it is enough to choose $W_1(t,x,x') = \alpha(t) - x$

and $W_2(t,x,x') = x - \beta(t)$. It is easy to check that W_1 satisfies all the assumptions of Theorem 1.6.1 with $g_1 \equiv 0$. The only condition that requires a little explanation is (iii), namely $D^+W_{1F}(t,x,x') \geq 0$ in the interior of D_1. For this, it is sufficient to prove that $W_1(t,x,x')$ does not attain a local maximum in the interior of D_1. Arguing as in the proof of Theorem 1.3.1, we see that it is necessary that $\alpha(t_1) = x(t_1)$ at some $t_1 \in (a,b)$ in order for W_1 to attain a local maximum. However, for $t = t_1$, we obtain $DW'_{1F}(t_1,x(t_1), x'(t_1)) > 0$ as in the proof of Theorem 1.3.1 which is impossible at a local maximum of W_1. A similar reasoning holds for W_2 and thus the conditions of Theorem 1.6.1 are verified.

The next theorem offers a general set of conditions to ensure that $|x'(t)|$ is bounded.

THEOREM 1.6.2. Assume that $\alpha,\beta \in C^{(1)}[J,R]$ and $\alpha(t) \leq \beta(t)$ on J. Define the sets $E_1 = J \times [x: \alpha(t) \leq x \leq \beta(t)] \times [x': x' \geq 0]$ and $E_2 = J \times [x: \alpha(t) \leq x \leq \beta(t)] \times [x'; x' \leq 0]$. Suppose that there exist Lyapunov functions V_i, $i = 1,2,3,4$, that are locally Lipschitzian in (x,x') such that

(i) $V_i \in C[E_1,R]$ and $V_i(t,x,x') \to \infty$ as $x' \to \infty$ uniformly on $J \times [x: \alpha(t) \leq x \leq \beta(t)]$, $i = 1,2$;

(ii) $D^+V_{1f}(t,x,x') \leq g_1(t,V_1(t,x,x'))$ and $D^+V_{2f}(t,x,x') \geq g_2(t,V_2(t,x,x'))$ on E_1;

(iii) $V_i \in C[E_2,R]$ and $V_i(t,x,x') \to \infty$ as $x' \to -\infty$ uniformly on $J \times [x: \alpha(t) \leq x \leq \beta(t)]$, $i = 3,4$;

(iv) on E_2, $D^+V_{3f}(t,x,x') \leq g_3(t,V_3(t,x,x'))$ and $D^+V_{4f}(t,x,x.) \geq g_4(t,V_4(t,x,x'))$;

(v) $V_i(t,x,x') \leq \psi_i(|x'|)$, $i = 1,2,3,4$, where $\psi_i(r) > 0$ for $r > 0$ are continuous functions.

Suppose further that

(vi) $g_i \in C[J \times R, R]$, $i = 1,2,3,4$ and for $t_0 \in J$, $r_0 \in R$, all solutions of $r' = g_i(t,r)$, $r(t_0) = r_0$, $i = 1,3$ exist on $[t_0, b]$, and all solutions of $r' = g_i(t,r)$, $r(t_0) = r_0$, $i = 2,4$ exist on $[a, t_0]$.

Then there exists an $N > 0$ such that every solution of (1.6.1) with $\alpha(t) \leq x(t) \leq \beta(t)$ on J satisfies $|x'(t)| \leq N$ on J.

Proof: Let

$$(b - a)\lambda = \max\left[|\alpha(b) - \beta(a)|, |\beta(b) - \alpha(a)|\right].$$

Define for $t \in J$

$$L_i = \max_{\substack{t_0 \in J \\ t \geq t_0}} \left|r_i\big(t, t_0, \psi_i(\lambda)\big)\right|, \qquad i = 1,3,$$

and

$$L_i = \max_{\substack{t_0 \in J \\ t \leq t_0}} \left|r_i\big(t, t_0, \psi_i(\lambda)\big)\right|, \qquad i = 2,4,$$

where $r_i\big(t, t_0, \psi_i(\lambda)\big)$, $i = 1,3$ $(i = 2,4)$ is the right (left) maximal solution of $r' = g_i(t,r)$, $r(t_0) = \psi_i(\lambda)$. Then condition (vi) guarantees that the L_i's are finite. By (i) and (iii), there exist $N_i > 0$, $i = 1,2,3,4$ such that

$$V_i(t, x, N_i) > L_i, \qquad i = 1,2, \tag{1.6.5}$$

and

$$V_i(t, x, -N_i) > L_i, \qquad i = 3,4.$$

Let $N > \max[N_i, \lambda]$, $i = 1,2,3,4$. We claim that $|x'(t)| \leq N$ on J where $x(t)$ is any solution of (1.6.1) with $\alpha(t) \leq x(t) \leq \beta(t)$ on J.

There exists a $t_0 \in (a,b)$ such that $x(b) - x(a) = x'(t_0)(b-a)$

and therefore it follows that $|x'(t_0)| \leq \lambda < N$. There are four cases to be considered depending upon whether there exists a t_2 such that $x'(t_2) = N$ or $x'(t_2) = -N$ and whether $t_2 > t_0$ or $t_2 < t_0$. Assume, for example, $x'(t_2) = N$ and $t_2 > t_0$. Then there exists a t_1 with $t_0 \leq t_1 < t_2$ such that $x'(t_1) = \lambda$ and $\lambda < x'(t) < N$ for $t \in (t_1, t_2)$. From (v) we obtain $V_1(t_1, x(t_1), x'(t_1)) \leq \psi_1(\lambda)$ and from (ii) we arrive at the estimate

$$(1.6.6) \quad V_1(t, x(t), x'(t)) \leq r_1(t, t_1, \psi_1(\lambda)), \qquad t \in [t_1, t_2],$$

using the theory of differential inequalities. Relations (1.6.5) and (1.6.6) lead us to the contradiction

$$L_1 < V_1(t_2, x(t_2), x'(t_2)) \leq L_1$$

because of the definitions of L_1 and N.

On the basis of the preceding argument, we can arrive at a similar contradiction in the remaining three cases. We omit the details. The proof is therefore complete.

The result we have just proved offers a general set of sufficient conditions which imply that $|x'(t)|$ is bounded and thus relaxes the more stringent Nagumo's condition assumed in Theorem 1.4.1. In fact, the intent of the following exercise is to clarify this advantage further.

EXERCISE 1.6.1. Let $h_i \in C[R_+, (0,\infty)]$, $f \in C[J \times R \times R, R]$, and

$$\int^{\infty} \frac{s\, ds}{h_i(s)} = \infty, \qquad i = 1,2,3,4.$$

Suppose that $\alpha, \beta \in C[J, R]$ with $\alpha(t) \leq \beta(t)$ on J. For $t \in J$ and $\alpha(t) \leq x \leq \beta(t)$, assume that

$$-h_2(x') \leq f(t,x,x') \leq h_1(x') \qquad \text{if} \quad x' \geq 0,$$

$$-h_3(-x') \leq f(t,x,x') \leq h_4(-x') \qquad \text{if} \quad x' \leq 0.$$

Show that for any solution $x \in C^{(2)}[J,R]$ of (1.6.1) with $\alpha(t) \leq x(t) \leq \beta(t)$ on J, there exists an $N > 0$ depending only on α, β, h_i such that $|x'(t)| \leq N$ on J.

Hint: Under the assumptions show one can construct four Lyapunov functions satisfying the conditions of Theorem 1.6.2. For example set $V_1 = \exp\left[t - x + \int_\lambda^{x'} s\, ds/h_1(s)\right]$ on E_1 and show the assumptions are verified with $g_1(t,r) = r$ and similarly construct V_2, V_3, and V_4.

We may combine Theorems 1.6.1 and 1.6.2 to obtain the existence of solutions (1.5.1). The proof is similar to Theorem 1.5.1. We merely state the following theorem which is a generalization of Theorem 1.5.1.

THEOREM 1.6.3. Let $\alpha, \beta \in C^{(1)}[J,R]$ with $\alpha(t) \leq \beta(t)$ on J. Assume that there exist two Lyapunov functions W_1, W_2 satisfying the hypotheses of Theorem 1.6.1. Suppose further that there exist four Lyapunov functions V_i, $i = 1,2,3,4$ such that they obey the hypotheses of Theorem 1.6.2. Then for any $\alpha(a) \leq c \leq \beta(a)$, $\alpha(b) \leq d \leq \beta(b)$ the BVP (1.5.1) has a solution $x \in C^{(2)}[J,R]$ with $\alpha(t) \leq x(t) \leq \beta(t)$ and $|x'(t)| \leq N$ on J.

1.7 EXISTENCE ON INFINITE INTERVALS

On the basis of Theorems 1.4.1 and 1.5.1 it is possible to obtain the existence of solutions on infinite intervals.

THEOREM 1.7.1. Assume that for each $b > a$, $f(t,x,x')$ satisfies Nagumo's condition on $[a,b]$ relative to the pair $\alpha, \beta \in C^{(1)}\left[[a,\infty),R\right]$ with $\alpha(t) \leq \beta(t)$ on $[a,\infty)$. Suppose also that α, β are lower and upper solutions of (1.6.1) on $[a,\infty)$, respectively. Then for any $\alpha(a) \leq c \leq \beta(a)$ the BVP

$$(1.7.1) \qquad x'' = f(t,x,x'), \qquad x(a) = c,$$

has a solution $x \in C^{(2)}[[a,\infty),R]$ such that $\alpha(t) \leq x(t) \leq \beta(t)$ on $[a,\infty)$.

Proof: By Theorem 1.5.1, it follows that for each $n \geq 1$ there is a solution $x_n \in C^{(2)}[[a,a+n],R]$ such that $x_n(a) = c$, $x_n(a+n) = \beta(a+n)$, and $\alpha(t) \leq x_n(t) \leq \beta(t)$ on $[a,a+n]$. Furthermore, there is an $N_n > 0$ such that $|x'(t)| \leq N_n$ on $[a,a+n]$ for any solution satisfying $\alpha(t) \leq x(t) \leq \beta(t)$ on $[a,a+n]$. Thus for any fixed $n \geq 1$, $x_m(t)$ is a solution on $[a,a+n]$ verifying $\alpha(t) \leq x_m(t) \leq \beta(t)$ and $|x_m'(t)| \leq N_n$ on $[a,a+n]$ for all $m \geq n$. Consequently, for $m \geq n$ the sequences $\{x_m(t)\}$, $\{x_m'(t)\}$ are both uniformly bounded and equicontinuous on $[a,a+n]$. Then, employing the standard diagonalization arguments, we obtain a subsequence which converges uniformly on all compact subintervals of $[a,\infty)$ to a solution $x(t)$. The desired solution of (1.7.1) is precisely this $x(t)$. Hence the proof is complete.

Essentially a similar proof may be given to the following,

THEOREM 1.7.2. Assume that $f(t,x,x')$ satisfies Nagumo's condition on $[-a,a]$ for each $a > 0$ with respect to the pair $\alpha,\beta \in C^{(1)}[R,R]$ with $\alpha(t) \leq \beta(t)$ on R. Suppose also that α,β are lower and upper solutions of (1.6.1) on R. Then there is a solution of (1.6.1) on R such that $\alpha(t) \leq x(t) \leq \beta(t)$ on R.

We shall merely state below a result analogous to Theorem 1.7.1 in terms of Lyapunov-like functions.

THEOREM 1.7.3. Let $\alpha,\beta \in C^{(1)}[[a,\infty),R]$ with $\alpha(t) \leq \beta(t)$ on $[a,\infty)$ and let $f \in C[[a,\infty)\times R\times R,R]$. Assume there exist two Lyapunov functions $W_1(t,x,x')$, $W_2(t,x,x')$ obeying the hypotheses of Theorem 1.6.1 where we now replace $[a,b]$ with $[a,\infty)$. Assume also that there exist two Lyapunov functions V_2, V_4 satisfying the corresponding properties of Theorem 1.6.2 with $[a,b]$ replaced by $[a,\infty)$. More precisely suppose that

(i) $V_2(t,x,x') \to \infty$ as $x' \to \infty$ uniformly on compact subsets of $[a,\infty)\times[x\colon \alpha(t) \leq x \leq \beta(t)]$;

(ii) $D^+V_{2f}(t,x,x') \geq g_2(t,V_2(t,x,x'))$ for $t \in [a,\infty)$, $\alpha(t) \leq x \leq \beta(t)$ and $x' \geq 0$;

(iii) $V_4(t,x,x') \to \infty$ as $x' \to -\infty$ uniformly on compact subsets of $[a,\infty)\times[x\colon \alpha(t) \leq x \leq \beta(t)]$;

(iv) $D^+V_{4f}(t,x,x') \geq g_4(t,V_4(t,x,x'))$ for $t \in [a,\infty)$, $\alpha(t) \leq x \leq \beta(t)$ and $x' \leq 0$;

(v) there exist ψ_i, $i = 2,4$ where $\psi_i \in C[[a,\infty)\times R,R]$ and $\psi_i(t,u)$ is increasing in t for each u such that for each $T \geq a$, $V_i(t,x,x') \leq \psi_i(T,|x'|)$ for $t \in [a,T]$ and $\alpha(t) \leq x \leq \beta(t)$;

(vi) $g_i \in C[[a,\infty)\times R,R]$, $i = 2,4$ such that for each $t_0 \in [a,\infty)$ and $r_0 \geq 0$ all solutions of $r' = g_i(t,r)$, $r(t_0) = r_0$ exist on $[a,t_0]$.

Then for any c satisfying $\alpha(a) \leq c \leq \beta(a)$ there exists a solution of (1.7.1) such that $\alpha(t) \leq x(t) \leq \beta(t)$ on $[a,\infty)$.

Clearly a theorem analogous to Theorem 1.7.2 may be formulated, which we leave as an exercise.

1.8 SUPER- AND SUBFUNCTIONS

Here we define subfunctions and superfunctions relative to the solutions of $x'' = f(t,x,x')$ and discuss necessary and sufficient conditions for such functions to be, respectively, lower and upper solutions. To avoid repetition, most of the results will be stated only in terms of lower solutions and subfunctions. When it becomes necessary later to refer to a result concerning superfunctions and upper solutions, we shall simply refer to the subfunction statement of the result.

DEFINITION 1.8.1. A function $\varphi(t)$ is said to be a <u>subfunction</u> relative to solutions of $x'' = f(t,x,x')$ on an interval J if for any $[t_1,t_2] \subset J$ and for any solution $x \in C^{(2)}\big[[t_1,t_2],R\big]$, $x(t_i) \geq \varphi(t_i)$, $i = 1,2$ implies $x(t) \geq \varphi(t)$ on $[t_1,t_2]$. A <u>superfunction</u> may be defined similarly by reversing the respective inequalities.

We have immediately the following result.

THEOREM 1.8.1. Assume that $\varphi \in C[J,R] \cap C^{(1)}[J^0,R]$ is a subfunction on J with respect to the solution of

$$x'' = f(t,x,x') \tag{1.8.1}$$

where $f \in C[J \times R \times R,R]$. Then φ is a lower solution of (1.8.1) on J.

<u>Proof</u>: Let $t_0 \in J^0$ and $h \geq 0$, $k \geq 0$, $h + k > 0$ be sufficiently small. Then by Theorem 1.1.2 the BVP

$$x'' = f(t,x,x'), \qquad x(t_0 - k) = \varphi(t_0 - k), \quad x(t_0 + h) = \varphi(t_0 + h),$$

has a solution $x \in C^{(2)}\big[[t_0 - k, t_0 + h],R\big]$. Since φ is a subfunction, we readily obtain

$$\frac{\varphi'(t_0 + h) - \varphi'(t_0 - k)}{h + k} \geq \frac{x'(t_0 + h) - x'(t_0 - k)}{h + k}$$

$$= x''(s) = f\big(s,x(s),x'(s)\big)$$

for some $t_0 - k < s < t_0 + h$. The continuity of f together with Theorem 1.1.2, implies that

$$f\big(s,x(s),x'(s)\big) \to f\big(t_0,x(t_0),x'(t_0)\big) \quad \text{as} \quad h+k \to 0.$$

Thus

$$\lim_{h+k\to 0} \frac{\varphi'(t_0+h)-\varphi'(t_0-k)}{h+k} \geq f\big(t_0,\varphi(t_0),\varphi'(t_0)\big)$$

and, in particular,

$$D\varphi'(t_0) \geq f\big(t_0,\varphi(t_0),\varphi'(t_0)\big)$$

from which it follows that φ is a lower solution relative to (1.8.1) on J.

Assuming only the continuity of f, it will not be possible in general to show that a lower function is a subfunction. Since a solution is a lower solution, if lower solutions are subfunctions, then solutions are subfunctions. From the definition of subfunctions it would then follow that, if a BVP on an interval has a solution, that solution is unique. Hence stronger assumptions other than continuity of f are required to conclude that a lower solution is a subfunction. A set of sufficient conditions is given in the next theorem.

THEOREM 1.8.2. Assume that

(i) $f \in C[J\times R\times R,R]$ and $f(t,x,y)$ is nondecreasing in x for each (t,y);

(ii) either f satisfies a one-sided Lipschitz condition in y on each compact subset of $J\times R\times R$ or solutions of initial value problems for (1.8.1) are unique.

Then a lower solution on J is a subfunction on J relative to solutions of (1.8.1).

The proof is a direct consequence of Theorems 1.2.2 and

1.2.3.

In the preceding result conditions are imposed on $f(t,x,y)$ which are sufficient to imply that lower solutions are subfunctions and, therefore, that solutions of BVP's, when they exist, are unique. In the next result we take the uniqueness of BVP's as one of the hypotheses.

THEOREM 1.8.3. Assume that each initial value problem for (1.8.1) has a solution which extends throughout $[a,b]$. Suppose further that solutions of boundary value problems

$$x'' = f(t,x,x'), \qquad x(t_1) = x_1, \quad x(t_2) = x_2, \quad [t_1,t_2] \subset J,$$

when they exist, are unique. Then, if $I \subset J$ and $\alpha \in C^{(1)}[I,R]$ is a lower solution on I, α is a subfunction on I.

<u>Proof</u>: Assume that α is not a subfunction on I. Then there is an interval $[c,d] \subset I$ and solution $x_0 \in C^{(2)}\big[[c,d],R\big]$ such that $x_0(c) = \alpha(c)$, $x_0(d) = \alpha(d)$, and $x_0(t) < \alpha(t)$ on (c,d). We now define $F(t,x,x')$ on $[c,d]\times R\times R$ by

$$F(t,x,x') = \begin{cases} f(t,x,x'), & x \geq \alpha(t), \\ f\big(t,\alpha(t),x'\big) - \big(\alpha(t) - x\big), & x < \alpha(t). \end{cases}$$

Since $F(t,x,x')$ is continuous on $[c,d]\times R\times R$ and $\alpha \in C^{(1)}\big[[c,d],R\big]$, it follows from Theorem 1.1.2 that there is a $\delta > 0$ such that $[t_1,t_2] \subset [c,d]$ and $t_2 - t_1 \leq \delta$ implies the BVP

$$x'' = F(t,x,x'), \qquad x(t_1) = \alpha(t_1), \quad x(t_2) = \alpha(t_2),$$

has a solution $x \in C^{(2)}\big[[t_1,t_2],R\big]$. Using the fact that α is a lower solution, we can show that $\alpha(t) \leq x(t)$ on $[t_1,t_2]$ with the same type of argument as used in the proof of Theorem 1.3.1. As a result, for $[t_1,t_2] \subset [c,d]$ and

$t_2 - t_1 \leq \delta$, the BVP

$$x'' = f(t,x,x'), \qquad x(t_1) = \alpha(t_1), \quad x(t_2) = \alpha(t_2),$$

has a solution $x \in C^{(2)}\big[[t_1,t_2],R\big]$ with $\alpha(t) \leq x(t)$ on $[t_1,t_2]$. Thus α is a subfunction "in the small."

Clearly $d - c > \delta$; for otherwise there would be a solution $x(t)$ with $x(c) = \alpha(c)$, $x(d) = \alpha(d)$, and $x(t) \geq \alpha(t)$ on $[c,d]$. This solution would be distinct from $x_0(t)$ contradicting the assumption concerning the uniqueness of solutions of BVP's. Now for each positive integer n, let $P(n)$ be the proposition that there exists an interval $[c_n,d_n] \subset [c,d]$ with $0 < d_n - c_n \leq d - c - (n-1)\delta$ and a solution $x_n \in C^{(2)}\big[[c_n,d_n],R\big]$ such that $x_n(c_n) = \alpha(c_n)$, $x_n(d_n) = \alpha(t)$ on (c_n,d_n). Evidently $P(1)$ is true with $[c_1,d_1] = [c,d]$ and $x_1(t) = x_0(t)$. Assume $P(k)$ is true. Then $d_k - c_k > \delta$, otherwise we would obtain a contradiction of $x_k(t)$ being the distinct solution with boundary values $\alpha(c_k)$ and $\alpha(d_k)$. Let $z_1(t)$ be the solution of the BVP

$$x'' = f(t,x,x'), \qquad x(c_k) = \alpha(c_k), \quad x(c_k + \delta) = \alpha(c_k + \delta).$$

Since each initial value problem has a solution extending throughout J, there is a solution $z_2(t)$ on $[c_k,d_k]$ such that $z_2(t) \equiv z_1(t)$ on $[c_k,c_k + \delta]$. Suppose that $P(k+1)$ is not true. Then we must have $z_2(t) \geq \alpha(t)$ on $[c_k + \delta,d_k]$. Also we must have $z_2(d_k) > \alpha(d_k)$. If $d_k - c_k - \delta \leq \delta$, the BVP

$$x'' = f(t,x,x'), \qquad x(c_k + \delta) = \alpha(c_k + \delta), \quad x(d_k) = \alpha(d_k)$$

has a solution $z_3 \in C^{(2)}\big[[c_k + \delta,d_k],R\big]$ with $\alpha(t) \leq z_3(t)$. Then, since $z_3(d_k) < z_2(d_k)$ and solutions of BVP's are unique

$$\alpha(t) \leq z_3(t) \leq z_2(t), \qquad t \in [c_k + \delta,d_k].$$

This implies

$$z_3(c_k+\delta) = z_2(c_k+\delta) \qquad \text{and} \qquad z_3'(c_k+\delta) = z_2'(c_k+\delta).$$

Consequently, $u(t)$, defined by

$$u(t) = \begin{cases} z_1(t) & \text{on } [c_k, c_k+\delta], \\ z_3(t) & \text{on } (c_k+\delta, d_k], \end{cases}$$

is of class $C^{(2)}\big[[c_k,d_k],R\big]$ and is a solution on $[c_k,d_k]$ with $u(c_k) = x_k(c_k)$, $u(d_k) = x_k(d_k)$. However, $u(t) \not\equiv x_k(t)$ on $[c_k,d_k]$ and this contradicts the uniqueness of solutions of BVP's. We conclude that $d_k - c_k > \delta$. This being the case, the BVP

$$x'' = f(t,x,x'), \qquad x(c_k+\delta) = \alpha(c_k+\delta), \quad x(c_k+2\delta) = \alpha(c_k+2\delta),$$

has a solution $z_4 \in C^{(2)}\big[[c_k+\delta, c_k+2\delta],R\big]$ with

$$\alpha(t) \le z_4(t) \le z_2(t) \quad \text{on} \quad [c_k+\delta, c_k+2\delta].$$

This again assures us that $z_4(c_k+\delta) = z_2(c_k+\delta)$ and $z_4'(c_k+\delta) = z_2'(c_k+\delta)$. Hence

$$v(t) = \begin{cases} z_1(t) & \text{on } [c_k, c_k+\delta], \\ z_4(t) & \text{on } (c_k+\delta, c_k+2\delta], \end{cases}$$

is such that $v \in C^{(2)}\big[[c_k, c_k+2\delta],R\big]$ and is a solution of BVP

$$x'' = f(t,x,x'), \qquad x(c_k) = \alpha(c_k), \quad x(c_k+2\delta) = \alpha(c_k+2\delta).$$

This solution $v(t)$ has an extension $z_5(t)$ on J. Since $P(k+1)$ is assumed to be false, we must have $z_5(t) \ge \alpha(t)$ on $[c_k+2\delta, d_k]$ and $z_5(d_k) > \alpha(d_k)$. In this way the foregoing arguments can be repeated and the assumption that

$P(k+1)$ is false permits us to work our way across the interval $[c_k, d_k]$ by subintervals of length δ until we obtain a solution $w \in C^{(2)}\big[[c_k, d_k], R\big]$ with $w(t) \geq \alpha(t)$ on $[c_k, d_k]$, $w(c_k) = \alpha(c_k) = x_k(c_k)$ and $w(d_k) = \alpha(d_k) = x_k(d_k)$. Then $w(t) \not\equiv x_k(t)$ on $[c_k, d_k]$ which again is a contradiction to uniqueness of solutions of BVP's.

We therefore conclude that, if $P(k)$ is true, then $P(k+1)$ is true. Thus $P(n)$ is true for all $n \geq 1$, which leads to the contradiction $0 < d - c - (n-1)\delta$ for all $n \geq 1$. Hence α is a subfunction on I and the proof is complete.

EXERCISE 1.8.1. Show that in Theorem 1.8.3 the hypothesis that each initial value problem for (1.8.1) has a solution which extends throughout J can be replaced by Nagumo's condition.

1.9 PROPERTIES OF SUBFUNCTIONS

Before proceeding to outline the Perron's method, it will be necessary to make a more detailed examination of the properties of sub- and superfunctions. This will be undertaken in this section. Again most results will be stated in terms of subfunctions and the obvious analogs for superfunctions will be omitted.

THEOREM 1.9.1. If φ is a subfunction on an interval $I \subset J$, then φ has right and left limits in the extended reals at each point in I^0 and has appropriate one-sided limits at the end points of I.

Proof: Clearly it is sufficient to consider one case. Let $t_0 \in I^0$ and suppose that $\varphi(t_0^-) = \lim_{t \to t_0^-} \varphi(t)$ does not exist in the extended reals. Then there exist real numbers α, β such that

$$\liminf_{t\to t_0^-} \varphi(t) < \alpha < \beta < \limsup_{t\to t_0^-} \varphi(t).$$

Let $\{t_n\}$, $\{s_n\}$ be strictly increasing sequences in I such that $t_n < s_n < t_{n+1}$ for $n \geq 1$, $\lim t_n = \lim s_n = t_0$,

$$\lim \varphi(t_n) = \limsup_{t\to t_0^-} \varphi(t) \quad \text{and} \quad \lim \varphi(s_n) = \liminf_{t\to t_0^-} \varphi(t).$$

Taking $\varepsilon = \frac{1}{4}(\beta - \alpha)$, it follows from Theorem 1.1.2 that there is a $\delta > 0$ such that for any $[t_1,t_2] \subset [t_1,t_0]$ with $t_2 - t_1 \leq \delta$ the BVP

$$\text{(1.9.1)} \qquad x'' = f(t,x,x'), \qquad x(t_1) = x(t_2) = \tfrac{1}{2}(\alpha + \beta),$$

has a solution $x \in C^{(2)}[[t_1,t_2],R]$ satisfying $|x(t) - \frac{1}{2}(\alpha + \beta)| < \varepsilon$ on $[t_1,t_2]$. Let n be a fixed positive integer sufficiently large so that $s_{n+1} - s_n \leq \delta$, $\varphi(s_n) < \alpha$, $\varphi(s_{n+1}) < \alpha$, and $\varphi(t_{n+1}) > \beta$. This implies that there is a solution of (1.9.1) with $[t_1,t_2] = [s_n,s_{n+1}]$ and $|x(t) - \frac{1}{2}(\alpha + \beta)| < \varepsilon$ on $[s_n,s_{n+1}]$. Since φ is a subfunction on I and $\varphi(s_n) < x(s_n)$, $\varphi(s_{n+1}) < x(s_{n+1})$, it must be the case that $\varphi(t_{n+1}) \leq x(t_{n+1})$. However,

$$x(t_{n+1}) < \tfrac{1}{2}(\alpha + \beta) + \varepsilon < \beta < \varphi(t_{n+1}),$$

a contradiction from which we conclude $\varphi(t_0^-)$ exists in the extended reals. The proof is complete.

COROLLARY 1.9.1. If φ is a bounded subfunction on $I \subset J$, then φ has at most a countable number of discontinuities on I. At each $t_0 \in I^0$, $\varphi(t_0) \leq \max[\varphi(t_0^+),\varphi(t_0^-)]$.

Proof: The first assertion is a classical result that follows from the fact that $\varphi(t_0^+)$ and $\varphi(t_0^-)$ exist at each

$t_0 \in I^0$. By Theorem 1.1.2 and the fact that φ is a subfunction, the second assertion readily follows.

We consider next the differentiability of subfunctions. For a function $h(t)$ with a finite right limit $h(t_0^+)$ at t_0, we define

$$Dh(t_0^+) = \lim_{t \to t_0^+} \frac{h(t) - h(t_0^+)}{t - t_0}$$

provided the limit exists. Analogously we define $Dh(t_0^-)$.

THEOREM 1.9.2. If φ is a bounded subfunction on $I \subset J$, then $D\varphi(t_0^+)$, $D\varphi(t_0^-)$ exist in the extended reals for each $t_0 \in I^0$. The appropriate one-sided derivatives exist at finite end points of I.

Proof: As in Theorem 1.9.1 we shall consider only one case. Suppose that $t_0 \in I^0$ and that

$$\liminf_{t \to t_0^+} \frac{\varphi(t) - \varphi(t_0^+)}{t - t_0} < \limsup_{t \to t_0^+} \frac{\varphi(t) - \varphi(t_0)}{t - t_0}.$$

Let m be a real number strictly between these two limits. Then the initial value problem

$$x'' = f(t, x, x'), \qquad x(t_0) = \varphi(t_0^+), \quad x'(t_0) = m$$

has a solution $x \in C^{(2)}[[t_0, t_0 + \delta], R]$ for some $\delta > 0$. As

$$\lim_{t \to t_0^+} \frac{x(t) - x(t_0)}{t - t_0} = m,$$

there exist t_1, t_2, t_3 such that $t_0 < t_1 < t_2 < t_3 < t_0 + \delta$,

$$\frac{\varphi(t) - \varphi(t_0^+)}{t - t_0} < \frac{x(t) - x(t_0)}{t - t_0} \quad \text{at} \quad t = t_1 \quad \text{and} \quad t = t_3,$$

and

$$\frac{\varphi(t_2) - \varphi(t_0^+)}{t_2 - t_0} > \frac{x(t_2) - x(t_0)}{t_2 - t_0} .$$

It follows that $\varphi(t_1) < x(t_1)$, $\varphi(t_3) < x(t_3)$, and $\varphi(t_2) > x(t_2)$. Since φ is a subfunction this is a contradiction and therefore we are done.

The classical results in the theory of real variables give as a consequence the following corollary.

COROLLARY 1.9.2. If φ is a bounded subfunction on $I \subset J$, then φ has a finite derivative almost everywhere on I.

LEMMA 1.9.1. If φ is a subfunction on $I \subset J$ and is bounded above on each compact subinterval of I, then $y(t) = \limsup_{s \to t} \varphi(s)$ is a subfunction on I.

Proof: Let $[t_1, t_2] \subset I$ and let $x(t)$ be a solution of (1.8.1) with $y(t_1) \le x(t_1)$ and $y(t_2) \le x(t_2)$. Then $\varphi(t_i) \le y(t_i) \le x(t_i)$, $i = 1,2$, and, because $\varphi(t)$ is a subfunction, $\varphi(t) \le x(t)$ on $[t_1, t_2]$. We therefore have for each $t_1 < t < t_2$, $y(t) = \limsup_{s \to t} \varphi(s) \le x(t)$. Hence $y(t) \le x(t)$ on $[t_1, t_2]$ and $y(t)$ is a subfunction on I.

We now proceed to discuss some lattice properties of subfunctions.

THEOREM 1.9.3. Assume that the set of subfunctions

$\{\varphi_\alpha : \alpha \in A\}$ on the interval $I \subset J$ is bounded above at each point of I. Then $\varphi_0(t) = \sup_{\alpha \in A} \varphi_\alpha(t)$ is a subfunction on I.

Proof: Suppose that $[t_1, t_2] \subset I$ and $x \in C^{(2)}[[t_1, t_2], R]$ is a solution with $\varphi_0(t) \leq x(t)$ at $t = t_1, t_2$. By the definition of φ_0, we then have $\varphi_\alpha(t) \leq x(t)$ at $t = t_1, t_2$ for each $\alpha \in A$. Since each φ_α is a subfunction on I, it follows that $\varphi_\alpha(t) \leq x(t)$ on $[t_1, t_2]$ for each $\alpha \in A$. This implies $\varphi_0(t) \leq x(t)$ on $[t_1, t_2]$ and $\varphi_0(t)$ is a subfunction on I.

THEOREM 1.9.4. Let φ be a subfunction on an interval $I \subset J$ and φ_1 a subfunction on I_1 with $I_1 = \overline{I}_1 \cap I$. Suppose further that $\varphi_1(t) \leq \varphi(t)$ at finite end points of I_1 which are contained in I. Then φ_2 defined by

$$\varphi_2(t) = \begin{cases} \max[\varphi_1(t), \varphi(t)], & t \in I_1, \\ \varphi(t), & t \in I - I_1, \end{cases}$$

is a subfunction on I.

Proof: By hypothesis $\varphi_2(t) = \varphi(t)$ is a subfunction on $I - I_1$ and, by Theorem 1.9.3, $\varphi_2(t)$ is a subfunction on I_1. Consequently, we need only to show that we have the correct behavior on intervals $[t_1, t_2] \subset I$ which are not contained in either I_1 or $I - I_1$. Again let us consider just one case since the arguments are similar. Assume $t_1 \in I_1$, $t_2 \in I - I_1$, and t_3, $t_1 < t_3 < t_2$, is the right end point of I_1. Suppose that $x \in C^{(2)}[[t_1, t_2], R]$ is a solution on $[t_1, t_2]$ such that $\varphi_2(t_1) \leq x(t_1)$ and $\varphi_2(t_2) \leq x(t_2)$. Then $\varphi(t_1) \leq \varphi_2(t_1) \leq x(t_1)$ and $\varphi(t_2) = \varphi_2(t_2) \leq x(t_2)$, and since φ

is a subfunction on I, we obtain $\varphi(t) \leq x(t)$ on $[t_1,t_2]$. In particular, $\varphi_2(t) = \varphi(t) \leq x(t)$ on $(t_3,t_2]$. Also $\varphi_1(t_3) \leq \varphi(t_3) \leq x(t_3)$ and $\varphi_1(t_1) \leq \varphi_2(t_1) \leq x(t_1)$, hence $\varphi_1(t) \leq x(t)$ on $[t_1,t_3]$ because $\varphi_1(t)$ is a subfunction on I_1. We thus infer that $\varphi_2(t) = \max[\varphi(t), \varphi_1(t)] \leq x(t)$ on $[t_1,t_3]$. Putting these things together we see that $\varphi_2(t) \leq x(t)$ on $[t_1,t_2]$. The other possibilities are dealt with in a similar way and we conclude that $\varphi_2(t)$ is a subfunction on I.

THEOREM 1.9.5. Suppose that

(i) $f \in C[J \times R \times R, R]$, $f(t,x,x')$ is nondecreasing in x for each (t,x') and

(ii) that solutions of initial value problems are unique.

Then, if $\psi(t)$ is an upper solution on $[t_1,t_2] \subset J$ and $\varphi(t)$ is a bounded subfunction on $[t_1,t_2]$ with $\varphi(t_1^+) \leq \psi(t_1)$ and $\varphi(t_2^-) \leq \psi(t_2)$, we have $\varphi(t) \leq \psi(t)$ on (t_1,t_2).

Proof: Suppose that $\varphi(t) > \psi(t)$ at some points of (t_1,t_2). Observe that it is enough to consider the case where φ is upper semicontinuous on (t_1,t_2). To see this, let $\bar{\varphi}(t) = \limsup_{s\to t} \varphi(s)$ on (t_1,t_2). Then by Lemma 1.9.1, $\bar{\varphi}(t)$ is a subfunction on (t_1,t_2). By Corollary 1.9.1, $\bar{\varphi}(t_1^+) = \varphi(t_1^+) \leq \psi(t_1)$ and $\bar{\varphi}(t_2^-$ $\varphi(t_2^-) \leq \psi(t_2)$. Moreover, $\varphi(t) > \psi(t)$ at some points in (t_1,t_2) implies $\bar{\varphi}(t) > \psi(t)$ at some points in (t_1,t_2). As a result we may assume $\varphi(t)$ is upper semicontinuous on (t_1,t_2). This being the case, $\varphi(t) > \psi(t)$ at some points in (t_1,t_2) and $\varphi(t_1^+) \leq \psi(t_1)$, $\varphi(t_2^-) \leq \psi(t_2)$ implies $\varphi(t) - \psi(t)$ has a positive maximum M which is assumed on a compact set $E \subset (t_1,t_2)$. Let $t_0 = \text{lub } E$. Then there exists a $\delta > 0$ and $\varepsilon > 0$ such that

$$[t_0 - \delta,\ t_0 + \delta] \subset (t_1,t_2), \qquad 0 < \varepsilon < \psi(t_0 + \varepsilon) + M - \varphi(t_0 + \delta),$$

and such that the BVP

$$x'' = f(t,x,x'), \qquad y(t_0 - \delta) = \psi(t_0 - \delta) + M,$$
$$y(t_0 + \delta) = \psi(t_0 + \delta) + M - \varepsilon$$

has a solution $x \in C^{(2)}\big[[t_0 - \delta,\ t_0 + \delta], R\big]$. This is assured by Theorem 1.1.2. Since $\psi(t) + M$ is an upper solution, we obtain by Theorem 1.2.3, $x(t) \leq \psi(t) + M$ on $(t_0 - \delta,\ t_0 + \delta)$. However,

$$\varphi(t_0 - \delta) \leq \psi(t_0 - \delta) + M = x(t_0 - \delta)$$

and

$$\varphi(t_0 + \delta) < \psi(t_0 + \delta) + M - \varepsilon = x(t_0 + \delta),$$

which yields $\varphi(t_0) \leq x(t_0)$, because of the fact $\varphi(t)$ is a subfunction on (t_1,t_2). This contradicts $t_0 \in E$ and $\varphi(t_0) = \psi(t_0) + M$. We therefore conclude that $\varphi(t) \leq \psi(t)$ on (t_1,t_2).

EXERCISE 1.9.1. Prove that the assertion of Theorem 1.9.5 remains valid if in place of the assumption "that solutions of initial value problems are unique" we suppose that $f(t,x,x')$ satisfies a Lipschitz condition with respect to x' on each compact subset of $J \times R \times R$.

We shall now discuss properties of bounded functions that are subfunctions and superfunctions simultaneously. We will need a well-known result concerning solutions of initial value problems which we merely state.

LEMMA 1.9.2. If $(t_0,x_0,x_0') \in J \times R \times R$, there exist $\delta > 0$, $M_1 > 0$, and $M_2 > 0$ such that every solution of the initial value problem $x'' = f(t,x,x')$, $x(t_0) = x_0$, $x'(t_0) = x_0'$ is defined on $I_\delta = [t_0 - \delta,\ t_0 + \delta] \cap J$. Moreover,

$|x(t)| \le M_1$, $|x'(t)| \le M_2$ on I_δ for all solutions $x(t)$.

THEOREM 1.9.6. Assume that $f(t,x,x')$ is such that $C^{(2)}$ solutions of boundary value problems, when they exist, are unique. Suppose that $z(t)$ is bounded on each compact subset $I \subset J$ and that $z(t)$ is simultaneously a subfunction and a superfunction on I. Then $z(t)$ is a solution of $x'' = f(t,x,x')$ on an open subset of I the complement of which has measure zero. Furthermore, if $t_0 \in I^0$ is a point of continuity of $z(t)$ at which $z(t)$ does not have a finite derivative, then either

$$Dz(t_0^+) = Dz(t_0^-) = +\infty \qquad \text{or} \qquad Dz(t_0^+) = Dz(t_0^-) = -\infty.$$

If $z(t_0^+) > z(t_0^-)$, $Dz(t_0^+) = Dz(t_0^-) = +\infty$ and $z(t_0^+) < z(t_0^-)$, $Dz(t_0^+) = Dz(t_0^-) = -\infty$.

Proof: By Corollary 1.9.2, $z(t)$ has a finite derivative almost everywhere on I. If $t_0 \in I^0$ is a point at which $z(t)$ has a finite derivative, there is a $\delta > 0$ such that $[t_0 - \delta, t_0 + \delta] \subset I$, $|z(t)| \le |z(t_0)| + 1$ on $[t_0 - \delta, t_0 + \delta]$, and

$$|(z(t_0 + \eta) - z(t_0 - \eta))/2\eta| \le |z'(t_0)| + 1, \qquad 0 < \eta \le \delta.$$

It then follows from Theorem 1.1.2 that there is a $\delta_1 > 0$, $0 < \delta_1 \le \delta$, such that the BVP

$$x'' = f(t,x,x'), \qquad x(t_0 \pm \delta_1) = z(t_0 \pm \delta_1),$$

has a solution $x \in C^{(2)}\big[[t_0 - \delta_1, t_0 + \delta_1],R\big]$. Since $z(t)$ is at the same time a subfunction and a superfunction on I, $z(t) \equiv x(t)$ on $[t_0 - \delta_1, t_0 + \delta_1]$. We conclude that $z(t)$ is a solution of $x'' = f(t,x,x')$ on an open subset of I the complement of which has measure zero.

Next let $t_0 \in I^0$ be a point of continuity of $z(t)$ at which $z(t)$ does not have a finite derivative. By Theorem 1.9.2 $Dz(t_0^+)$ and $Dz(t_0^-)$ both exist in extended reals. If both are finite, then by the same argument as used above there is an interval around t_0 in which $z(t)$ is a solution. This contradicts the assumption that $z(t)$ does not have a finite derivative at t_0. Consequently, at least one of $Dz(t_0^+)$, $Dz(t_0^-)$ is infinite. To be specific, suppose $Dz(t_0^+) = +\infty$ and $Dz(t_0^-) \neq +\infty$. Then there exist numbers $\delta > 0$ and N such that

$$z(t) > w(t) = z(t_0) + N(t - t_0) \qquad \text{on} \quad t_0 - \delta \leq t < t_0.$$

By Theorem 1.1.2, there is a δ_1, $0 < \delta_1 \leq \delta$, such that the BVP

$$x'' = f(t,x,x'), \qquad x(t_0 - \delta_1) = w(t_0 - \delta_1), \quad x(t_0) = z(t_0),$$

has a solution $x_1 \in C^{(2)}\big[[t_0 - \delta_1, t_0], R\big]$. By Lemma 1.9.2, there is a $\delta_2 > 0$ such that $x_1(t)$ can be extended to be a solution on $[t_0 - \delta_1, t_0 + \delta_2]$ and such that all solutions of the initial value problem

$$x'' = f(t,x,x'), \qquad x(t_0) = x_1(t_0), \quad x'(t_0) = x_1'(t_0),$$

exist on $[t_0, t_0 + \delta_2]$ and satisfy $|x(t) - x(t_0)| \leq M(t - t_0)$ on $[t_0, t_0 + \delta_2]$, where $M = |x_1'(t_0)| + 1$. Again applying Theorem 1.1.2 we infer that there is an η, $0 < \eta \leq \min[\delta_1, \delta_2]$, such that for $0 < \delta_3 \leq \eta$ the BVP

$$X'' = f(t,x,x'), \qquad x(t_0 - \delta_3) = x_1(t_0 - \delta_3),$$
$$x(t_0 + \delta_3) = x_1(t_0) + (M + \varepsilon)\delta_3$$

has a solution $x_2 \in C^{(2)}\big[[t_0 - \delta_3, t_0 + \delta_3], R\big]$ for some fixed $\varepsilon > 0$. As $Dz(t_0^+) = +\infty$, we can assume that δ_3 is chosen so that $0 < \delta_3 \leq \eta$ and $x_1(t_0) + (M+\varepsilon)\delta_3 < z(t_0 + \delta_3)$. Then

$$x_1(t_0 + \delta_3) < x_2(t_0 + \delta_3) < z(t_0 + \delta_3)$$

and

$$x_1(t_0 - \delta_3) = x_2(t_0 - \delta_3) \leq z(t_0 - \delta_3).$$

The last inequality follows from the fact that $z(t)$ is a superfunction and $x_1(t_0) = z(t_0)$, $x_1(t_0 - \delta_1) < z(t_0 - \delta_1)$. Since $z(t)$ is a superfunction, we deduce from the above inequalities that $x_2(t) \leq z(t)$ on $[t_0 - \delta_3, t_0 + \delta_3]$. From the same inequalities and the fact that the solutions of BVP's, when they exist, are unique, we also infer that $x_1(t) \leq x_2(t)$ on $[t_0 - \delta_3, t_0 + \delta_3]$. Thus $x_1(t_0) = z(t_0) = x_2(t_0)$ and $x_1'(t_0) = x_2'(t_0)$, hence, $x_2(t)$ is a solution of the initial value problem with the initial conditions $x(t_0) = x_1(t_0)$, $x'(t_0) = x_1'(t_0)$. However,

$$|x_2(t_0 + \delta_3) - x_2(t_0)| = (M+\varepsilon)\delta_3$$

which contradicts the fact that all solutions of this initial value problem satisfy $|x(t) - x_1(t_0)| \leq M(t - t_0)$ on $[t_0, t_0 + \delta_2]$. We are forced to conclude that $Dz(t_0^-) = +\infty$. By similar arguments using the fact that $z(t)$ is also a subfunction, the other statements regarding the behavior of $z(t)$ at a point of continuity can be established.

We consider now the behavior of $z(t)$ at points of discontinuity. If $t_0 \in I^0$ is a point of discontinuity of $z(t)$, then by Theorem 1.9.1, $z(t_0^+)$ and $z(t_0^-)$ both exist and are finite since $z(t)$ is bounded on each compact interval of I. Furthermore, $z(t_0^+) \neq z(t_0^-)$ since, by Corollary 1.9.1,

$$\min\left[z(t_0^+), z(t_0^-)\right] \le z(t_0) \le \max\left[z(t_0^+), z(t_0^-)\right].$$

Assume that $z(t_0^+) > z(t_0^-)$ and that $Dz(t_0^+) \neq +\infty$. Then there is a $\delta > 0$ and an N such that $[t_0, t_0 + \delta] \subset I$ and $z(t) < w(t) = z(t_0^+) + N(t - t_0)$ on $t_0 < t \le t_0 + \delta$. By Theorem 1.1.2 there is a δ_1, $0 < \delta_1 \le \delta$, such that the BVP

$$x'' = f(t,x,x'), \qquad x(t_0) = z(t_0^+), \quad x(t_0 + \delta_1) = w(t_0 + \delta_1),$$

has a solution $x_1 \in C^{(2)}[[t_0, t_0 + \delta_1], R]$. Since

$$x_1(t_0) = z(t_0^+) \ge z(t_0), \qquad x_1(t_0 + \delta_1) = w(t_0 + \delta_1) > z(t_0 + \delta_1),$$

and $z(t)$ is a subfunction, $x_1(t) \ge z(t)$ on $[t_0, t_0 + \delta_1]$. Now proceeding as in the previous paragraph and using the fact that $z(t)$ is a subfunction, we can obtain a solution of the initial value problem

$$x'' = f(t,x,x'), \qquad x(t_0) = x_1(t_0), \quad x'(t_0) = x_1'(t_0),$$

the graph of which is not contained in the sector to the left of t_0 in which such solutions must be. From this contradiction, we conclude that $Dz(t_0^+) = +\infty$. The other assertions concerning derivatives at points of discontinuity of $z(t)$ are dealt with in a similar way. The proof of the theorem is now complete.

1.10 PERRON'S METHOD

Employing the properties of sub- and superfunctions and the existence "in the small" theorem, we consider the existence in the large for the boundary value problems by Perron's method.

DEFINITION 1.10.1. A bounded real-valued function

defined on J is said to be an <u>underfunction</u> with respect to the BVP

$$(1.10.1) \qquad x'' = f(t,x,x'), \qquad x(a) = A, \quad x(b) = B,$$

where $f \in C[J \times R \times R, R]$, in case $\varphi(a) \leq A$, $\varphi(b) \leq B$, and φ is a subfunction on J relative to (1.8.1). An <u>overfunction</u> is defined similarly in an obvious way.

THEOREM 1.10.1. Assume that solutions of BVP's for $x'' = f(t,x,x')$ on subintervals of J, when they exist, are unique. Suppose that there exist both an underfunction φ_0 and an overfunction ψ_0 relative to BVP (1.10.1) such that $\varphi_0(t) \leq \psi_0(t)$ on J. Let Φ be the set of all underfunctions φ such that $\varphi(t) \leq \psi_0(t)$ on J. Then $z(t) = \sup_{\varphi \in \Phi} \varphi(t)$ is simultaneously a sub- and superfunction on J.

<u>Proof</u>: It follows from Theorem 1.9.3 that $z(t)$ is a subfunction on J. Suppose now that $z(t)$ is not a superfunction on J. Then there is a subinterval $[t_1, t_2] \subset J$ and a solution $x \in C^{(2)}[[t_1,t_2],R]$ such that $x(t_1) \leq z(t_1)$, $x(t_2) \leq z(t_2)$, but $x(t) > z(t)$ at some points of (t_1, t_2). We define $z_1(t)$ on J by

$$z_1(t) = \begin{cases} \max[x(t), z(t)] & \text{on } [t_1, t_2] \\ z(t) & \text{on } J - [t_1, t_2]. \end{cases}$$

Then by Theorem 1.9.4, $z_1(t)$ is a subfunction on J and

$$z_1(a) = z(a) \leq A, \qquad z_1(b) = z(b) \leq B.$$

Moreover

$$x(t_1) \leq z(t_1) \leq \psi_0(t_1), \qquad x(t_2) \leq z(t_2) \leq \psi_0(t_2),$$

and ψ_0 a superfunction implies that $x(t) \leq \psi_0(t)$ on $[t_1,t_2]$. Consequently, we have $z_1(t) \leq \psi_0(t)$ on J. Hence, we infer that $z_1 \in \Phi$ and $z_1(t) \leq z(t)$. However, $z_1(t) = x(t) > z(t)$ at some points in (t_1,t_2), a contradiction which proves $z(t)$ is a superfunction on J. This proves the theorem.

DEFINITION 1.10.2. The function $z(t)$ obtained in the preceding theorem depends on the BVP (1.10.1) and on the overfunction ψ_0. We shall say that $z(t)$ is a <u>generalized solution</u> of the BVP (1.10.1) and we shall designate it by $z(t;\psi_0)$.

Notice that since $z(t;\psi_0)$ is a subfunction and a superfunction at the same time, the assertions made in Theorem 1.9.6 apply to $z(t;\psi_0)$. We therefore have to consider the behavior of $z(t;\psi_0)$ at the end points of $J = [a,b]$. We discuss this in the following theorem.

THEOREM 1.10.2. Assume that the hypotheses of Theorem 1.10.1 are satisfied and let $z(t) = z(t;\psi_0)$ be the corresponding generalized solution of (1.10.1). Then $z(a) = A$. If $Dz(a+) \neq +\infty$, $z(a+) \leq z(a)$. If $z(a+) < A$, $Dz(a+) = -\infty$. Hence, if $Dz(a+)$ is finite, $z(a+) = z(a) = A$. Similar assertions are true at $t = b$.

<u>Proof</u>: Observe that if $\varphi(t)$ is a subfunction on J, the function $\varphi^*(t)$ defined by $\varphi^*(t) = \varphi(t)$ on $(a,b]$ and $\varphi^*(a) = \varphi(a) + c$, $c > 0$, is also a subfunction. This, together with the definition of $z(t)$, yields $z(a) = A$.

Assume now that $Dz(a+) \neq +\infty$ and $z(a+) > z(a)$. Then there is a $\delta, 0 < \delta < b - a$, and an N such that

$$z(t) < w(t) = z(a+) + N(t - a), \qquad a < t \leq a + \delta.$$

By Theorem 1.1.2 it follows that for $0 < \varepsilon < z(a+) - z(a)$ and $0 < \delta_1 \leq \delta$ sufficiently small, the BVP

$$x'' = f(t,x,x'), \qquad x(a) = z(a+) - \varepsilon, \quad x(a+\delta_1) = w(a+\delta_1),$$

has a solution $x \in C^{(2)}\big[[a,a+\delta_1],R\big]$. Since $z(t)$ is a subfunction, we have $z(t) \leq x(t)$ on $[a,a+\delta_1]$ which yields

$$z(a+) \leq x(a+) = x(a) = z(a+) - \varepsilon.$$

This contradiction proves that $Dz(a+) \neq +\infty$ implies $z(a+) \leq z(a)$.

Finally, suppose that $z(a+) < A$ and $Dz(a+) \neq -\infty$. Then using an analogous argument as above and the fact that $z(t)$ is also a superfunction on J, we can conclude that $z(a+) < A$ implies $Dz(a+) = -\infty$.

If $Dz(a+)$ is finite, then combining the assertions of this theorem we arrive at $A = z(a) \geq z(a+) \geq A$. The proof is complete.

From the preceding results it is clear that Perron's method of studying the BVP (1.10.1) can be separated into two parts. The first part deals with the problem of establishing the existence of an overfunction ψ_0 and an underfunction φ_0 such that $\varphi_0(t) \leq \psi_0(t)$ on J. The second part consists of finding conditions under which the generalized solution $z(t;\psi_0)$ is of class $C^{(2)}[J,R]$ and is a solution on J. In view of Theorems 1.9.6, 1.10.1, and 1.10.2, accomplishing this is equivalent to showing that $Dz(t+)$ is finite on $[a,b)$ and $Dz(t-)$ is finite on $(a,b]$.

Let us first give sufficient conditions for the existence of under- and overfunctions.

LEMMA 1.10.1. Let $f \in C[J \times R \times R, R]$ and $f(t,x,x')$ be nondecreasing in x for fixed (t,x'). Assume that $f(t,x,x')$

is such that lower and upper solutions of the differential equation $x'' = f(t,x,x')$ are sub- and superfunctions, respectively. Then,

(a) if $u \in C^{(2)}[J,R]$ is a solution of $x'' = f(t,x,x')$ on J, there exist overfunctions and underfunctions relative to any BVP on J;

(b) if there is a $k > 0$ such that

$$|f(t,0,x') - f(t,0,0)| \leq k|x'|, \qquad t \in J, \quad x' \in R, \tag{1.10.2}$$

there exist overfunctions and underfunctions relative to any BVP on J.

Proof: For any sufficiently large $M > 0$, $\psi_0(t) = u(t) + M$ is an overfunction, $\varphi_0(t) = u(t) - M$ is an underfunction, and $\varphi_0(t) \leq \psi_0(t)$ on J. This proves (a).

To prove (b), we let $M = \max|f(t,0,0)|$ on J and consider the solution $w(t)$ of BVP

$$w'' = -kw' - M, \qquad w(a) = 0, \quad w'(b) = 0.$$

Then $w(t) \geq 0$ and $w'(t) \geq 0$ on J. Hence, by (1.10.2),

$$w'' = -kw' - M = -k|w'| - M \leq f(t,0,w') - f(t,0,0) - M.$$

Thus $w'' \leq f(t,0,w') \leq f(t,w,w')$ on J and we assert that, for a given BVP, $\psi_0(t) = w(t) + M$ will be an overfunction provided $M > 0$ is sufficiently large.

Similarly, let $v(t)$ be the solution of BVP

$$v'' = -kv' + M, \qquad v(a) = 0, \quad v'(b) = 0.$$

Then $v(t) \leq 0$ and $v'(t) \leq 0$ on J, from which it follows that

$$v'' = k|v'| + M \geq f(t,0,v') - f(t,0,0) + M \geq f(t,0,v')$$

on J. As a result, $v'' \geq f(t,v,v')$ on J and, for $M > 0$ sufficiently large, $\varphi_0(t) = v(t) - M$ is an underfunction relative to a given BVP. Obviously, we have $\varphi_0(t) \leq \psi_0(t)$ on J. This completes the proof of the lemma.

We are now ready to state our main result.

THEOREM 1.10.3. Assume that $f \in C[J \times R \times R, R]$, $f(t,x,x')$ is nondecreasing in x for each (t,x'), and that solutions of initial value problems are unique. In addition, suppose that there is a $k > 0$ such that $|f(t,0,x') - f(t,0,0)| \leq k|x'|$ for $t \in J$ and $x' \in R$. Then for any BVP on J with an associated overfunction $\psi_0(t)$, the generalized solution $z(t) = z(t;\psi_0) \in C^{(2)}[J^0,R]$ and $z'' = f(t,z,z')$ on J^0.

Proof: By Lemma 1.10,1, we readily obtain the existence of an underfunction φ_0 and an overfunction ψ_0 with respect to BVP (1.10.1) such that $\varphi_0(t) \leq \psi_0(t)$ on J. Consequently, the generalized solution $z(t) = z(t;\psi_0)$ is defined. Moreover, the hypotheses imply that solutions of BVP's when they exist, are unique and, therefore, the conclusions of Theorem 1.9.6 apply to $z(t)$. Thus it suffices to show that $Dz(t_0^+)$, $Dz(t_0^-)$ are finite for every $t_0 \in J^0$. Let $t_0 \in J^0$ and suppose that $z(t_0^+) \geq z(t_0^-)$. The alternative case can be dealt with in a similar way and will not be treated.

We break the discussion into two parts. First assume that $z(t_0^+) \geq 0$. Let $\psi(t)$ be a solution of $\psi'' = -k\psi' - M$, where $M = \max_J |f(t,0,0)|$, with $\psi(t_0) = 0$, $\psi'(t) \geq 0$ on $[t_0,b]$, and $\psi(b) \geq z(b^-)$, which can be computed. Then as in Lemma 1.10.1, $\psi_1(t) = \psi(t) + z(t_0^+)$ is an upper solution on $[t_0,b]$ with $z(t_0^+) \leq \psi_1(t_0)$ and $z(b^-) \leq \psi_1(b)$. It follows from Theorem 1.9.5 that $z(t) \leq \psi_1(t)$ on (t_0,b), which implies that $Dz(t_0^+) \leq \psi_1'(t_0) < \infty$. Applying Theorem 1.9.6, we assert that $z(t)$ is continuous at t_0. Now let $\psi(t)$ be a solution of $\psi'' = k\psi' - M$ on $[a,t_0]$ such that

$\psi(t_0) = 0$, $\psi(a) \geq z(a+)$, and $\psi'(t) \leq 0$ on $[a,t_0]$. Then again by Theorem 1.9.5, we have $z(t) \leq \psi(t) + z(t_0)$, which implies $Dz(t_0^-) \geq \psi'(t_0) > -\infty$. We infer that in this case $z(t)$ has a finite derivative at t_0.

Finally, suppose that $z(t_0^+) < 0$ and let $\varphi_1(t)$ be a solution of $\varphi'' = k\varphi' + M$ on $[a,t_0]$ such that $\varphi(a) \leq z(a+)$, $\varphi(t_0) = 0$, and $\varphi'(t) \geq 0$ on $[a,t_0]$. Then, by Theorem 1.9.5, $z(t) \geq \varphi_1(t) - z(t_0^-)$ on (a,t_0), which assures us that $Dz(t_0^-) \leq \varphi_1'(t_0) < \infty$. It follows from Theorem 1.9.6 that $z(t)$ is continuous at t_0. Similarly we can show that $Dz(t_0^+) > -\infty$. Thus we again conclude that $z(t)$ has a finite derivative at t_0.

Since $z(t)$ has a finite derivative at each point of J^0, it follows from Theorem 1.9.6 that $z \in C^{(2)}[J^0,R]$ and is a solution of $z'' = f(t,z,z')$ on J^0. This completes the proof.

The following exercise is instructive.

EXERCISE 1.10.1. Under the hypotheses of Theorem 1.10.3, show that the BVP $x'' = f(t,x,x')$, $x(a) = x(b) = 0$ has a unique solution $x(t)$ such that

$$|x(t)| < M/k^2 | e^{k(b-a)} - e^{\frac{1}{2}k(b-a)} - \tfrac{1}{2}k(b-a)|,$$

$$|x'(t)| \leq M/k | e^{k(b-a)} - 1|$$

on J, where M,k as in Theorem 1.10.3.

<u>Hints</u>: If $\psi_1(t),\psi_2(t)$ are the solutions of BVP's $\psi_1'' = -k\psi_1 - M$, $\psi_1(a) = 0$, $\psi_1'(b) = 0$, $\psi_2'' = k\psi_2 - M$, $\psi_2(b) = 0$, $\psi_2'(a) = 0$, show that $\psi_0(t) = \min[\psi_1(t),\psi_2(t)]$ is an over-function. Similarly obtain an underfunction $\varphi_0(t) \leq \psi_0(t)$ on J. Using Theorems 1.10.2 and 1.10.3, proceed to show that $\varphi_0(t) \leq z(t;\psi_0) \leq \psi_0(t)$ on J and $z(t;\psi_0)$ is a solution of

the desired BVP which is unique. Compute φ_0, ψ_0 and establish the bound on $|x(t)|$; show that $\max_J \psi_0(t) = -\min_J \varphi_0(t)$. If $t_0 \in J^0$ and $x(t_0) \geq 0$, let $\psi_1(t;t_0)$, $\psi_2(t;t_0)$ be the respective solutions of BVP's

$$\psi'' = -k\psi' - M, \qquad \psi(t_0) = 0, \quad \psi'(b) = 0;$$

$$\psi'' = k\psi' - M, \qquad \psi(t_0) = 0, \quad \psi'(a) = 0.$$

Prove that $\psi_2'(t_0;t_0) \leq x'(t_0) \leq \psi_1'(t_0;t_0)$. Similarly, if $x(t_0) < 0$, obtain analogous relations. Since the functions involved and their derivatives can be computed, deduce the desired bound on $|x'(t)|$.

1.11 MODIFIED VECTOR FUNCTION

Some of the results considered so far for scalar second-order differential equations will now be extended, in the sections that follow, to finite systems of second-order differential equations. To avoid repetition, let us agree on the following: the subscripts i,j range over the integers $1,2,\ldots,n$; vectorial inequalities will be used freely with the understanding that the same inequalities hold between their respective components.

Let us consider a system of second-order differential equations written in the vector form

$$x'' = f(t,x,x'), \tag{1.11.1}$$

where $f \in C[J \times R^n \times R^n, R^n]$. We shall next define modified vector functions in different ways.

DEFINITION 1.11.1. Let $\alpha, \beta \in C^{(1)}[J,R^n]$ with $\alpha(t) \leq \beta(t)$ on J and let a constant vector $c > 0$ be such that $|\alpha'(t)|$, $|\beta'(t)| < c$ on J. Then define

$$F^*(t,x,x') = f(t,x,\bar{x}')$$

where

$$\bar{x}'_j = \begin{cases} c_j & \text{if} \quad x'_j > c_j, \\ x'_j & \text{if} \quad -c_j \le x'_j \le c_j, \\ -c_j & \text{if} \quad x'_j < -c_j, \end{cases}$$

$$F_i(t,x,x') = \begin{cases} F_i^*((t,\bar{x},x') + (x_i - \beta_i(t))/1+x_i^2 & \text{if } x_i > \beta_i(t), \\ F_i^*(t,\bar{x},x') & \text{if } \alpha_i(t) \le x_i \le \beta_i(t) \\ F_i^*((t,\bar{x},x') + (x_i - \alpha_i(t))/1+x_i^2 & \text{if } x_i < \alpha_i(t), \end{cases}$$

where

$$\bar{x}_j = \begin{cases} \beta_j(t) & \text{if} \quad x_j > \beta_j(t), \\ x_j & \text{if} \quad \alpha_j(t) \le x_j \le \beta_j(t), \\ \alpha_j(t) & \text{if} \quad x_j < \alpha_j(t). \end{cases}$$

The function $F(t,x,x')$ will be called the <u>modification</u> of $f(t,x,x')$ relative to the triple α,β,c. It is clear from the definition that $F(t,x,x')$ is continuous and bounded on $J \times R^n \times R^n$. Also note that $|\bar{x}'| \le c$ and $\alpha(t) \le \bar{x} \le \beta(t)$ on J.

DEFINITION 1.11.2. Let $\alpha,\beta \in C[J,R^n]$ with $\alpha(t) \le \beta(t)$ on J and let $\varphi,\psi \in C[\Delta,R^n]$ such that $\varphi(t,x) \le \psi(t,x)$ on Δ, where, as before, $\Delta = [(t,x)\colon \alpha(t) \le x \le \beta(t),\ t \in J]$. Then define

$$F(t,x,x') = f(t,\bar{x},\bar{x}'),$$

where

$$\bar{x}_j = \begin{cases} \beta_j(t) & \text{if} \quad x_j > \beta_j(t), \\ x_j & \text{if} \quad \alpha_j(t) \le x_j \le \beta_j(t), \\ \alpha_j(t) & \text{if} \quad x_j < \alpha_j(t), \end{cases}$$

and

$$\bar{x}_j' = \begin{cases} \psi_j(t,x) & \text{if } x_j' > \psi_j(t,x), \\ x_j' & \text{if } \varphi_j(t,x) \le x_j' \le \psi_j(t,x), \\ \varphi_j(t,x) & \text{if } x_j' < \varphi_j(t,x). \end{cases}$$

We shall call the function $F(t,x,x')$, the modified function of $f(t,x,x')$ associated with α,β,φ, and ψ. Again, it is easily seen that $F(t,x,x')$ is continuous and bounded on $J \times R^n \times R^n$. Furthermore, $\alpha(t) \le \bar{x} \le \beta(t)$ and $\varphi(t,x) \le \bar{x}' \le \psi(t,x)$.

DEFINITION 1.11.3. Let ρ, $N > 0$ be given and let

$$E = [(t,x,x'): t \in J,\ \|x\| < \rho,\ x' \in R^n].$$

Let $\delta \in C[R^+,R^+]$ be such that $\delta(s) = 1$, $0 < \delta(s) < 1$, and $\delta(s) = 0$ accordingly as $0 \le s \le N$, $N < s < 2N$, and $s > 2N$ respectively. Then define

$$F(t,x,x') = \begin{cases} \delta(\|x'\|)f(t,x,x') & \text{on} \quad E \\ (\rho/\|x\|)F(t, \rho x/\|x\|,x') & \text{for} \quad \|x\| > \rho. \end{cases}$$

We shall say that $F(t,x,x')$ is a modified function of $f(t,x,x')$ relative to ρ,δ. Clearly the function $F(t,x,x')$ is continuous and bounded on $J \times R^n \times R^n$.

A result analogous to Theorem 1.3.1 will be proved next.

THEOREM 1.11.1. Assume that $\alpha,\beta \in C[J,R^n] \cap C^{(1)}[J^0,R^n]$ with $\alpha(t) \le \beta(t)$ on J and for $t \in J^0$. Then

$$D_-\alpha_i'(t) \ge f_i\big(t,\alpha(t),a(t,i)\big),$$

$$D^-\beta_i(t) \le f_i\big(t,\beta(t),b(t,i)\big),$$

where

$$a(t,i) = (-c_1,\dots,-c_{i-1},\alpha_i'(t),-c_{i+1},\dots,-c_n),$$

$$b(t,i) = \left(c_1, \ldots, c_{i-1}, \beta_i'(t), c_{i+1}, \ldots, c_n \right),$$

and $|\alpha'(t)|, |\beta'(t)| < c$. Suppose further that $f \in [J \times R^n \times R^n, R^n]$ and $f(t,x,y)$ is quasimonotone nonincreasing in x,y, that is, for fixed (t,x), $f_i(t,x,y)$ is nonincreasing in y_j for $j \neq i$ and for fixed (t,y), $f_i(t,x,y)$ is nonincreasing in x_j for $j \neq i$. Then, if $F(t,x,x')$ is the modification of $f(t,x,x')$ associated with α,β,c according to Definition 1.11.1, and $\alpha(a) \leq \gamma \leq \beta(a)$, $\alpha(b) \leq \delta \leq \beta(b)$, the BVP

$$x'' = F(t,x,x'), \qquad x(a) = \gamma, \quad x(b) = \delta, \tag{1.11.2}$$

has a solution $x \in C^{(2)}[J,R^n]$ satisfying $\alpha(t) \leq x(t) \leq \beta(t)$ on J.

Proof: By Corollary 1.1.1, the BVP (1.11.2) has a solution. Hence we need only to show that $\alpha(t) \leq x(t) \leq \beta(t)$ on J. We will show that $\alpha(t) \leq x(t)$ on J, since the proof of $x(t) \leq \beta(t)$ on J follows similarly.

Suppose it is not true that $\alpha(t) \leq x(t)$ on J. Then there exists an index k and an interval $[t_1,t_2] \subset (a,b)$ such that

$$x_k(t_1) = \alpha_k(t_1), \qquad x_k(t_2) = \alpha_k(t_2),$$

and

$$x_k(t) < \alpha_k(t), \qquad t \in (t_1,t_2).$$

Thus $\alpha_k(t) - x_k(t)$ has a maximum at some $t_0 \in (t_1,t_2)$ and $x_k'(t_0) = \alpha_k'(t_0)$. Hence $|x_k'(t_0)| < c_k$ and consequently, we have

$$\begin{aligned}
D_-\alpha_k'(t_0) - x_k''(t_0) &\geq f_k\big(t_0,\alpha(t_0),a(t_0,k)\big) - F_k\big(t_0,x(t_0),x'(t_0)\big) \\
&\geq f_k\big(t_0,\alpha(t_0),a(t_0,k)\big) - f_k\big(t_0,\bar{x}(t_0),\bar{x}'(t_0)\big) \\
&\quad + \frac{\big(\alpha_k(t_0) - x_k(t_0)\big)}{1 + x_k^2(t_0)} \\
&\geq f_k\big(t_0,\alpha(t_0),a(t_0,k)\big) - f_k\big(t_0,\alpha(t_0),a(t_0,k)\big) \\
&\quad + \frac{\big(\alpha_k(t_0) - x_k(t_0)\big)}{1 + x_k^2(t_0)} \\
&> 0,
\end{aligned}$$

in view of the definition of $F(t,x,x')$ and the quasimonotone nonincreasing character of $f(t,x,x')$ in x,x'. This is impossible at a maximum of $\alpha_k(t) - x_k(t)$ and hence the proof.

The quasimonotone nonincreasing nature of $f(t,x,x')$ in x,x' assumed in Theorem 1.11.1 becomes superfluous if the assumptions concerning α,β are made stronger. This we state in the following exericse.

EXERCISE 1.11.1. Suppose that $\alpha,\beta \in C[J,R^n] \cap C^{(1)}[J^0,R^n]$ with $\alpha(t) \leq \beta(t)$ on J and for $t \in J^0$, $D_-\alpha_i'(t) \geq f_i\big(t,A(t,i),A'(t,i)\big)$, $D^-\beta_i'(t) \leq f_i\big(t,B(t,i),B'(t,i)\big)$, where

$$\begin{aligned}
A(t,i) &= \big(x_1,\ldots,x_{i-1},\alpha_i(t),x_{i+1},\ldots,x_n\big), \\
A'(t,i) &= \big(x_1',\ldots,x_{i-1}',\alpha_i'(t),x_{i+1}',\ldots,x_n'\big), \\
B(t,i) &= \big(x_1,\ldots,x_{i-1},\beta_i(t),x_{i+1},\ldots,x_n\big), \\
B'(t,i) &= \big(x_1',\ldots,x_{i-1}',\beta_i'(t),x_{i+1}',\ldots,x_n'\big),
\end{aligned}$$

provided $\alpha_j(t) \leq x_j \leq \beta_j(t)$ and $-c_j \leq x_j' \leq c_j$ for $j \neq i$, c being any vector satisfying $|\alpha'(t)|, |\beta'(t)| < c$. Show that the BVP (1.11.2) such that $\alpha(a) \leq \gamma \leq \beta(a)$, $\alpha(b) \leq \delta \leq \beta(b)$ has a solution $x \in C^{(2)}[J,R^n]$ satisfying

$\alpha(t) \leq x(t) \leq \beta(t)$ on J.

REMARK 1.11.1. From the foregoing discussion, a natural question that arises is whether a result analogous to Theorem 1.11.1 remains true if α, β are defined as lower and upper solutions, in a natural way, relative to the vector differential equation $x'' = f(t,x,x')$. This can be done by defining a modified function which has a jump discontinuity along the upper and lower solutions.

EXERCISE 1.11.2. Let $x \in C^{(2)}[J,R^n]$ be a solution of $x'' = F(t,x,x')$, where F is the modified function of $f \in C[J \times R^n \times R^n, R^n]$ relative to ρ, δ as in Definition 1.11.3, such that $\|x(a)\|, \|x(b)\| \leq \rho$. Suppose that there exists a $V \in C[D,R^+]$ such that $V(t,x,y)$ is locally Lipschitzian in (x,y), $V(t,x,y) = 0$ if $\|x\| = \rho$, $V(t,x,y) > 0$ if $\|x\| > \rho$ and $D^+V_F(t,x,x') \geq 0$ in the interior of D, D being the set $D = \big[[a,b] \times [x\colon \|x\| \geq \rho] \times R^n\big]$. Show that $\|x(t)\| \leq \rho$ on $[a,b]$.

1.12 NAGUMO'S CONDITION (CONTINUED)

In remark 1.4.1, we noted that Nagumo's condition is not sufficient to conclude $\|x'(t)\|$ is bounded, if $f \in C[J \times R^n \times R^n, R^n]$. Nonetheless, we can make conclusions on the arc length of the curve $\|x(t)\|$, that is $\int \|x'(s)\|\, ds$, as the next lemma shows.

LEMMA 1.12.1. Assume that $f \in C[J \times R^n \times R^n, R^n]$ and $h \in C[R^+, (0,\infty)]$. Suppose that

$$\|f(t,x,x')\| \leq h\big(\|x'\|\big), \qquad (t,x,x') \in J \times R^n \times R^n, \tag{1.12.1}$$

where $\int_0^\infty s\, ds/h(s) = \infty$. Then, there exists a $\gamma\colon R^+ \to R^+$

such that

$$\|x'(t)\| \leq \gamma(\theta_1), \qquad t \in J,$$

where $x \in C^{(2)}[J,R^n]$ is any solution of (1.11.1) and $\theta_1 = \int_a^b \|x'(s)\|\, ds$.

Proof: Since $\int_0^\infty s\, ds/h(s) = \infty$, there exists a function $\gamma\colon R^+ \to R^+$ satisfying

$$\int_{\theta/(b-a)}^{\gamma(\theta)} \frac{s\, ds}{h(s)} = \theta, \qquad \theta \in R^+. \tag{1.12.2}$$

From assumption (1.12.1), we obtain

$$\|x'(s)\| \geq \|x'(s)\| \frac{\|x''(s)\|}{h(\|x'(s)\|)} \geq \left| \frac{x'(s) \cdot x''(s)}{h(\|x'(s)\|)} \right| .$$

Hence if $a \leq v,\ w \leq b$,

$$\theta_1 \geq \left| \int_v^w \frac{x'(s) \cdot x''(s)}{h(\|x'(s)\|)}\, ds \right| = \left| \int_{\|x'(v)\|}^{\|x'(w)\|} \frac{s\, ds}{h(s)} \right| \tag{1.12.3}$$

by the change of variables $\|x'(s)\| \to s$; the two integrals are equal (and zero) for $v = w$, and the derivatives with respect to w exist almost everywhere and are equal. It follows that the integrals are equal since $\|x'(w)\|$ is absolutely continuous. Choose v and w such that $\|x'(v)\| = \theta_1/(b-a)$ (which is possible by the mean value theorem) and $\|x'(w)\| = \max \|x'(s)\|$ on $a \leq s \leq b$. The desired result then follows from (1.12.2) and (1.12.3) and the proof is complete.

It is, however, possible to extend Theorem 1.4.1 to deal with the vector situation.

THEOREM 1.12.1. Suppose that the following assumptions hold:

(i) $f \in C[J \times R^n \times R^n, R^n]$, $\alpha, \beta \in C[J, R^n]$ such that $\alpha(t) \leq \beta(t)$ on J and for $t \in J$, $\alpha(t) \leq x \leq \beta(t)$, $x' \in R^n$,

$$|f_i(t,x,x')| \leq h_i(|x_i'|), \qquad i = 1,\ldots,n, \tag{1.12.4}$$

(ii) $h_i \in C[R^+, (0,\infty)]$ and satisfies

$$\int_{\lambda}^{\infty} \frac{s\,ds}{h_i(s)} > \max_J \beta_i(t) - \min_J \alpha_i(t), \tag{1.12.5}$$

where

$$\lambda(b - a) = \max\left[|\alpha(a) - \beta(a)|, |\alpha(b) - \beta(a)|\right].$$

Then, for any solution $x \in C^{(2)}[J,R^n]$ of (1.11.1) with $\alpha(t) \leq x(t) \leq \beta(t)$ on J there exists a constant vector $N \geq 0$, depending only on α, β, h, such that

$$|x'(t)| \leq N \quad \text{on } J. \tag{1.12.6}$$

<u>Proof</u>: In view of (1.12.5), there exists a constant vector $N > 0$ such that

$$\int_{\lambda}^{N_i} \frac{s\,ds}{h_i(s)} > \max_J \beta_i(t) - \min_J \alpha_i(t).$$

If (1.12.6) is not true, we may assume, without loss of generality, that there is an index k such that for some $t \in J$, $|x_k(t)| > N_k$. However, by the mean value theorem, we have

$$x_k(b) - x_k(a) = x_k'(t_0)(b - a),$$

where $t_0 = t_0(k) \in (a,b)$ and consequently, it follows that $|x_k'(t_0)| \leq \lambda_k < N_k$. The remainder of the proof is almost identical to the proof of Theorem 1.4.1 and we therefore leave it to the reader.

The next result is an extension of Theorem 1.4.3 to the present situation.

THEOREM 1.12.2. Suppose that

(i) $\alpha, \beta \in C^{(1)}[J, R^n]$ with $\alpha(t) \le \beta(t)$ on J;

(ii) $\varphi, \psi \in C^{(1)}[\Delta, R^n]$ with $\varphi(t,x) \le \psi(t,x)$ and for $(t,x) \in \Delta$, where $\Delta = [(t,x): \alpha(t) \le x \le \beta(t),\ t \in J]$,

$$\varphi_t(t,x) + \varphi_x(t,x) \cdot \varphi(t,x) \le f\big(t,x,\varphi(t,x)\big),$$

$$\psi_t(t,x) + \psi_x(t,x) \cdot \psi(t,x) \ge f\big(t,x,\psi(t,x)\big);$$

(iii) $f \in C[J \times R^n \times R^n, R^n]$ and $\varphi\big(a,x(a)\big) \le x'(a) \le \psi\big(a,x(a)\big)$.

Then, for any solution $x \in C^{(2)}[J, R^n]$ of (1.11.1) such that $\alpha(t) \le x(t) \le \beta(t)$ on J, we have

$$\varphi\big(t,x(t)\big) \le x'(t) \le \psi\big(t,x(t)\big), \qquad t \in J. \tag{1.12.7}$$

<u>Proof</u>: Let $x(t)$ be any solution of (1.11.1) satisfying $\varphi\big(a,x(a)\big) \le x'(a) \le \psi\big(a,x(a)\big)$.

We shall only prove that $\varphi\big(t,x(t)\big) \le x'(t)$ on J, since the proof of the other case in (1.12.7) is similar. Assume that this is not true. Then there exists an index k and a $t_0 \in (a,b]$ such that $x_k'(t_0) < \varphi_k\big(t_0, x(t_0)\big)$. Set

$$z_k(t) = \Big[\varphi_k\big(t,x(t)\big) - x_k'(t)\Big] \exp\Big[\int_t^{t_0} H_k(s)\, ds\Big],$$

where

$$H_k(t) = \varphi_{kx}\big(t,x(t)\big) \cdot \Big[\varphi\big(t,x(t)\big) - x'(t)\Big] \Big/ \Big[\varphi_k\big(t,x(t)\big) - x_k'(t)\Big]$$

on $(t_1, t_0]$. Such an interval $(t_1, t_0]$ exists where $\varphi_k\big(t,x(t)\big) - x_k'(t) > 0$. It then follows that

$$\exp\left[\int_t^{t_0} H_k(s)\,ds\right] z_k'(t) = \varphi_{kt}\big(t,x(t)\big)$$

$$+\ \varphi_{kx}\big(t,x(t)\big)\cdot\varphi\big(t,x(t)\big) - x_k''(t)$$

on $(t_1,t_0]$. This together with assumption (ii) yields $z_k'(t) \le 0$ on $(t_1,t_0]$ and hence $z_k(t)$ is nonincreasing as t increases. Thus $z_k(t_1) > 0$. This argument can be extended to conclude that $z_k(a) > 0$ which implies that $\varphi_k\big(a,x(a)\big) > x_k'(a)$, a contradiction to assumption (iii). This proves $\varphi\big(t,x(t)\big) \le x'(t)$ on J and the proof is complete.

Another result that may be useful in some situations is the following.

THEOREM 1.12.3. Suppose that

(i) $\alpha,\beta \in C^{(1)}[J,R^n]$ with $\alpha(t) \le \beta(t)$ on J and $\alpha(a) = \beta(a)$;

(ii) $h \in C[J\times R_+^n, R_+^n]$, $h(t,u)$ is quasimonotone nondecreasing in u, that is, $h_i(t,u)$ is nondecreasing in u_j, $j \ne i$ for each $t \in J$, and the maximal solution of

$$u' = h(t,u), \qquad u(a) = \lambda, \tag{1.12.8}$$

where $\lambda = \max\big[|\alpha'(a)|, |\beta'(a)|\big]$, exists on $[a,b]$;

(iii) for $t \in J$, $\alpha(t) \le x \le \beta(t)$,

$$|f(t,x,x')| \le h\big(t,|x'|\big).$$

Then for any solution $x \in C^{(2)}[J,R^n]$ of (1.11.1) with $\alpha(t) \le x(t) \le \beta(t)$ on J, there exists a constant vector $N > 0$, depending only on α,β,h such that

$$|x'(t)| \le N, \qquad t \in J.$$

<u>Proof</u>: Let $x(t)$ be any solution of (1.11.1) with $\alpha(t) \le x(t) \le \beta(t)$ on J. This, in view of the assumption that $\alpha(a) = \beta(a)$, implies that $\alpha'(a) \le x'(a) \le \beta'(a)$. Define $m(t) = |x'(t)|$. Then $m(a) \le \lambda$ and, by (iii),

$$D^+m(t) \le |x''(t)| = |f(t,x(t),x'(t))| \le h(t,m(t)), \qquad t \in J.$$

Consequently, by the theory of differential inequalities, we have

$$m(t) \le r(t,a,\lambda), \qquad t \in J,$$

where $r(t,a,\lambda)$ is the maximal solution of (1.12.8) which exists on $[a,b]$. Let $N > 0$ be a vector such that $r(t,a,\lambda) \le N$, $t \in J$. Then the stated conclusion follows immediately.

One can also deduce a bound on $\|x'(t)\|$ by employing a Lyapunov-like method. This is the content of the next two theorems.

THEOREM 1.12.4. Assume that

(i) $V \in C[D_1,R^+]$, $V(t,x,y)$ is locally Lipschitzian in x,y, where $D_1 = J \times [x: \|x\| \le \rho] \times R^n$, $b_1,b_2 \in C[R^+,R^+]$, $b_1(u) \to \infty$ as $u \to \infty$ and

$$b_1(\|y\|) \le V(t,x,y) \le b_2(\|y\|), \qquad (t,x,y) \in D_1;$$

(ii) $$D^+V(t,x,x') = \limsup_{h\to 0^+} (1/h)\left[V(t+h,x+hx', x'+hf(t,x,x')) - V(t,x,x')\right] \le g(t,V(t,x,x')) \text{ for } (t,x,x') \in \operatorname{Int} D_1;$$

(iii) $g \in C[J \times R^+,R]$ and the maximal solution $r(t,a,\lambda)$ of

$$u' = g(t,u), \qquad u(a) = \lambda \ge 0,$$

exists on $[a,b]$.

Then, for any solution $x \in C^{(2)}[J,R^n]$ of (1.11.1) such that $\|x(t)\| \le \rho$ and $\|x'(a)\| \le \rho_0$, there exists an $N > 0$ depending only on ρ, ρ_0, g such that

$$\|x'(t)\| \le N \quad \text{on } J.$$

Proof: Let $x(t)$ be any solution of (1.11.1) such that $\|x(t)\| \le \rho$ and $\|x'(a)\| \le \rho_0$. Define $m(t) = V(t,x(t),x'(t))$. Then by (i), we obtain $m(a) \le b_2(\|x'(a)\|) \le b_2(\rho_0)$. Choose $\lambda = b_2(\rho_0)$. Also, because of the assumptions on V, it follows that

$$D^+m(t) \le g(t,m(t)), \qquad t \in (a,b]$$

and consequently, by the theory of differential inequalities, we obtain

$$m(t) \le r(t,a,\lambda), \qquad t \in J.$$

By (i), this in turn yields

$$b_1(\|x'(t)\|) \le V(t,x(t),x'(t)) = m(t) \le r(t,a,\lambda), \qquad t \in J.$$

Since $r(t,a,\lambda)$ exists on $[a,b]$, there is a $M > 0$ such that $r(t,a,\lambda) \le M$ on $[a,b]$. Furthermore, as $b_1(u) \to \infty$ as $u \to \infty$, there exists an $N > 0$ such that $M \le b(N)$. These considerations imply that $\|x'(t)\| \le N$ on J. Clearly N depends only on ρ, ρ_0, and g. This completes the proof.

THEOREM 1.12.5. Suppose that there exists a $V \in C[E,R^+]$, where $E = [[a,b] \times [x\colon \|x\| \le \rho] \times R^n]$, such that $V(t,x,y)$ is locally Lipschitzian in (x,y), $V(a,x(a),x'(a)) = 0$, $V(t,x,x') \ge (t-a)b(\|x'\|)$, where $b \in C[R^+,R^+]$ with $b(r) \to \infty$ as $r \to \infty$, and $D^+V_f(t,x,x') \le L$ in the interior of E. Then for any solution $x \in C^{(2)}[[a,b],R^n]$ of (1.11.1) such that $\|x(t)\| \le \rho$ on $[a,b]$, there exists an $N > 0$ satisfying $\|x'(t)\| \le N$ on $[a,b]$.

Proof: By assumption on $b(r)$, there exists an $N > 0$ such that $b(r) > L$ if $r \geq N$. Also, we have by the theory of differential inequalities

$$V\big(t,x(t),x'(t)\big) \leq V\big(a,x(a),x'(a)\big) + L(t-a), \qquad t \in [a,b].$$

If we now suppose that for some $t_2 \in (a,b]$, $\|x(t_2)\| \geq N$, then we are lead to the contradiction

$$0 < (t_2 - a)\big[b(\|x'(t_2)\|) - L\big] \leq V\big(a,x(a),x'(a)\big) = 0.$$

Hence the conclusion of the theorem is true.

1.13 EXISTENCE IN THE LARGE FOR SYSTEMS

We shall first prove an existence result analogous to Theorem 1.5.1 for the BVP

$$(1.13.1) \qquad x'' = f(t,x,x'), \qquad x(a) = A, \quad x(b) = B,$$

where $f \in C[J \times R^n \times R^n, R^n]$, $A,B \in R^n$.

THEOREM 1.13.1. Suppose that the following conditions hold:

(i) $\alpha,\beta \in C[J,R^n] \cap C^{(1)}[J^0,R^n]$ such that $\alpha(t) \leq \beta(t)$ on J and for $t \in J^0$,

$$D_-\alpha_i'(t) \geq f_i\big(t,\alpha(t),a(t,i)\big), \qquad D^-\beta'(t) \leq f_i\big(t,\beta(t),b(t,i)\big),$$

where

$$a(t,i) = \big(-c_1,\ldots,-c_{i-1},\alpha_i'(t),-c_{i+1},\ldots,-c_n\big),$$

$$b(t,i) = \big(c_1,\ldots,c_{i-1},\beta_i'(t),c_{i+1},\ldots,c_n\big),$$

and $|\alpha'(t)|, |\beta'(t)| < c$;

(ii) $f(t,x,y)$ is quasimonotone nonincreasing in x,y as in the statement of Theorem 1.11.1 and for $t \in J$, $\alpha(t) \leq x \leq \beta(t)$ and $x' \in R^n$,

$$|f_i(t,x,x')| \le h_i(|x_i'|);$$

(iii) $h_i \in C[R^+,(0,\infty)]$ and h_i satisfies (1.12.5).

Then for any $\alpha(a) \le A \le \beta(a)$, $\alpha(b) \le B \le \beta(b)$, the BVP (1.13.1) has a solution $x \in C^{(2)}[J,R^n]$ such that $\alpha(t) \le x(t) \le \beta(t)$ on J. Moreover there exists a constant vector $N_0 > 0$ such that $|x'(t)| \le N_0$ on J.

Proof: By hypothesis (ii), we can choose a vector $N > 0$, depending only on α,β,h, as in Theorem 1.12.1. Let N_0 be such that

$$N_0 > \max\Big[N, \max_J |\alpha'(t)|, \max_J |\beta'(t)|\Big].$$

Choose $c > N_0$. Let $F(t,x,x')$ be the modification function as defined in Definition 1.11.1 relative to the triple α,β,c. Then by Theorem 1.11.1 it follows that the BVP

$$x'' = F(t,x,x'), \qquad x(a) = A, \quad x(b) = B,$$

has a solution $x \in C^{(2)}[J,R^n]$ with $\alpha(t) \le x(t) \le \beta(t)$ on J. Hence we have

$$x'' = F(t,x,x') = f(t,x,\bar{x}')$$

in view of the definition of F. Consequently, we obtain, by (ii),

$$|f_i(t,x,\bar{x}')| \le h_i(|\bar{x}_i'|)$$

whenever $\alpha(t) \le x \le \beta(t)$, $t \in J$. Recalling that $|\bar{x}_i'| < c_i$ and proceeding as in the proof of Theorem 1.12.1, it is easily shown that $|x'(t)| \le N_0$ on J_0. This then implies that $F(t,x,x') = f(t,x,x')$ and therefore $x(t)$ is actually a solution of the desired BVP (1.13.1) completing the proof.

Based on Theorem 1.12.2, it is possible to exhibit

another existence theorem which is an extension of Theorem 1.5.5.

THEOREM 1.13.2. Assume that

(i) $\alpha,\beta \in C[J,R^n] \cap C^{(1)}[J^0,R^n]$ with $\alpha(t) \leq \beta(t)$ on J, $\alpha(a) = \beta(a)$ and

$$D_-\alpha_i'(t) \geq f_i\big(t,\alpha(t),R(t,i)\big), \qquad D^-\beta_i'(t) \leq f_i\big(t,\beta(t),S(t,i)\big)$$

on J_0, where

$$R(t,i) = \big[\varphi_1\big(t,\alpha(t)\big),\ldots,\varphi_{i-1}\big(t,\alpha(t)\big),\alpha_i'(t),$$
$$\varphi_{i+1}\big(t,\alpha(t)\big),\ldots,\varphi_n\big(t,\alpha(t)\big)\big],$$
$$S(t,i) = \big[\psi_1\big(t,\beta(t)\big),\ldots,\psi_{i-1}\big(t,\beta(t)\big),\beta_i'(t),$$
$$\psi_{i+1}\big(t,\beta(t)\big),\ldots,\psi_n\big(t,\beta(t)\big)\big],$$

and $\varphi,\psi \in C^{(1)}[\Delta,R^n]$ such that $\varphi(t,x) \leq \psi(t,x)$ on Δ, where

$$\Delta = [(t,x)\colon \alpha(t) \leq x \leq \beta(t), t \in J];$$

(ii) $f(t,x,y)$ is quasimonotone nonincreasing in (x,y) as in the statement of Theorem 1.11.1;

(iii) for $t \in J$,

$$\varphi\big(t,\alpha(t)\big) \leq \alpha'(t) \leq \psi\big(t,\alpha(t)\big),$$
$$\varphi\big(t,\beta(t)\big) \leq \beta'(t) \leq \psi\big(t,\beta(t)\big),$$

and on Δ,

$$\varphi_t(t,x) + \varphi_x(t,x)\cdot\varphi(t,x) \leq f\big(t,x,\varphi(t,x)\big),$$
$$\psi_t(t,x) + \psi_x(t,x)\cdot\psi(t,x) \geq f\big(t,x,\psi(t,x)\big).$$

Then for any $\alpha(a) = A = \beta(a)$, $\alpha(b) \leq B \leq \beta(b)$, the BVP (1.13.1) has a solution $x \in C^{(2)}[J,R^n]$ such that

$$\big(t,x(t),x'(t)\big) \in \Omega,$$

Ω being the set

$$\Omega = [(t,x,x): (t,x) \in \Delta \text{ and } \varphi(t,x) \leq x' \leq \psi(t,x)].$$

Proof: Let $F(t,x,x')$ be a modified function of $f(t,x,x')$ associated with $\alpha,\beta,\varphi,\psi$ as in Definition 1.11.2 which is continuous and bounded on $J \times R^n \times R^n$ and therefore satisfies hypotheses (i) and (iii) of Theorem 1.13.1. Furthermore, in view of the assumptions relative to α',β' in (iii), it is easily verified that hypothesis (i) is true with respect to $F(t,x,x')$ in place of $f(t,x,x')$. This implies that condition (i) of Theorem 1.13.1 holds with φ,ψ instead of $-c,c$ respectively. As a result, it follows, by Theorem 1.13.1 that there exists a solution $x \in C^{(2)}[J,R^n]$ to the modified BVP

$$x'' = F(t,x,x'), \qquad x(a) = A, \quad x(b) = B,$$

such that $\alpha(t) \leq x(t) \leq \beta(t)$ on J, where $\alpha(a) = A = \beta(a)$, $\alpha(b) \leq B \leq \beta(b)$. We now apply Theorem 1.12.2 to assert that $x(t)$ is actually a solution of the BVP (1.13.1) satisfying $\big(t,x(t),x'(t)\big) \in \Omega$, $t \in J$. For this, it is necessary to check that all the hypotheses of Theorem 1.12.2 are satisfied which we leave to the reader since it is similar to the proof of Theorem 1.5.5. This completes the proof.

EXERCISE 1.13.1. Under the assumptions of Exercises 1.11.2 and Theorem 1.12.5, show that the BVP $x'' = f(t,x,x')$, $x(a) = A$, $x(b) = B$ has a solution.

1.14 FURTHER RESULTS FOR SYSTEMS

Let us consider the differential system

$$x'' = f(t,x,x'), \tag{1.14.1}$$

where $f \in C\big[[0,1] \times R^n \times R^n, R^n\big]$, subject to the boundary conditions

$$x(0) - A_0 x'(0) = 0, \tag{1.14.2}$$

$$x(1) + A_1 x'(1) = 0, \tag{1.14.3}$$

A_0, A_1 being $d \times d$ matrices.

Here we shall study the existence of solutions of the BVP (1.14.1) - (1.14.3) in a more general set up than the earlier results. We employ Lyapunov-like functions and the theory of differential inequalities in a slightly different way which throws much light on the underlying ideas. The technique is, of course, the modified function approach. The following lemma is very useful in our discussion.

LEMMA 1.14.1. Assume that

(i) $u \in C^{(2)}[[0,1],R^+]$, $g \in C[[0,1] \times R^+ \times R, R^-]$, $g(t,u,v)$ is nonincreasing in u for each (t,v) and

$$u'' \geq g(t,u,u'); \tag{1.14.4}$$

(ii) $u'(0) \geq 0$, $u'(1) \leq 0$, and $u(0) \leq \alpha u'(0)$ for some $\alpha \geq 0$;

(iii) $G \in C[[0,1] \times R^+, R]$ and there exists an $L > 0$ such that for $u \geq L$, $t \in [0,1]$,

$$(1/u)\, g(t,u,v) - (v/u)^2 \geq G(t, v/u), \tag{1.14.5}$$

and for any $\tau \in (0,1]$, the left maximal solution $r(t,\tau,0)$ of

$$z' = G(t,z), \qquad z(\tau) = 0, \tag{1.14.6}$$

satisfies the estimate $r(t,\tau,0) < \alpha_0$, $t \in [0,\tau]$, where $\alpha_0 = \min(\frac{1}{2}, 1/\alpha)$;

(iv) the left maximal solution $r(t,1,0)$ and the right minimal solution $\rho(t,0,0)$ of

$$v' = g(t,2L,v) \tag{1.14.7}$$

exists on $[0,1]$.

Then there exists a $B_0 > 0$ such that

$$(1.14.8) \quad u(t) \leq B_0 \quad \text{and} \quad |u'(t)| \leq B_0, \quad 0 \leq t \leq 1.$$

Proof: Assume that the maximum of $u(t)$ occurs at a point t_1. From conditions $u'(0) \geq 0$ and $u'(1) \leq 0$, it follows that $u'(t_1) = 0$. Clearly $t_1 > 0$, for otherwise we would have $u(t_1) \leq \alpha u'(t_1) = 0$ and consequently $u(t) \equiv 0$.

We shall show that $u(t) \leq 2L$, $0 \leq t \leq 1$. If not, let $u(t_1) > 2L$. Define $t_0 = 0$, if $u(t) > L$ for $t \in [0,t_1]$. If not, define

$$t_0 = \sup\left[t \in [0,t_1]: u'(t) \geq \tfrac{1}{2}u(t)\right].$$

Since $u(t_1) > 2L$, by the mean value theorem, t_0 is well defined. It is then easily seen that

$$(1.14.9) \quad u'(t_0) \geq \alpha_0 u(t_0), \quad L \leq u(t), \quad t \in [t_0,t_1].$$

Setting $z(t) = u'(t)/u(t)$ for $t \in [t_0,t_1]$ and using assumption (1.14.5) we readily obtain

$$z'(t) \geq G\big(t,z(t)\big), \quad t \in [t_0,t_1].$$

Notice that $z(t_0) = 0$ and $z(t_0) \geq \alpha_0 > 0$. By the theory of differential inequalities, we then infer that

$$z(t) \leq r\big(t,t_1,z(t_1)\big), \quad t \in [t_0,t_1],$$

where $r\big(t,t_1,z(t_1)\big)$ is the left maximal solution of (1.14.6) with $\tau = t_1$. Since $z(t_1) = 0$, we see that $r(t,t_1,0) < \alpha_0$ on $[t_0,t_1]$, and as a result, we are lead to the contradiction

$$\alpha_0 \leq z(t_0) \leq r(t_0,t_1,0) < \alpha_0.$$

This proves that $u(t) \leq 2L$ on $[0,1]$.

Using this inequality and the nonincreasing nature of $g(t,u,v)$ in u, we obtain

$$u'' \geq g(t,2L,u').$$

Again, using the fact that $u'(0) \geq 0$, $u'(1) \leq 0$, and the theory of differential inequalities, we have

$$u'(t) \leq r(t,1,0), \qquad 0 \leq t \leq 1,$$

$$u'(t) > \rho(t,0,0), \qquad 0 \leq t \leq 1,$$

where $r(t,1,0)$, $\rho(t,0,0)$ are, respectively, left maximal and right minimal solutions of (1.14.7) which are assumed to exist on $[0,1]$. Thus, we can find a $B > 0$ such that $|u'(t)| \leq B$, for $0 \leq t \leq 1$, where

$$B = \max\left[\left|\max_{0\leq t\leq 1} r(t,1,0)\right|, \left|\min_{0\leq t\leq 1} \rho(t,0,0)\right|\right].$$

The conclusion of the lemma now follows by choosing $B_0 = \max[2L,B]$. The proof is complete.

COROLLARY 1.14.1. The functions $g(t,u,v) = -k[1+(2u)^{\frac{1}{2}} + |v|]$, $k > 0$, $G(t,z) = -(a+k|z|+z^2)$, $V(t,x) = \frac{1}{2}\|x\|^2$ are admissible in Lemma 1.14.1, where $La = k\left(1 + (1/2h) + hk\right)$, $h = \frac{1}{2}\alpha_0 e^{-(k+1)}$, provided $a < \alpha_0 k e^{-(k+1)}$.

Our aim is to prove the following result.

THEOREM 1.14.1. Assume that

(a) $f \in C\left[[0,1]\times R^n \times R^n, R^n\right]$ and A_0, A_1 are positive definite or identically zero;

(b) $V \in C^{(2)}\left[[0,1]\times R^n, R^+\right]$, $V(t,x)$ is positive definite, $g \in C\left[[0,1]\times R^+ \times R, R^-\right]$, $g(t,u,v)$ is nonincreasing in u for each (t,v); and for $(t,x,x') \in [0,1]\times R^n \times R^n$,

$$V''_f(t,x) \geq g\left(t,V(t,x),V'(t,x)\right) + \sigma\|f(t,x,x')\|, \quad \sigma > 0, \tag{1.14.10}$$

(1.14.11) $U(t,x,x') \equiv V_{tt}(t,x) + 2V_{tx}(t,x)\cdot x' + V_{xx}(t,x)x'\cdot x' \geq 0,$

where

$$V'(t,x) = V_t(t,x) + V_x(t,x)\cdot x',$$

$$V_f''(t,x) = V_{tt}(t,x) + 2V_{tx}(t,x)\cdot x' + V_{xx}(t,x)x'\cdot x' + V_x(t,x)\cdot f(t,x,x');$$

(c) The boundary conditions (1.14.2), (1.14.3) imply, for some $\alpha \geq 0$, that

(1.14.12) $V'(0,x(0)) \geq 0, \quad V'(1,x(1)) \leq 0$ and $V(0,x(0)) \leq \alpha V'(0,x(0));$

(d) $G \in C[[0,1]R^+,R]$ and there exists an $L > 0$ such that for $u \geq L$, $t \in [0,1]$

$$(1/u)g(t,u,v) - (v/u)^2 \geq G(t,v/u)$$

and for any $\tau \in (0,1]$, the left maximal solution $r(t,\tau,0)$ of

$$z' = G(t,z), \qquad z(\tau) = 0$$

satisfies the inequality $r(t,\tau,0) < \alpha_0$, $t \in [0,\tau]$, where $\alpha_0 = \min(\frac{1}{2},1/\alpha)$,

(e) the left maximal solution $r(t,1,0)$ and the right minimal solution $\rho(t,0,0)$ of

$$v' = g(t,2L,v)$$

exist on $[0,1]$.

Then there exists a solution $x \in C^{(2)}[[0,1],R^d]$ of the boundary value problem (1.14.1) - (1.14.3).

<u>Proof</u>: Define the function $\delta(u,v)$ on $R^+ \times R^+$ as

$$(1.14.13)\quad \delta(u,v) = \begin{cases} 1, & 0 \le u, \quad v \le B, \\ (1 + B - v), & 0 \le u \le B \le v \le B+1, \\ (1 + B - u), & 0 \le v \le B \le u \le B+1, \\ (1 + B - u)(1 + B - v), & B \le u, \quad v \le B+1, \\ 0, & \text{otherwise}. \end{cases}$$

Next define the modified function F of f on $[0,1] \times R^d \times R^d$ by

$$(1.14.14)\qquad F(t,x,x') = \delta\big(\|x\|, \|x'\|\big) f(t,x,x').$$

Clearly the function F is continuous and bounded on $[0,1] \times R^d \times R^d$. Hence there exists a solution $x \in C^{(2)}\big[[0,1], R^d\big]$ of the boundary value problem

$$(1.14.15)\qquad x'' = F(t,x,x'), \qquad x(0) - Ax'(0) = 0, \quad x(1) + A_1 x'(1) = 0.$$

Set $m(t) = V\big(t,x(t)\big)$ so that, because of assumption (c), we have the relations

$$(1.14.16)\quad m'(0) \ge 0, \qquad m'(1) \le 0, \qquad \text{and} \qquad m(0) \le \alpha m'(0).$$

Since

$$(1.14.17)\qquad V''_F(t,x) = \delta\big(\|x\|, \|x'\|\big) V''_f(t,x) + \big[1 - \delta\big(\|x\|, \|x'\|\big)\big] U(t,x,x'),$$

we obtain, in view of

$$0 \le \delta(u,v) \le 1, \qquad g(t,u,v) \le 0, \qquad U(t,x,x') \ge 0,$$

and assumption (1.14.10), the inequality

$$V''_F(t,x) \ge g\big(t, V(t,x), V'(t,x)\big) + \sigma\|F(t,x,x')\|$$

which leads to the further inequality

$$(1.14.18)\qquad m''(t) \ge g\big(t, m(t), m'(t)\big) + \sigma\|x''(t)\|.$$

Hence, by Lemma 1.14.1, it follows that there exists a $B_0 > 0$ such that

$$m(t) \le B_0 \quad \text{and} \quad |m'(t)| \le B_0, \quad 0 \le t \le 1.$$

As a result, setting

$$-N = [\min g(t,u,v):\ 0 \le t \le 1,\ u \le B_0,\ |v| \le B_0],$$

we obtain from (1.14.18)

$$m''(t) \ge -N + \sigma\|x''(t)\|.$$

Thus, for $0 \le s \le t \le 1$,

$$2B_0 \ge m'(t) - m'(s) \ge -N(t-s) + \sigma\left\| \int_s^t x''(\xi)\, d\xi \right\|$$

$$\ge -N + \sigma\|x'(t) - x'(s)\|.$$

Integrating this from 0 to 1 gives

$$\frac{(2B_0 + N)}{\sigma} \ge \int_0^1 \|x'(t) - x'(\xi)\|\, d\xi$$

$$\ge \left\| \int_0^1 \big(x'(t) - x'(\xi)\big)\, d\xi \right\|$$

$$\ge \|x'(t)\| - \|x(1)\| - \|x(0)\|.$$

Since $V(t,x)$ is assumed to be positive definite, it follows, from the estimate $V\big(t,x(t)\big) = m(t) \le B_0$, $0 \le t \le 1$, that $\|x(t)\| \le B^*$, $0 \le t \le 1$. Consequently, we deduce that

$$\|x(t)\| \le 2B^* + \frac{(2B_0 + N)}{\sigma} \equiv B, \quad 0 \le t \le 1.$$

Eivdently, this implies that

$$\text{(1.14.19)} \qquad \|x(t)\| \le B \quad \text{and} \quad \|x'(t)\| \le B, \quad 0 \le t \le 1,$$

This, in view of the definition of F, assures us that $x(t)$ is actually a solution of the boundary value problem (1.14.)-(1.14.3). The proof is complete.

If f satisfies Nagumo's condition, assumption (1.14.10) may be weakened as the next theorem shows.

THEOREM 1.14.2. Let the hypotheses of Theorem 1.14.1 hold except that inequalities (1.14.10) and (1.14.11) are replaced by

$$V_f''(t,x) \geq g\big(t,V(t,x),V'(t,x)\big) + \sigma\|x'\|, \qquad \sigma > 0, \tag{1.14.20}$$

$$U(t,x,x') + 1 \geq \|x'\|. \tag{1.14.21}$$

Suppose that $\|f(t,x,x')\| \leq h\,\|x'\|$ for $(t,x,x') \in [0,1]\times R^n \times R^n$, where $h \in C[R^+,(0,\infty)]$ and $\int_0^\infty s\,ds/h(s) = \infty$. Then there exists a solution $x \in C^{(2)}\big[[0,1],R^n\big]$ of the boundary value problem (1.14.1) - (1.14.3).

<u>Proof</u>: We proceed exactly as in the proof of Theorem 1.14.1 until we arrive at inequalities (1.14.16).

Consider first the case when $\sigma \geq 1$. Then, in view of (1.14.17), relations (1.14.20) and (1.14.21) yield the inequality

$$\begin{aligned} V_F''(t,x) &\geq g\big(t,V(t,x),V'(t,x)\big) + \sigma\delta\|x'\| \\ &\quad + (1-\delta)\big[\|x'\| - 1\big] \\ &\geq g\big(t,V(t,x),V'(t,x)\big) - 1 + \|x'\|[\delta(\sigma-1)+1] \\ &\geq g\big(t,V(t,x),V'(t,x)\big) - 1 + \|x'\|, \end{aligned} \tag{1.14.22}$$

using the facts $g \leq 0$, $0 \leq \delta \leq 1$. Here we have used δ for $\delta(u,v)$. If, on the other hand, $0 < \sigma < 1$, noting that (1.14.21) implies $\sigma\|x'\| \leq 1 + U(t,x,x')$, we obtain

$$\begin{aligned} V_F''(t,x) &\geq g\big(t,V(t,x),V'(t,x)\big) + \sigma\delta\|x'\| \\ &\quad + (1-\delta)\big[\sigma\|x'\| - 1\big] \\ &\geq g\big(t,V(t,x),V'(t,x)\big) - 1 + \|x'\|[\sigma\delta + (1-\delta)\sigma] \\ &\geq g\big(t,V(t,x),V'(t,x)\big) - 1 + \sigma\|x'\|. \end{aligned} \tag{1.14.23}$$

Since by Lemma 1.14.1 we have

$$m(t) \le B_0, \qquad |m'(t)| \le B_0, \qquad 0 \le t \le 1,$$

inequalities (1.14.22), (1.14.23) lead to

$$m''(t) \ge -(N+1) + \|x'(t)\|$$

and

$$m''(t) \ge -(N+1) + \sigma\|x'(t)\|,$$

respectively, where, as before,

$$-N = [\min g(t,u,v)\colon 0 \le t \le 1,\ u \le B_0,\ |v| \le B_0].$$

Let $\theta(t) = \int_0^t \|x'(s)\|\, ds$. Then, the preceding inequalities give

$$2B_0 \ge m'(1) - m'(0) \ge \int_0^1 [-(N+1) + \|x'(s)\|]\, ds \ge -(N+1) + \theta(1)$$

or

$$2B_0 \ge m'(1) - m'(0) \ge \int_0^1 [-(N+1) + \sigma\|x'(s)\|]\, ds \ge -(N+1) + \sigma\theta(1).$$

It then follows that $\theta(1) \le (2B_0 + N + 1)/\sigma \equiv M$ in any case. From Lemma 1.12.1, we have

$$\|x'(t)\| \le \gamma(\theta(1)) \le \gamma(M), \qquad 0 \le t \le 1.$$

Letting $B = \max[B^*, \gamma(M)]$, we obtain (1.14.19) which concludes the proof as before.

Finally, for later use, we shall state a uniqueness result leaving the proof as an exercise.

THEOREM 1.14.3. Assume that $f \in C[[0,1] \times R^n \times R^n, R^n]$, $V \in C^{(2)}[[0,1] \times R^n \times R^n, R^+]$, $V(t,x,y) \equiv 0$ if and only if $x = y$ and

$$\begin{aligned} V''(t,x,y) \equiv\ & V_{tt}(t,x,y) + 2\big(V_{tx}(t,x,y)\cdot x' + V_{ty}(t,x,y)\cdot y' \\ & + V_{xy}(t,x,y)x'\cdot y'\big) + V_{xx}(t,x,y)x'\cdot x' + V_{yy}(t,x,y)y'\cdot y' \\ & + V_x(t,x,y)\cdot f(t,x,x') + V_y(t,x,y)\cdot f(t,y,y') > 0 \end{aligned}$$

provided $V(t,x,y) \neq 0$ and $V'(t,x,y) = 0$, where

$$V'(t,x,y) = V_t(t,x,y) + V_x(t,x,y)\cdot x' + V_y(t,x,y)\cdot y'.$$

Then the BVP (1.14.1), $x(0) = x_0$ and $x(1) = x_1$ has at most one solution.

1.15 NOTES AND COMMENTS

For the existence results in the small contained in Section 1.1, see Hartman [3], Jackson [2], and Bailey, et al. [3]. The results of Sections 1.2 and 1.3 are taken from Jackson [2] except Theorems 1.2.4 and 1.3.2 which are based on the work of Schmitt [5]. Theorem 1.4.1, the Nagumo's condition, is taken from Jackson [2], while Theorems 1.4.2 and 1.4.3 are adapted from Schmitt [2] and Ako [1], respectively (see also Hartman [3]). Theorem 1.5.1 is taken from Jackson [2]. Theorems 1.5.2, 1.5.4 are due to Erbe [1] and Theorem 1.5.5 is due to Schmitt [2]. The results of Section 1.6 are due to Bernfeld et al. [1]. Exercise 1.6.1 contains the work of Schrader [2]. See George and Sutton [1] for the use of Lyapunov-like functions. The contents of Sections 1.7 - 1.10 are taken from Jackson [2] where other related references may be found. The work contained in Sections 1.11 - 1.13 is due to Bernfeld et al. [4] except Lemma 1.12.1 which is due to Lasota and Yorke [13]. The definition of the modified function as given in Definition 1.11.3 may be found in Hartman [3] where a number of results for second-order systems are given. The contents of Section 1.14 are due to Bernfeld et al. [3]. For further results in this direction, see Hartman [5].

For related results, see Knobloch [1,2], Schrader [5,6], Jackson and Schrader [1], Schmitt [3], Moyer [1], Gaines [3,4], Halikov [1], Gudkov and Lepin [1], Gudkov [2], and Mamedov [1].

Chapter 2
SHOOTING TYPE METHODS

2.0 INTRODUCTION

This chapter is essentially divided into two parts in which the shooting method serves as the underlying technique. The first part is concerned with the question as to whether uniqueness of solutions of boundary value problems implies the existence of solutions. Although the interdependence between uniqueness and existence is complicated for nonlinear equations, it can be formulated in simple terms for second-order differential equations. We examine this problem under linear and nonlinear boundary conditions and show that the results obtained are the best possible.

The second part is devoted to an important method known as the "angular function technique." A number of results concerning existence, uniqueness, and criteria for existence of a finite or infinite number of solutions are studied using this method as a tool. Employing Lyapunov-like functions and the theory of differential inequalities, this technique is extended to cover nonlinear boundary conditions and systems of differential equations.

2.1 UNIQUENESS IMPLIES EXISTENCE

For linear differential equations, it is well known that the uniqueness of solutions implies the existence of solutions. For nonlinear equations the interdependence between uniqueness and existence is much more complicated. However, in the case of second-order differential equations this relationship can surprisingly be formulated in simple terms.

An extremely useful technique in handling these questions for second-order boundary value problems is the so-called shooting method. The idea of this method is to fix one initial value and to allow the slope at the initial point to vary through the real numbers. From the connectedness of the solution funnel and the uniqueness of BVP's, it is then possible to show that the values of the solutions at the final point cover the real line. In what follows we plan to illustrate this important technique in a detailed manner. Our discussion depends on the following variation of Kneser's theorem.

LEMMA 2.1.1. Consider the initial value problem

$$x' = F(t,x), \qquad x(t_0) = x_0,$$

where $F \in C[R \times R^n, R^n]$. Let S be any compact, connected set. Assume all solutions $x(t,t_0,x_0)$ exist on $[t_0,t_1]$. Then

$$\bigcup_{\substack{x_0 \in S \\ t \in [t_0,t_1]}} x(t,t_0,x_0)$$

is a compact, connected set.

Our first result will be concerned with a system of two first-order equations subject to simple boundary conditions.

Consider the differential system

$$x' = f(t,x) \tag{2.1.1}$$

with the boundary conditions

$$x_1(t_1) = c_1, \qquad x_1(t_2) = c_2, \tag{2.1.2}$$

where $f \in C[R \times R^2, R^2]$. For this problem we have the following result.

THEOREM 2.1.1. Let $J = (a,b]$, $-\infty \leq a < b < \infty$, and $J^0 = (a,b)$. Assume

(i) $f_1(t,x_1,x_2)$ is an increasing function of x_2 for fixed (t,x_1) satisfying $f_1(t,x_1,x_2) \to \pm\infty$ as $x_2 \to \pm\infty$ uniformly on compact sets in $(a,b) \times R$;

(ii) all solutions of (2.1.1) exist on J;

(iii) there exists at most one solution of (2.1.1), (2.1.2) for all $t_1, t_2 \in J^0$ and all c_1, c_2.

Then every BVP (2.1.1), (2.1.2) has exactly one solution if $t_1 \in J^0$, $t_2 \in J$.

Proof: We may assume without loss of generality $t_2 = b$. Let t_1, c_1 be given where $a < t_1 < b$. Let $x(t) = (x_1(t), x_2(t))$ be a solution of (2.1.1) satisfying the initial condition

$$x(t_1) = (c_1, r). \tag{2.1.3}$$

Then define

$$S_k = [x(b)\colon x(t) \text{ is a solution of } (2.1.1) \text{ satisfying } (2.1.3) \text{ for each } |r| \leq k].$$

By Lemma 2.1.1, S_k is connected. Let $\tilde{S}_k$ be the projection of S_k on the x_1 axis. Hence $\tilde{S}_k$ is an interval which is nondecreasing in k. Define $\tilde{S} = \bigcup_{k=1}^{\infty} \tilde{S}_k$; then $\tilde{S}$ is an interval. To prove the stated result, it is sufficient to show $\tilde{S} = (-\infty,\infty)$. If this is not true, suppose that $\tilde{S}$ is bounded above. A similar proof holds if $\tilde{S}$ is bounded below. Then $\tilde{S} \subset (-\infty, M)$ for some $M > 0$. Let $x_n(t)$ be a sequence of solutions of (2.1.1) satisfying

$$x_n(t_1) = (c_1, n), \qquad n = 1,2,\ldots . \tag{2.1.4}$$

If $s_1, s_2, \ldots, s_n$ is a sequence of t-values on a compact subinterval of J^0, then

$$\|x_n(s_n)\| \to \infty \quad \text{as} \quad n \to \infty, \tag{2.1.5}$$

for otherwise, from the well-known convergence theorems a subsequence of $\{x_n(t)\}$ converges uniformly to a solution of (2.1.1), on all compact subsets of J^0, a contradiction to (2.1.4).

Let $x_{n,1}(t)$ be the first component of $x_n(t)$. Observe that $x_{n+1,1}(t) > x_{n,1}(t)$ and $x_{n+1,1}(s) < x_{n,1}(s)$ for $t \in (t_1, \delta]$ and $s \in [t_1 - \delta, t_1)$ for sufficiently small δ. However, from (iii) we immediately obtain

$$x_{n+1,1}(s) < x_{n,1}(s), \qquad x_{n+1,1}(t) > x_{n,1}(t) \tag{2.1.6}$$

for $a < s < t_1 < t < b$.

We claim that the sequence $x_{n,1}(t)$ is not uniformly bounded on any subinterval of J^0. If we assume not, then there exists a $C > 0$ and an interval $[a_1, b_1]$ such that $|x_{n,1}(t)| \leq C$ for $t \in [a_1, b_1]$. From the mean value theorem there exists $r_n \in [a_1, b_1]$ such that $|x'_{n,1}(r_n)| \leq 2C/(b_1 - a_1)$. Then condition (i) implies $\|x_n(r_n)\|$ is bounded for all n. However, this contradicts (2.1.5). Hence, on any subintervals of $[t_1 - \delta_m, t_1], [t_1, t_1 + \delta_m]$ the sequence $\{x_{n,1}(t)\}$ is not uniformly bounded. Letting $\delta_m \to 0$ we see there exist points v_n, w_n where

$$v_n < t_1, \qquad v_n \to t_1, \qquad w_n > t_1, \qquad w_n \to t_1$$

such that, in view of (2.1.6),

$$x_{n,1}(v_n) \to -\infty \quad \text{and} \quad x_{n,1}(w_n) \to +\infty \tag{2.1.7}$$

as $n \to \infty$.

Let $x^*(t)$ be any solution of (2.1.1) satisfying

$x_1^*(b) = M$; hence, since $S \in (-\infty, M), x_1^*(b) > x_{n,1}(b)$ for all n. From (2.1.7), for large n

$$x_{n,1}(v_n) < x_1^*(v_n), \qquad x_{n,1}(w_n) > x_1^*(w_n).$$

This implies $x_{n,1}(t)$ and $x_1^*(t)$ intersect twice, contradicting assumption (iii). This completes the proof of Theorem 2.1.1.

COROLLARY 2.1.1. Consider the BVP

$$x'' = h(t,x,x'), \tag{2.1.8}$$

$$x(t_1) = c_1, \quad x(t_2) = c_2, \quad t_1 \in (a,b), \quad t_2 \in (a,b], \tag{2.1.9}$$

where $h \in C[J \times R \times R, R]$, $J = (a,b]$. Assume that

(i) all solutions of (2.1.8) exist on $(a,b]$;

(ii) there exists at most one solution of (2.1.8), (2.1.9) for $t_1, t_2 \in (a,b]$, $c_1, c_2 \in R$.

Then there exists exactly one solution of (2.1.8) and (2.1.9).

<u>Proof</u>: The BVP (2.1.8), (2.1.9) can be written as

$$x_1' = x_2, \qquad x_2' = h(t, x_1, x_2),$$
$$x_1(t_1) = c_1, \qquad x_1(t_2) = c_2.$$

We see immediately that the conditions in Theorem 2.1.1 are satisfied. Hence Corollary 2.1.1 follows.

EXERCISE 2.1.1. Consider the BVP (2.1.8), (2.1.9) and suppose that $h(t,x,x')$ is strictly increasing in x. Assume that all solutions of (2.1.8) exist on $(a,b]$. Then there exists exactly one solution of (2.1.8), (2.1.9) for all $t_1, t_2 \in (a,b]$ and all $c_1, c_2 \in R$.

<u>Hint</u>: Show that there exists at most one solution by

showing that the difference of any two solutions cannot attain a positive maximum or negative minimum. Then use Corollary 2.1.1.

EXERCISE 2.1.2. Assume $h(t,x,x')$ satisfies

$$|h(t,\bar{x},\bar{y}) - h(t,x,y)| \le q(t)|x - \bar{x}| + p(t)|y - \bar{y}|,$$

where

$$\int_a^b q(t) \exp\left(\tfrac{1}{2} \int_a^b p(t)\, dt\right) < 4.$$

Then (2.1.8), (2.1.9) has a unique solution.

Hint: Show that the inequality $|x''| \le q(t)|x| + p(t)|x'|$ with boundary conditions $x(t_1) = 0$, $x(t_2) = 0$ has only the zero solution, where $|t_2 - t_1| < b - a$. Then use Corollary 2.1.1.

Observe that in Theorem 2.1.1 we obtained the existence of a solution on $(a,b]$. In the following example we show that we cannot substitute $[a,b]$ for $(a,b]$.

EXAMPLE 2.1.1. The implicit equation

$$\varphi + (p/2) \tan^{-1} \varphi = q, \qquad p > -2,$$

has a unique solution $\varphi(p,q)$. The family of all solutions of

$$\text{(2.1.10)} \qquad x'' = -x + \tfrac{1}{2} \tan^{-1} \varphi(\sin t, x \sin t + x' \cos t)$$

can be represented by

$$x(t) = A \cos t + B \sin t + \tfrac{1}{2} \tan^{-1} B,$$

where A and B are arbitrary constants. Consider the boundary conditions

$$\text{(2.1.11)} \quad x(t_1) = x_1 \quad \text{and} \quad x(t_2) = x_2, \qquad t_1, t_2 \in [0,\pi].$$

Then

$$x_i = A \cos t_i + B \sin t_i + \tfrac{1}{2} \tan^{-1} B, \qquad i = 1,2.$$

Eliminating A, we obtain

$$B \sin(t_2 - t_1) + \tfrac{1}{2}(\cos t_1 - \cos t_2) \tan^{-1} B = x_2 \cos t_1 - x_1 \cos t_2. \tag{2.1.12}$$

We now claim that there exists at most one solution of (2.1.10), (2.1.11). If $0 \leq t_1 < t_2 < \pi$ or $0 < t_1 < t_2 \leq \pi$, then

$$\sin(t_2 - t_1) > 0 \qquad \text{and} \qquad \cos t_2 - \cos t_1 < 0.$$

Thus from (2.1.12), we see that, for any x_1, x_2, t_1, t_2, B and A are uniquely determined. Thus solutions of (2.1.10), (2.1.11) are uniquely determined. For the case $t_1 = 0$, $t_2 = \pi$, (2.1.12) reduces to

$$\tan^{-1} B = x_1 + x_2. \tag{2.1.13}$$

Once again uniqueness of solutions follows. Thus on $[0,\pi]$ the conditions of Corollary 2.1.1 hold. However, observe (2.1.13) implies there exists no solution, if $|x_1 + x_2| = \pi/2$.

EXERCISE 2.1.3. Show that solutions of

$$y'' = -y + \arctan y, \tag{2.1.14}$$

$$y(t_1) = c_1, \qquad y(t_2) = c_2, \tag{2.1.15}$$

when they exist, are unique for $t_1, t_2 \in [0,\pi]$. Then show there exists no solution of (2.1.14) satisfying

$$y(0) = 0, \qquad y(\pi) = 3\pi. \tag{2.1.16}$$

<u>Hint</u>: Assume there exist two solutions $y_1(t)$, $y_2(t)$ of

(2.1.14), (2.1.15) for $t_1, t_2 \in (0,\pi)$. Show that $w(t) \equiv y_1(t) - y_2(t)$ satisfies $w''(t) \geq -w(t)$ and prove that $w(t)$ is a subfunction of $y'' = -y$. If $[t_1, t_2]$ is not a proper subinterval of $[0,\pi]$, then show that $w(0) > u(0)$, $w(\pi) > u(\pi)$, where $u'' = -u$ and $w(\pi/2) = u(\pi/2)$, $w'(\pi/2) = u'(\pi/2)$. Prove that $u(t)$ has two zeros on $[0,\pi]$ which leads to a contradiction. To show there exists no solution of (2.1.15), (2.1.16), prove that the solution $v(t)$ of

$$v'' = -v + \pi, \qquad v(0) = 0, \qquad v'(0) = m + 1,$$

is a subfunction on $[0,\pi]$ with respect to (2.1.14) for any m. Argue that $y(\pi) < v(\pi) = 2\pi$, where y satisfies (2.1.14), $y(0) = 0$, $y'(0) = m$.

REMARK 2.1.1. Interestingly enough we can, under certain conditions on $h(t,x,x')$, derive the results of Corollary 2.1.1 for the case when $J = [a,b]$. It is sufficient, for example, that h satisfy a Lipschitz condition in both x and x'. This follows by considering the linear equation derived from the Lipschitz condition, and using the fact that the uniqueness of the BVP for the linear equation on all subintervals of $[a,b]$ implies the existence of unique solutions of BVP's for $x'' = h(t,x,x')$ on all subintervals of $[a,b]$. Moreover, an estimate of $[a,b]$ can be derived in terms of hyperbolic functions.

2.2 GENERAL LINEAR BOUNDARY CONDITIONS

The question as to whether uniqueness of solutions of (2.1.8) with the general linear boundary conditions

$$(2.2.1) \qquad a_1 x(t_1) + a_2 x'(t_1) = c_1, \qquad b_1 x(t_2) + b_2 x'(t_2) = c_2$$

implies existence is naturally more difficult since it depends upon the coefficients a_1, a_2, b_1, b_2.

We shall show, as a consequence of our next result, that Corollary 2.1.1 holds for (2.1.8) and (2.2.1) whenever $a_2b_2 = 0$.

THEOREM 2.2.1. Let J, x, f be as in Theorem 2.1.1 except that condition (iii) is replaced by the following:

(iv) there is at most one solution of (2.1.1) satisfying

$$(2.2.2) \qquad x_1(t_1) = c_1 \quad \text{and} \quad x_2(t_2) = c_2, \qquad t_1 \neq t_2,$$

for any $t_1, t_2 \in J^0$ and $c_1, c_2 \in R$.

Then

(a) every boundary value problem (2.1.1), (2.1.2) has a unique solution;

(b) every boundary value problem (2.1.1), (2.2.2) has a unique solution.

Proof: In order to show (a), we must show condition (iii) of Theorem 2.1.1 holds. Let $x(t)$, $y(t)$ be distinct solutions of (2.1.1) satisfying $x_1(t_1) = y_1(t_1)$ for some $t_1 \in J^0$. It suffices to show that $x_1(t) \neq y_1(t)$ for $t_1 \neq t \in J^0$. We show this for $t_1 < t < b$ as a similar argument holds for $t \in (a, t_1)$. Without loss of generality, we may assume from (iv) that $y_2(t) > x_2(t)$ for $t \in (t_1, b)$. Then either $y_2(t_1) = x_2(t_1)$ or $y_2(t_1) > x_2(t_1)$. If $y_2(t_1) = x_2(t_1)$, then, by (iv), $y_1(t) \neq x_1(t)$ for $t_1 \neq t \in J^0$. If $y_2(t_1) > x_2(t_1)$, then $y_1'(t_1) > x_1'(t_1)$ by (i), so that there exists a $\delta > 0$ such that $y_1(t) > x_1(t)$ for $t \in (t_1, t_1 + \delta)$. Assume that there exists a least $\tau, t_1 < \tau < b$ such that $x_1(\tau) = y_1(\tau)$. However, then $x_1'(\tau) \geq y_1'(\tau)$, which implies $x_2(\tau) \geq y_2(\tau)$. This contradiction shows that $x_1(t) \neq y_1(t)$ on (t_1, b) and thus completes the proof of part (a).

The proof of part (b) is similar to that of Theorem 2.1.1. Assume without loss of generality that $t_2 = b$.

Define S_k as in Theorem 2.1.1 and let $\hat{S}_k$ be the projection of S_k on the x_2 axis. Let $\hat{S} = \bigcup_{k=1}^{\infty} \hat{S}_k$; then $\hat{S}$ is an interval. It suffices to show $\hat{S}$ is the real line. Suppose $\hat{S}$ is bounded above (a similar proof holds if $\hat{S}$ is bounded below). We may assume that there exists $M > 0$ such that $\hat{S} \subset (-\infty, M)$ but $(-\infty, M] \not\subset \hat{S}$. As before, let $x_n(t)$ be a sequence of solutions of (2.1.1) satisfying

$$x_n(t_1) = (c_1, n), \qquad n = 1, 2, \ldots . \tag{2.2.3}$$

Then there exist sequences v_n, w_n such that (2.1.7) holds [this follows since (a) is the same as (iii) in Theorem 2.1.1].

From (2.2.3) there exists a sequence $\{z_n\}$ such that $t_1 < z_n \to t_1$ and $x_{n2}(z_n) \to \infty$ as $n \to \infty$. Let $x^*(t)$ be any solution of (2.1.1) satisfying $x_2^*(b) = M$. From the fact that $\hat{S} \subset (-\infty, M)$, $x_2^*(b) > x_{n2}(b)$ for all n, but for large n $x_2^*(z_n) < x_{n2}(z_n)$. Thus there exists z_n' such that

$$x_2^*(z_n') = x_{n2}(z_n'), \qquad z_n < z_n' < b. \tag{2.2.4}$$

From (2.1.7) $x_1^*(w_n) < x_{n1}(w_n)$, $x_1^*(v_n) > x_{n1}(v_n)$ for large n so that there exists z_n'' such that

$$x_1^*(z_n'') = x_{n1}(z_n''), \qquad v_n < z_n'' < w_n. \tag{2.2.5}$$

If $z_n' \neq z_n''$, then (2.2.4) and (2.2.5) contradict (iv). Suppose $z_n' = z_n''$; then define $\bar{x}(t) = x_n(t)$ for $a < t \leq z_n'$, and $\bar{x}(t) = x^*(t)$ for $z_n' < t \leq b$. Then $\bar{x}(t)$ is a solution of (2.1.1) such that $\bar{x}_1(t_1) = x_{n1}(t_1) = c_1$ and $\bar{x}_2(b) = x_2^*(b) = M$, so that $(-\infty, M] \in \hat{S}$, a contradiction. This completes the proof of Theorem 2.2.2.

COROLLARY 2.2.1. Consider the BVP (2.1.8), (2.2.1) and

assume $a_2 b_2 = 0$. Suppose all solutions of (2.1.8) exist on $(a,b]$ and there exists at most one solution of (2.1.8), (2.2.1) for every $t_1, t_2 \in (a,b)$. Then there exists

(a) a unique solution of (2.1.8), (2.1.9), and

(b) a unique solution of (2.1.8), (2.2.1).

Proof: Without loss of generality, assume $a_2 = 0$, $b_2 = 1$ and $a_1 = 1$. Then (2.1.8), (2.2.1) can be written as the first-order BVP for $x = (x_1, x_1' + b_1 x_1)$ in which

$$x_1' = x_2 - b_1 x_1, \qquad x_2' = h(t, x_1, x_2 - b_1 x_1) + b_1(x_2 - b_1 x_1),$$

$$x_1(t_1) = c_1, \qquad x_2(t_2) = c_2.$$

We therefore see that the hypotheses of Theorem 2.2.1 hold and the result follows.

EXERCISE 2.2.1. Assume in (2.1.8) that the function $h(t,x,y) \in C^{(1)}[(a,b) \times R^2, R]$ satisfies the inequalities $|h_x| \leq M$, $|h_y| \leq K$. If $|b - a| < \int_{-1/b_2}^{\infty} ((u^2 + K|u| + M)^{-1}) du$ for $a_1 = 1$, $a_2 = 0$, $b_1 = 1$, $b_2 \neq 0$, then the BVP (2.1.8), (2.2.1) has a unique solution.

Hint: Prove (2.1.8), (2.2.1) has at most one solution. This can be done by assuming that there exist solutions x_1, x_2 and by setting $x = x_1 - x_2$ obtain $|x''| \leq M|x| + K|x'|$, $x(t_1) = 0$, $x(t_2) + b_2 x'(t_2) = 0$. Then letting $w = x'/x$, deduce $|w'| \leq w^2 + K|w| + M$, $w(t_1 - 0) = +\infty$ and $w(t_2) = -1/b_2$. By the theory of differential inequalities conclude that $w(t) < u(t)$, where $u(t)$ is the solution of

$$-u' = u^2 + K|u| + M, \qquad u(t_2) = -1/b_2.$$

Finally obtain a contradiction.

In the next two examples, we show that Corollary 2.2.1 does not necessarily hold without the assumption that $a_2 b_2 = 0$.

EXAMPLE 2.2.1. The equation

$$x'' = \frac{2t}{1+t^2} x' - \frac{2}{1+t^2} x, \qquad -\infty < t < +\infty \tag{2.2.6}$$

has the general solution $x(t) = A(t^2 - 1) + Bt$. Hence there exists one solution satisfying the boundary conditions $x'(t_1) = c_1, x'(t_2) = c_2$. Observe $x_1(t) \equiv 0$ and $x_2(t) \equiv t^2 - 1$ are two different solutions satisfying $x(-1) = 0$, $x(1) = 0$. Moreover, there exists no solution satisfying $x(-1) = 0$, $x(1) = 1$. Thus (a) in Corollary 2.2.1 is violated.

EXAMPLE 2.2.2. Let φ be defined as in Example 2.1.1. Consider the differential equation

$$x'' = e^t \varphi(2e^{-t}, x'e^{-t}) \tag{$*$}$$

with the boundary conditions

$$x'(t_1) - x(t_1) = r_1, \qquad x'(t_2) - x(t_2) = r_2.$$

The family of solutions of $(*)$ is given by

$$x(t) = A + Be^t + t \tan^{-1} B,$$

where A and B are arbitrary constants. The boundary conditions lead to

$$(1 - t_i) \tan^{-1} B - A = r_i \qquad (i = 1,2).$$

Eliminating A we obtain

$$(t_1 - t_2) \tan^{-1} B = r_2 - r_1$$

and thus there exists at most one solution of the BVP. However, if

$$|r_2 - r_1| \geq (\pi/2)\ |t_2 - t_1|,$$

there exists no solution. Hence (b) in Corollary 2.2.1 does not hold.

When $a_2 b_2 \neq 0$ we can rewrite (2.2.1) in the form

$$(2.2.7) \qquad rx(t_1) + x'(t_1) = c_1, \qquad sx(t_2) + x'(t_2) = c_2.$$

We now present a result in which uniqueness of (2.1.8) and (2.2.7) implies existence under certain added conditions. Consider the ordering on R^2 defined as

$$(a_1,b_1) \leq (a_0,b_0) \Leftrightarrow \begin{cases} a_0 < b_0 \quad \text{and} \quad a_1 \leq a_0 \quad \text{and} \quad b_1 \geq b_0, \\ \text{or} \\ a_0 > b_0 \quad \text{and} \quad a_1 \geq a_0 \quad \text{and} \quad b_1 \leq b_0. \end{cases}$$

Write $(a_1,b_1) < (a_0,b_0)$ when $(a_1,b_1) \leq (a_0,b_0)$ and $(a_1,b_1) \neq (a_0,b_0)$. Problem (2.1.8) and (2.2.7) will be denoted by $(h;r,s)$.

THEOREM 2.2.2. Assume all solutions of (2.1.8) exist on (a,b) and are uniquely determined by their initial values. Assume problem $(h;r_0,s_0)$ is unique for any $t_1,t_2 \in (a,b)$, then there exists a unique solution of $(h;r_1,s_1)$ whenever $(r_1,s_1) < (r_0,s_0)$.

Proof: Assume that

$$r_0 < s_0, \qquad r_1 = r_0, \qquad s_1 > s_0.$$

The other cases have similar arguments and therefore we shall consider this case only. Fix t_1, t_2, and c_1 and let $x(t,v)$ be the solution of (2.1.8) satisfying

$$x(t_1,v) = v, \qquad x'(t_1,v) = c_1 - r_0 v.$$

Define

$$T(t,v) = x'(t,v) + s_0 x(t,v).$$

From the uniqueness of $(h;r_0,s_0)$, $v \to T(t,v)$ is a one-to-one mapping for $t \geq t_1$. Define

$$\Delta T/\Delta v \equiv \big(T(t,v_2) - T(t,v_1)\big)/(v_2 - v_1).$$

Then $\Delta T/\Delta v|_{t=t_1} = -r_0 + s_0 > 0$. Thus $\Delta T/\Delta v$ is positive for all $t \geq t_1$. Then, since

$$\Delta T/\Delta v = \Delta x'/\Delta v + s_0 \, \Delta x/\Delta v > 0,$$

we see that

$$\Delta x/\Delta v|_{t=t_2} > \Delta x/\Delta v|_{t=t_1} \exp\big(-s_0(t_2 - t_1)\big) = \exp\big(-s_0(t_2 - t_1)\big).$$

Setting

$$S(t,v) = x'(t,v) + s_1 x(t,v) = T(t,v) + (s_1 - s_0)x(t,v),$$

we have

$$\Delta S/\Delta v|_{t=t_2} = \Delta T/\Delta v|_{t=t_2} + (s_1 - s_0)\, \Delta x/\Delta v|_{t=t_2}$$
$$> (s_1 - s_0) \exp\big(-s_0(t_2 - t_1)\big).$$

Hence $v \to S(t_2,v)$ is a one-to-one mapping of R onto itself. This is equivalent to $(h;r_1,s_1)$ having exactly one solution and thus the proof is complete.

COROLLARY 2.2.2. Assume $h(t,x,y)$ is strictly increasing in x and that solutions of (2.1.8) exist on (a,b) and are uniquely determined by initial conditions. Then, for $r < 0$, $s > 0$, problem (2.1.8), (2.2.7) has a unique solution.

Proof: To prove this, choose r_0, s_0 such that $r < r_0 < 0$, $0 < s_0 < s$. Hence $(r_0,s_0) > (r,s)$ and by Theorem 2.2.2, it suffices to show $(h;r_0,s_0)$ is unique.

Assume there exist two solutions x_1 and x_2 and let $x = x_1 - x_2$. Since $x(t) \not\equiv 0$, $x(t)$ attains either its positive maximum or negative minimum on $[t_1,t_2]$. Assume that x attains its positive maximum (the case in which x attains its negative minimum uses a similar argument). If the maximum occurs at t_1, then $x(t_1) > 0$, $x'(t_1) \le 0$, and the first boundary condition $r_0x(t_1) + x'(t_1) = 0$ cannot hold. If the maximum occurs at t_2, then $x(t_2) > 0$, $x'(t_2) \ge 0$, and the second condition $s_0x(t_2) + x'(t_2) = 0$ cannot hold. If the maximum occurs at $t_0 \in (t_1,t_2)$, then

$$x(t_0) > 0, \qquad x'(t_0) = 0, \qquad x''(t_0) \le 0;$$

however, by the monotonicity of $h(t,x,x')$, we have

$$x''(t_0) = h\big(t_0,x_1(t_0),x_1'(t_0)\big) - h\big(t_0,x_2(t_0),x_2'(t_0)\big) > 0,$$

a contradiction.

In the next exercise, we show that the strict inequality $(r_1,s_1) < (r_0,s_0)$ cannot be weakened to the inequality $(r_1,s_1) \le (r_0,s_0)$. This will provide another example in which $a_2b_2 \ne 0$ and uniqueness of the boundary value problem does not imply existence.

EXERCISE 2.2.2. Construct an example in which the hypotheses of Theorem 2.2.2 hold with $(r_1,s_1) < (r_0,s_0)$ replaced by the weaker condition $(r_1,s_1) \le (r_0,s_0)$ such that there exists no solution of the problem (2.1.8), (2.2.7).

Hint: Let $\varphi(p,q)$ be the solution of the equation $\varphi + p \arctan \varphi = q$. Consider the differential equation

$x'' = F''\varphi(1/F', x'/F')$ on the interval $(-1,2)$, where F will be determined. Let $w = x'$; then w satisfies the general solution

$$w'(t) = BF'(t) + \arctan B.$$

Solve for $x(t)$ in terms of $w(t)$ and substitute the boundary conditions, where we may assume $s = 1$, $r > 1$. Then obtain conditions on $F(t_1)$, $F'(t_1)$, $F(t_2)$, $F'(t_2)$ such that uniqueness holds (see Example 2.1.1) and such that $F'(t) > 0$ on $(-1,2)$. This leads to

$$F(t) = \begin{cases} 1 - e^{-r(t+1)}, & t \in [-1,0], \\ F_2(t), & t \in [0,1], \\ 1 + \left(F_2(1) - 1\right)e^{1-t}, & t \in [1,2], \end{cases}$$

where F_2 satisfies

$$F_2 + g(t)F_2' = 1, \qquad F_2(0) = 1 - e^{-r},$$

in which $g \in C^{(1)}\left[[0,1],R\right]$ such that $g(0) = 1/r'$ $g(1) = 1$, $1/r \leq g(t) \leq 1$ for $0 \leq t \leq 1$, and $g'(0) = g'(1) = 0$. Show that for $t_1 = -\frac{1}{2}$, $t_2 = 3/2$ and for c_1, c_2 chosen so that $|(rc_2 - c_1)/(2r - 1)| \geq \pi/2$, there exists no solution of (2.1.8) and (2.2.7). (Use ideas in Example 2.1.1.)

2.3 WEAKER UNIQUENESS CONDITIONS

It is natural to ask whether it is possible to weaken the uniqueness condition of the BVP (2.1.8), (2.1.2) by not demanding that this condition hold for all $t_1, t_2 \in [a, b + \varepsilon)$ $\left(\text{or } t_1, t_2 \in (a - \varepsilon, b]\right)$. In this section, we plan to discuss how some of the previous results remain true even when the uniqueness assumptions are weakened for $t_1 = a$ and for $t_2 \in (b - \varepsilon, b + \varepsilon)$ for some $\varepsilon > 0$.

THEOREM 2.3.1. Let $h \in C\left[[a, b + \varepsilon) \times R^2, R\right]$ for some

$\varepsilon > 0$ and suppose that all solutions of initial value problems for (2.1.1) exist on $[a,b+\varepsilon)$. Assume that for a fixed c_1 there exists at most one solution on $[a,t_2]$ of the BVP (2.1.1) and

$$(2.3.1) \qquad x(a) = c_1, \quad x(t_2) = c_2,$$

for all $c_2 \in R$ and $t_2 \in (b-\varepsilon, b+\varepsilon)$. Then there exists exactly one solution of (2.1.1) and

$$(2.3.2) \qquad x(a) = c_1, \quad x(b) = c_2$$

for each $c_2 \in R$.

Proof: For each $m \in R$ define

$$A_m = \big[x(t,m) \in C^{(2)}\big[[a,b+\varepsilon)\times R,R\big]: x(t,m) \text{ is a solution of (2.1.1) with } x(a) = c_1 \text{ and } x'(a) = m\big].$$

Define the mapping $\Gamma(m) = [x(b,m)]$; Γ maps points into sets and an application of Lemma 2.1.1 yields that Γ maps compact, connected sets into compact, connected sets. If $m_1 \neq m_2$, then $\Gamma(m_1) \cap \Gamma(m_2) = \emptyset$, because of the uniqueness of solutions of BVP. We can thus define $\Gamma(m_1) \leq \Gamma(m_2)$ if and only if $\max \Gamma(m_1) \leq \max \Gamma(m_2)$.

We now show $\Gamma(m)$ is monotone. Assume, for example, that $m_1 < m_2 < m_3$ and $\Gamma(m_1) < \Gamma(m_3) < \Gamma(m_2)$. Let $x_i \in \Gamma(m_i)$. Since $x_1 < x_3 < x_2$ and $\Gamma([m_1,m_2])$ is connected, there exists $m^* \in [m_1,m_2]$ such that $x(b,m^*) = x_3$. Since $m^* \leq m_2 < m_3$, we have a contradiction to the uniqueness of the BVP. The other cases may be treated by a similar argument.

In order to complete the proof of Theorem 2.3.1, it is enough to show $\Gamma(R) = R$. We first prove that $\Gamma(R)$ is open. If not, there exists a $\psi \in \Gamma(R)$ that is not in the interior.

Hence there is an $\overline{m}$ such that $x(b,\overline{m}) = \psi$. By the monotonicity of Γ we can find $\tilde{m} > \overline{m}$ or $\tilde{m} < \overline{m}$ such that $\Gamma((\tilde{m},-\tilde{m}))$ contains ψ as an interior point. This is a contradiction.

It thus suffices to prove that $\Gamma(R)$ is neither bounded above nor below. Assume $\Gamma(R)$ is bounded above (a similar argument is true for the other case) and let $\eta = \sup \Gamma(R)$. Let $x_n(t,m_n)$ denote a sequence of solutions of (2.1.1), $(x_n(a,m_n)) = c_1, (x_n'(a,m_n)) = m_n$ such that $x_n(b,m_n) \to \eta$. For $c \in (b, b+\varepsilon)$ and from the uniqueness of BVP,

$$\frac{x_n(c,m_n) - x_n(b,m_n)}{c-b} \geq \frac{x_1(c,m_1) - x_n(b,m_n)}{c-b} \geq K$$

where $K = \min\{0, (x_1(c,m_1) - \eta)/(c-b)\}$.

We may assume that, for infinitely many n's, we have $x_n'(b,m_n) \leq 0$, because a similar argument can be made for $c \in (b-\varepsilon, b)$ in case $x_n'(b,m_n) \geq 0$ for infinitely many n's. We restrict our arguments to those n such that $x_n'(b,m_n) \leq 0$.

Let

$$S_n = [t\colon 0 \geq x_n'(t,m_n) \geq K,\ t \in [b,c]].$$

This set is nonempty by the mean value theorem. Let $s_n = \min S_n$; then $b \leq s_n \leq c$ and $K \leq x_n'(s_n,m_n) \leq 0$. Thus, without loss of generality, $s_n \to s_0$ and $x_n'(s_n,m_n) \to \overline{x}_0$. If $x_n'(b,m_n) = 0$, then $b = s_n$ and $x_n(s_n,m_n) < \eta$. If $x_n'(b,m_n) < 0$, then $x_n(t,m_n) < \eta$ for $t \in [b,t^*]$, where $x_n'(t^*,m_n) = 0$ and t^* is the first such point. Then $t^* \in S_n$ and therefore $s_n \in [b,t^*]$. Thus again $x_n(s_n,m_n) < \eta$. Since $x_n(s_n,m_n) \geq \min [x_1(t,m_n)\colon t \in [b,c]]$ (from the uniqueness of BVP), there exists a subsequence of $\{x_n(s_n,m_n)\}$ which we choose to call by the same labeling, such that $x_n(s_n,m_n) \to x_0$. Thus, $s_n \to s_0$, $x_n(s_n,m_n) \to x_0$, $x_n'(s_n,m_n) \to \overline{x}_0$.

Let $S(\delta)$ be a δ-ball in R^3 about $(s_0,x_0,\overline{x}_0)$, where $\delta < b + \varepsilon - s_0$. For n sufficiently large $\left(s_n, x_n(s_n,m_n), x_n'(s_n,m_n)\right) \in S(\delta)$. Since all solutions of (2.1.8) exist on $[a, b+\varepsilon)$, there exists $Q > 0$ such that $|x_n'(t,m_n)| < Q$ for $t \in [a, s_0 + \delta]$. Let $x(t,Q)$ and $x(t,-Q)$ be solutions of (2.1.8) with $x(a,\pm Q) = c_1$; then by the montonicity of $x(b,\cdot)$, either $x(b,-Q) > x_n(b,m_n)$ so that $x(b,-Q) \geq \eta$ or $x(b,Q) > x_n(b,m_n)$ which means $x(b,Q) \geq \eta$. This is a contradiction and the proof of Theorem 2.3.1 is complete.

COROLLARY 2.3.1. Under the conditions of Theorem 2.3.1, the BVP (2.1.8), (2.2.1) has a unique solution, if $a_2 b_2 = 0$.

We leave the proof as an exercise. The following hints may be helpful. Consider the boundary conditions

$$r_1 x(a) + x'(a) = c_1, \qquad x(b) = c_2. \tag{2.3.3}$$

For a fixed c_1, define

$$\Gamma(m) = [x(b,m)\text{: where } x(t,m) \text{ is a solution of (2.1.8) such that } x(a,m) = m \text{ and } x'(a,m) = c_1 - r_1 m].$$

Then using the same techniques as in Theorem 2.3.1, show $\Gamma(R) = R$.

By combining the techniques employed in Theorems 2.1.1 and 2.3.1, one can deduce the following generalization, the proof of which we leave to the reader.

THEOREM 2.3.2. Consider the BVP (2.1.1), (2.1.2) assuming conditions (i) and (ii) of Theorem 2.1.1. Suppose that the condition relative to the uniqueness of BVP, in the sense of Theorem 2.3.1 is true. Then there exists a unique solution of (2.1.1), (2.1.2).

2.4 NONLINEAR BOUNDARY CONDITIONS

We now present a result in which the boundary conditions are nonlinear. Under some additional hypotheses on $h(t,x,x')$, it is possible to deduce existence results for general linear boundary conditions for which $a_2b_2 \neq 0$. This we state as a corollary to our main result.

THEOREM 2.4.1. Consider the boundary value problem

$$x' = f_1(t,x,y), \qquad y' = f_2(t,x,y), \tag{2.4.1}$$

$$g_1\big(x(a),y(a)\big) = c_1, \tag{2.4.2}$$

$$g_2\big(x(b),y(b)\big) = c_2, \tag{2.4.3}$$

where $f_i \in C\big[[a,b+\varepsilon)\times R^2,R\big]$, $g_i \in C^{(1)}[R^2,R]$. For any fixed c_1, if we define

$$S_1 = [(x,y) \in R^2 : g_1(x,y) = c_1],$$

then S_1, when parameterized in terms of arc length, leaves every bounded region in R^2 in either direction. For every c, let $g_2^{-1}(c)$ be nonempty. Assume for $f = (f_1,f_2)$

$$g_2^2(x,y) + [\text{grad } g_2(x,y)\cdot f(x,y)]^2 \to \infty \tag{2.4.4}$$

as $x^2 + y^2 \to \infty$ uniformly for $t \in [a,b+\varepsilon)$, and that all solutions of (2.4.1) exist on $[a,b+\varepsilon)$. Then if there exists at most one solution of (2.4.1), (2.4.2) and $g_2\big(x(s),y(s)\big) = c$ for $s \in (b-\varepsilon,b+\varepsilon)$ and all $c \in R$, there exists exactly one solution of (2.4.1) - (2.4.3).

<u>Proof</u>: The idea of the proof is similar to that of the proof of Theorem 2.3.1. Let P_0 be an arbitrary point of $g_1(x,y) = c_1$ and parameterize S_1 by arc length with P_0

as the initial point. The points on S_1 will be denoted by the parameter. For any fixed $s \in [a, b+\varepsilon)$, define for $P \in S_1$

$$T_s(P) = \Big[\big(x(s,P),y(s,P)\big),\ \text{where}\ \big(x(t,P),y(t,P)\big)\ \text{denotes a solution of (2.4.1) satisfying}\ \big(x(a,P),y(a,P)\big) = P\Big],$$

and let

$$\Gamma_s(P) = \Big[g_2\big(x(s,P),y(s,P)\big) : \big(x(s,P),y(s,P)\big) \in T_s(P)\Big].$$

It is sufficient to show $\Gamma_b(R) = R$, to prove the theorem.

From Lemma 2.1.1, $T_s(A)$ is compact and connected, if A is compact and connected. Since $g_2(x,y)$ is continuous, $\Gamma_s(A)$ is compact and connected. It follows from a simple argument that $\Gamma_s(R)$ is connected.

By uniqueness of the BVP, $\Gamma_s(P_1) \cap \Gamma_s(P_2) = \emptyset$ for $s \in (b-\varepsilon, b+\varepsilon)$. Thus $\Gamma_s(P)$ is an interval or point. We now show that Γ_s is monotone for $s \in (b-\varepsilon, b+\varepsilon)$. Let $P_1 < P_2 < P_3$ (ordered by arc length) and assume $\Gamma_s(P_2) < \Gamma_s(P_1) < \Gamma_s(P_3)$ (the other cases follow similarly). Let $x_i \in \Gamma_s(P_i)$, $i = 1,2,3$. Since $\Gamma_s([P_2,P_3])$ is connected, there exists P^*, $P_2 \leq P^* \leq P_3$ such that $x_1 \in \Gamma_s(P^*)$. Since $P_1 < P_2 \leq P^*$, this contradicts uniqueness of solutions of boundary value problems.

Assume now $\Gamma_b(R)$ is not all of R and that $\Gamma_b(R)$ is bounded above [a similar argument holds if $\Gamma_b(R)$ is bounded below]. Let $\eta = \sup \Gamma_b(R)$. We claim $\eta \notin \Gamma_b(R)$. Suppose that $\eta \in \Gamma_b(R)$. Then there exists a solution of (2.4.1), $\big(\bar{x}(t),\bar{y}(t)\big)$ satisfying (2.4.2) and $g_2\big(x(b),y(b)\big) = \eta$. Let $P_1 = \big(\bar{x}(a),\bar{y}(a)\big)$ and choose P_2, P_3 such that $P_2 < P_1 < P_3$. Then from the monotonicity of Γ_b, either $\Gamma_b(P_2) > \Gamma_b(P_1)$ or $\Gamma_b(P_3) > \Gamma_b(P_1)$. Since $\eta \in \Gamma_b(P_1)$ this contradicts the maximality of η and thus $\eta \notin \Gamma_b(R)$.

There exists $z_n \in \Gamma_b(R)$ such that $z_n \uparrow \eta$. To each z_n there exists a unique point $P_n \in S_1$ and a unique solution of (2.4.1), $(x_n(t), y_n(t))$ through P_n at $t = a$ such that $g_2(x_n(b), y_n(b)) = z_n$. Also P_n is monotonic in n. Let $\psi_n(t) = g_2(x_n(t), y_n(t))$, $t \in [a, b+\varepsilon)$. Then

$$\psi_n'(t) = \partial g_2/\partial x \,(x_n(t), y_n(t)) f_1(t, x_n(t), y_n(t)) + \partial g_2/\partial y \,(x_n(t), y_n(t)) f_2(t, x_n(t), y_n(t)).$$

As in the proof of Theorem 2.3.1, we want to obtain a convergent subsequence of initial points (x_n, y_n). For fixed $t \in (b-\varepsilon, b+\varepsilon)$, $\psi_n(t)$ is monotone in n since $\{P_n\}$ is monotone in n and BVP's are unique. Also, either $\psi_n'(b) \geq 0$ or $\psi_n'(b) \leq 0$ for infinitely many n's. We may assume $\psi_n'(b) \leq 0$. A similar argument holds when $\psi_n'(b) \geq 0$. Assume also that $\psi_n(t)$ is monotone increasing in n for $t \in (b-\varepsilon, b+\varepsilon)$ [a similar argument is valid if $\psi_n(t)$ is monotone decreasing in n]. Let $t^* \in [b, b+\varepsilon)$. Then

$$(\psi_n(t^*) - \psi_n(b))/(t^* - b) \geq (\psi_1(t^*) - \psi_n(b))/(t^* - b) \geq K = \min[0, (\psi_1(t^*) - \eta)/(t^* - b)].$$

Define $S_n = [t\colon t \geq b,\ 0 \geq \psi_n'(t) \geq K]$. From the mean value theorem, S_n is not empty. Let $s_n = \min S_n$. Then $b \leq s_n \leq b + \varepsilon$ and $K \leq \psi_n'(s_n) \leq 0$. Thus, without loss of generality $s_n \to s_0$, $\psi_n'(s_n) \to \overline{\psi}_0$. If $\psi_n'(b) = 0$, then $b = s_n$. If $\psi_n'(b) < 0$, $\psi_n(t) < \eta$ until $\psi_n'(t^*) = 0$ for some $t^* > b$. However, then $t^* \in S_n$ and thus $\psi_n(s_n) < \eta$. From the monotonicity of ψ_n we have $\psi_1(s_n) < \psi_n(s_n)$. Since $\psi_n(s_n)$, $\psi_n'(s_n)$ are bounded condition (2.4.4) implies $\{x_n(s_n), y_n(s_n)\}$ lie in a bounded region.

Choose a convergent subsequence $s_n \to s_0$, $x_n(s_n) \to x_0$,

$y_n(s_n) \to y_0$. Let S_δ be a ball of radius δ about (s_0,x_0,y_0) where $\delta < b+\varepsilon - s_0$. For large n, $(s_n,x_n,y_n) \in S_\delta$. Since all solutions exist on $[a,s_0+\delta]$, we have $\big(x_n(a),y_n(a)\big) \in S_k$ for k sufficiently large. Let $P_n = \big(x_n(a),y_n(a)\big)$. Pick $Q_1,Q_2 \in S_1$ such that $\|Q_1\| > k$, $\|Q_2\| > k$, and $Q_1 > P_n > Q_2$. From the montonicity of Γ_b either $\Gamma_b(Q_1) > \Gamma_b(P_n)$ or $\Gamma_b(Q_2) > \Gamma_b(P_n)$ for all n. Thus either $\Gamma_b(Q_1) \geq \eta$ or $\Gamma_b(Q_2) \geq \eta$. This is a contradiction to the maximality of η and the proof follows.

COROLLARY 2.4.1. Consider the BVP

$$x'' = h(t,x,x'), \tag{2.4.5}$$

$$r_1x(a) + r_2x'(a) = c_1, \tag{2.4.6}$$

$$s_1x(b) + s_2x'(b) = c_2, \tag{2.4.7}$$

where $h \in C\big[[a,b+\varepsilon)\times R^2,R\big]$ and assume the BVP (2.4.5), (2.4.6) and $s_1x(t) + s_2x'(t) = c$ has at most one solution on $[a,t]$ for each $t \in (b-\varepsilon,b+\varepsilon)$ and all $c \in R$. If

$$[s_1x+s_2x']^2 + [s_1x'+s_2h(t,x,x')]^2 \to \infty \quad \text{as} \quad x^2+x'^2 \to \infty$$

uniformly for $t \in [b-\varepsilon,b+\varepsilon]$, then there exists exactly one solution of (2.4.5) - (2.4.7).

2.5 ANGULAR FUNCTION TECHNIQUE

In the preceding sections of this chapter, we used a shooting type method to prove the existence of solutions from uniqueness assumptions. In what follows, we wish to employ the angular function technique to study a number of results concerning existence and uniqueness in a unified way. In particular, we are interested in discussing the solutions of the differential equations

$$(2.5.1) \qquad x' = f(t,x,y), \qquad y' = g(t,x,y)$$

satisfying the boundary conditions

$$(2.5.2) \qquad x(a) \sin \alpha - y(a) \cos \alpha = 0,$$

$$(2.5.3) \qquad x(b) \sin \beta - y(b) \cos \beta = 0.$$

Here we assume $f,g \in C[[a,b] \times R^2, R]$, $0 \leq \alpha < \pi$ and $0 < \beta \leq \pi$. We may geometrically interpret the problem as finding a solution $z(t) = (x(t),y(t))$ of (2.5.1) which lies on the straight line $x \sin \alpha - y \cos \alpha = 0$ at $t = a$, and on the line $x \sin \beta - y \cos \beta = 0$ at $t = b$.

With respect to the solution $z(t) = (x(t),y(t))$ the polar angle $\varphi(t)$ is defined in the xy plane. The function $\varphi(t)$ is called the <u>angular function</u> of $z(t)$ and is well defined as long as $z(t)$ does not vanish. Thus the following formulas relate $\varphi(t)$ and $z(t) \equiv (x(t),y(t))$:

$$x(t) = \|z(t)\| \cos \varphi(t), \qquad y(t) = \|z(t)\| \sin \varphi(t),$$

$$\left(\|z(t)\| = (x^2(t) + y^2(t))^{\frac{1}{2}}\right),$$

$$\varphi(t) = \arctan \left(y(t)/x(t)\right),$$

and

$$\varphi'(t) = \frac{x(t)y'(t) - y(t)x'(t)}{x^2(t) + y^2(t)} .$$

Observe also that $z(t)$ is a solution of the BVP (2.5.1) - (2.5.3) if and only if its angular function $\varphi(t)$ satisfies

$$\varphi(a) = \alpha, \qquad \varphi(b) = \beta + k\pi$$

for some integer k.

2.6 FUNDAMENTAL LEMMAS

The following lemma is a version of Lemma 2.1.1.

LEMMA 2.6.1. Assume all solutions of (2.5.1) exist on $[a,b]$. Let S be any compact connected set in $[a,b] \times R^2$. Then the set of all solutions passing through S form a compact, connected set in R^2.

Consider now the initial conditions

$$x(t_0) = x_0, \qquad y(t_0) = y_0, \tag{2.6.1}$$

and denote the solutions of (2.5.1) and (2.6.1) as $z(t,z_0) = (x(t,x_0,y_0), y(t,x_0,y_0))$, where $z_0 = (x_0,y_0)$. Let $\varphi(t,z_0)$ denote the angular function of any solution of (2.5.1) satisfying (2.6.1). Observe that $\varphi(t,z_0)$, for each $t \in [a,b]$, is multivalued since solutions are not in general uniquely defined by the initial conditions. We now present a lemma similar to Lemma 2.6.1 for angular functions.

LEMMA 2.6.2. Let S be any compact connected set of R^2 such that there exists a line through the origin which does not intersect S. Assume that for all $z_0 \in S$, $z(t,z_0) \neq 0$ for all $t \in [a,b]$. Then

$$\bigcup_{z_0 \in S} \varphi(t,z_0)$$

forms a segment for each $t \in [a,b]$.

EXERCISE 2.6.1. Prove Lemma 2.6.2.

Hint: Assume that the result is not true and use Lemma 2.6.1 to obtain a contradiction.

LEMMA 2.6.3. There exist functions $m(u)$ and $M(u)$ such that

$$m(\|z_0\|) \leq \|z(t,z_0)\| \leq M(\|z_0\|),$$

where

$$\lim_{u \to +\infty} m(u) = +\infty.$$

Proof: Define

$$M(u) = \sup_{\substack{\|z_0\| \leq u \\ t \in [a,b]}} \|z(t,z_0)\|,$$

and

$$m(u) = \inf_{\substack{\|z_0\| \geq u \\ t \in [a,b]}} \|z(t,z_0)\|.$$

By Lemma 2.6.1, $M(u)$ and $m(u)$ are well defined. We now show $\lim_{u \to +\infty} m(u) = +\infty$. Let $r_0 > 0$ be given. Consider the solutions of (2.5.1) passing through the set $S = [(t,z): t \in [a,b], \|z\| \leq r_0]$. By Lemma 2.6.1 there exists a $u_0 > 0$ such that $\|z(t)\| \leq u_0$ where $z(t)$ passes through S. Hence if $\|z_0\| > u_0$, then $\|z(t,z_0)\| > r_0$ for all $t \in [a,b]$ and thus $m(u) \geq r_0$ for $u > u_0$. This proves the lemma.

We now state the following known result on the semi-continuity behavior of solution funnels.

LEMMA 2.6.4. Assume that solutions of

$$r' = F(t,r),$$

where $F \in C[[a,b] \times R^n, R^n]$, exist on $[a,b]$. For each $\varepsilon > 0$ there exists $\delta > 0$ such that for each solution $r_1(t)$ of

$$r' = F(t,r) + \delta(t),$$

where $\delta(t)$ is continuous and $\|\delta(t)\| < \delta$, there is a solution $r_2(t)$ of $r' = F(t,r)$, where $r_1(a) = r_2(a)$, satisfying

$$\|r_1(t) - r_2(t)\| < \varepsilon$$

for all $t \in [a,b]$.

We shall now compare (2.5.1) with

$$x' = \tilde{f}(t,x,y), \qquad y' = \tilde{g}(t,x,y), \tag{2.6.2}$$

where $\tilde{f},\tilde{g} \in C\big[[a,b]\times R^2,R\big]$. Assume also that the right-hand side of (2.6.2) are positively homogeneous, that is,

$$\tilde{f}(t,cx,cy) = c\tilde{f}(t,x,y), \qquad \tilde{g}(t,cx,cy) = c\tilde{g}(t,x,y) \tag{2.6.3}$$

for all $c \geq 0$. We write $\{f,g\} \geq \{\tilde{f},\tilde{g}\}$ if

$$xg(t,x,y) - yf(t,x,y) \geq x\tilde{g}(t,x,y) - y\tilde{f}(t,x,y).$$

We now present a comparison result.

LEMMA 2.6.5. Let solutions $z(t)$ and $\tilde{z}(t)$ of systems (2.5.1) and (2.6.2), respectively, be defined on $[a,b]$ and never vanish. Let $\varphi(t)$ and $\tilde{\varphi}(t)$ be the respective angular functions of the solutions and assume

$$\varphi(a) \geq \tilde{\varphi}(a), \qquad \{f,g\} \geq \{\tilde{f},\tilde{g}\}. \tag{2.6.4}$$

Then

$$\varphi(t) \geq \tilde{\varphi}_m(t) \qquad \text{for all } t \in [a,b],$$

where $\tilde{\varphi}_m(t)$ is the minimal solution of

$$\tilde{\varphi}'(t) = \tilde{g}(t,\cos\tilde{\varphi},\sin\tilde{\varphi})\cos\tilde{\varphi} - \tilde{f}(t,\cos\tilde{\varphi},\sin\tilde{\varphi})\sin\tilde{\varphi}. \tag{2.6.5}$$

<u>Proof</u>: Since

$$\varphi(t) = \arctan\big(y(t)/x(t)\big), \quad \tilde{\varphi}(t) = \arctan\big(\tilde{y}(t)/\tilde{x}(t)\big),$$

we have, using (2.6.3),

$$\varphi'(t) = \big(1/\|z(t)\|\big)\,\big[g\big(t,\|z(t)\|\cos\varphi,\ \|z(t)\|\sin\varphi\big)\cos\varphi - f\big(t,\ \|z(t)\|\cos\varphi,\ \|z(t)\|\sin\varphi\big)\sin\varphi\big]$$

and

$$\tilde{\varphi}'(t) = \tilde{g}(t, \cos\tilde{\varphi}, \sin\tilde{\varphi})\cos\tilde{\varphi} - \tilde{f}(t, \cos\tilde{\varphi}, \sin\tilde{\varphi})\sin\tilde{\varphi}$$
$$\equiv F(t, \tilde{\varphi}).$$

From (2.6.4), it follows that

$$\varphi(a) \geq \tilde{\varphi}(a) \qquad \text{and} \qquad \varphi'(t) \geq F\big(t, \varphi(t)\big).$$

The theory of differential inequalities then gives

$$\varphi(t) \geq \tilde{\varphi}_m(t) \qquad \text{for} \quad t \in [a,b].$$

The proof is complete.

If solutions of (2.5.1) and (2.6.2) are uniquely determined by initial conditions, then we may conclude that $\varphi(t) \geq \tilde{\varphi}(t)$. Observe also that $\tilde{f}(t,0,0) \equiv \tilde{g}(t,0,0) \equiv 0$ so that $z(t) \equiv 0$ is a solution of (2.6.2). Moreover, we notice, from the homogeneity, that $\|\tilde{f}(t,x,y)\| \leq A\,(x^2 + y^2)^{\frac{1}{2}}$, $\|\tilde{g}(t,x,y)\| \leq A\,(x^2 + y^2)^{\frac{1}{2}}$ for some $A > 0$. This implies that all solutions of (2.6.2) exist on $[a,b]$. From the calculations used in the previous lemma, we see that the angular function satisfies a differential equation independent of the solution, so it is consistent to discuss the angular function of a system with a given initial angle.

2.7 EXISTENCE

Before proving existence results for (2.5.1) - (2.5.3), it is necessary that certain restrictions be placed on the solutions of the comparison equation (2.6.2) which satisfy the boundary conditions (2.5.2) and (2.5.3). When solutions of (2.5.1) and (2.6.2) are uniquely determined by initial conditions, these restrictions reduce to the following: let $\tilde{\varphi}^+(t)$ and $\tilde{\varphi}^-(t)$ denote the angular functions of (2.6.2) satisfying the initial conditions $\tilde{\varphi}^+(a) = \alpha$, $\tilde{\varphi}^-(a) = \alpha + \pi$ and suppose that

$$\beta + k\pi < \tilde{\varphi}^{+}(b) < \beta + (k+1)\pi$$

and

$$\beta + (k+1)\pi < \tilde{\varphi}^{-}(b) < \beta + (k+2)\pi$$

for a certain integer k. This condition, in fact, implies that the only solution of the BVP (2.6.2), (2.5.2), (2.5.3) is $z(t) \equiv 0$.

Since we do not assume the uniqueness of solutions, the previous conditions will be in terms of minimal and maximal solutions.

Consider the auxiliary systems

$$x' = a(t,x,y), \qquad y' = b(t,x,y) \tag{2.7.1}$$

and

$$x' = c(t,x,y), \qquad y' = d(t,x,y), \tag{2.7.2}$$

where $a,b,c,d \in C\big[[a,b]\times R^2,R\big]$ and are positively homogeneous in x and y. Let $\theta(t)$ and $\psi(t)$ be the angular functions of (2.7.1) and (2.7.2). Then a computation similar to one in Lemma 2.6.5 yields

$$\theta'(t) = b(t,\cos\theta,\sin\theta)\cos\theta - a(t,\cos\theta,\sin\theta)\sin\theta, \tag{2.7.3}$$

and

$$\psi'(t) = d(t,\cos\psi,\sin\psi)\cos\psi - c(t,\cos\psi,\sin\psi)\sin\psi. \tag{2.7.4}$$

In order to avoid repetition, let us list below certain hypotheses that will be needed subsequently.

(H_1) The maximal solutions $\psi_M^{+}(t)$ and $\psi_M^{-}(t)$ of (2.7.4) satisfying $\psi_M^{+}(a) = \alpha$ and $\psi_M^{-}(a) = \alpha + \pi$ obey

$$\psi_M^{+}(b) < \beta + (k+1)\pi \qquad \text{and} \qquad \psi_M^{-}(b) < \beta + (k+2)\pi$$

for some integer k;

(H_2) The minimal solutions $\theta_m^+(t)$ and $\theta_m^-(t)$ of (2.7.3) satisfying $\theta_m^+(a) = \alpha$ and $\theta_m^-(a) = \alpha + \pi$ obey

$$\theta_m^+(b) > \beta + k\pi \quad \text{and} \quad \theta_m^-(b) > \beta + (k+1)\pi$$

for the same integer k as in (H_1);

(H_3) The following inequality holds between systems (2.5.1), (2.7.1), and (2.7.2):

$$\begin{aligned} b(t,x,y)x - a(t,x,y)y + \underline{\delta}(t,x,y) &\le g(t,x,y)x - f(t,x,y)y \\ &\le d(t,x,y)x - c(t,x,y)y + \overline{\delta}(t,x,y), \end{aligned}$$

where $\underline{\delta}, \overline{\delta} \in C\big[[a,b] \times R^2, R\big]$ and

$$\lim_{x^2+y^2 \to \infty} \frac{\underline{\delta}(t,x,y)}{x^2+y^2} = 0, \qquad \lim_{x^2+y^2 \to \infty} \frac{\overline{\delta}(t,x,y)}{x^2+y^2} = 0$$

uniformly for $t \in [a,b]$.

Observe that when solutions are unique (H_1) and (H_2) imply that the only solution of (2.7.1), (2.5.2), (2.5.3) and (2.7.2), (2.5.2), (2.5.3) is the trivial solution.

We now state and prove our main existence result.

THEOREM 2.7.1. Under the hypotheses (H_1) - (H_3), the BVP (2.5.1) - (2.5.3) admits a solution.

Proof: Let $z(t,c) = \big(x(t,c), y(t,c)\big)$ be any solution of (2.5.1) satisfying the initial condition

$$x(a) = c\cos\alpha, \qquad y(a) = c\sin\alpha. \tag{2.7.5}$$

Let $\varphi(t,c)$ be the angular function of $z(t,c)$ such that

$$\varphi(a,c) = \begin{cases} \alpha & \text{if } c > 0, \\ \alpha + \pi & \text{if } c < 0. \end{cases}$$

From (H_3), we obtain

$$(2.7.6)\quad \varphi'(t,c) \geq b(t,\cos\varphi,\sin\varphi)\cos\varphi - a(t,\cos\varphi,\sin\varphi)\sin\varphi + \delta_0(t),$$

where $\delta_0(t) \equiv \underline{\delta}(t,x(t,c),y(t,c))/(x^2(t,c)+y^2(t,c))$. Let $\tilde{\theta}_m^+(t)$, $\tilde{\theta}_m^-(t)$ be the minimal solutions of

$$(2.7.7)\quad \tilde{\theta}'(t) = b(t,\cos\tilde{\theta},\sin\tilde{\theta})\cos\tilde{\theta} - a(t,\cos\tilde{\theta},\sin\tilde{\theta})\sin\tilde{\theta} + \delta_0(t),$$

where $\tilde{\theta}_m^+(a) = \alpha$, $\tilde{\theta}_m^-(a) = \alpha + \pi$.

From (2.7.6) and the proof of Lemma 2.6.5, it follows that

$$(2.7.8)\qquad \varphi(t,c) \geq \begin{cases} \theta_m^+(t) & \text{if } c > 0, \\ \theta_m^-(t) & \text{if } c < 0. \end{cases}$$

In view of (H_2), we may pick $\varepsilon > 0$ so small that

$$(2.7.9)\quad \theta_m^+(b) > \beta + k\pi + \varepsilon, \qquad \theta_m^-(b) > \beta + (k+1)\pi + \varepsilon.$$

By Lemma 2.6.4 and (H_2), we may pick $\delta > 0$ so that there exist solutions $\theta^+(t)$, $\theta^-(t)$ of (2.7.3), where $\theta^+(a) = \alpha$, $\theta^-(a) = \alpha + \pi$, such that

$$(2.7.10)\qquad |\tilde{\theta}_m^+(t) - \theta^+(t)| < \varepsilon \qquad \text{and} \qquad |\theta_m^-(t) - \theta^-(t)| < \varepsilon,$$

whenever $|\delta_0(t)| < \delta$ for all $t \in [a,b]$. Because of (H_3), there exists a $\rho > 0$ satisfying $\delta_0(t) < \delta$ if $\|z(t,c)\| > \rho$. Moreover, Lemma 2.6.3 implies that there exists an $r_0 > 0$ such that, if $|c| \equiv \|z(a,c)\| > r_0$, then $\|z(t,c)\| > \rho$.

We thus obtain, using (2.7.8) - (2.7.10), that for $c \geq r_0$

$$(2.7.11)\quad \varphi(b,c) > \theta_m^+(b) \geq \theta^+(b) - \varepsilon > \theta_m^+(b) - \varepsilon > \beta + k\pi,$$

and for $c \leq -r_0$

$$\varphi(b,c) \geq \theta_m^-(b) \geq \theta^-(b) - \varepsilon > \theta_m^-(b) - \varepsilon > \beta + (k+1)\pi. \tag{2.7.12}$$

By applying a similar analysis to the maximal solutions of (2.7.4) and using (H_1), we infer the existence of $r_1 > 0$ such that, for $c \geq r_1$

$$\varphi(b,c) < \beta + (k+1)\pi, \tag{2.7.13}$$

and for

$$c \leq -r_1$$

$$\varphi(b,c) < \beta + (k+2)\pi. \tag{2.7.14}$$

Hence, for $c \geq r_3 \equiv \max(r_0, r_1)$,

$$\beta + k\pi < \varphi(b,c) < \beta + (k+1)\pi,$$

and for $c \leq -r_3$

$$\beta + (k+1)\pi < \varphi(b,c) < \beta + (k+2)\pi.$$

This asserts that the points $z(b,r_3)$ and $z(b,-r_3)$ lie on different sides of the straight line $x \sin\beta - y\cos\beta = 0$. Consider the set of solutions $\bigcup_{\|z\| \leq r_3} [z(t,c)]$. By Lemma 2.6.1, this set is connected in R^2 for each t. Hence there is $c^* \in (-r_3, r_3)$ such that $z(b,c^*)$ lies on the straight line $x \sin\beta - y\cos\beta = 0$. Thus $z(t,c^*)$ is a solution of the BVP. This concludes the proof of Theorem 2.7.1.

We leave the proof of the following theorem as an exercise since its proof is very similar to the proof of Theorem 2.7.1.

THEOREM 2.7.2. Consider the system (2.7.1) and assume (H_2) holds. Moreover suppose the maximal solutions $\theta_M^+(t)$ and $\theta_M^-(t)$ of (2.7.3) satisfying $\theta_M^+(a) = \alpha$ and $\theta_M^-(a) = \alpha + \pi$

obey $\theta_M^+(b) < \beta + (k+1)\pi$ and $\theta_M^-(b) < \beta + (k+2)\pi$. Then there exists a solution of

$$x' = a(t,x,y) + u(t,x,y), \qquad y' = b(t,x,y) + v(t,x,y) \tag{2.7.15}$$

satisfying the boundary conditions (2.5.2), (2.5.3) provided

$$\lim_{|x|+|y|\to\infty} \frac{|u(t,x,y)| + |v(t,x,y)|}{|x| + |y|} = 0.$$

EXERCISE 2.7.1. Prove Theorem 2.7.2.

COROLLARY 2.7.1. Under the hypotheses of Theorem 2.7.1 (Theorem 2.7.2) there exists solutions of (2.5.1) [(2.7.15)] satisfying the boundary conditions

$$\begin{aligned} x(a)\sin\alpha - y(a)\cos\alpha &= c_1, \\ x(b)\sin\beta - y(b)\cos\beta &= c_2 \end{aligned} \tag{2.7.16}$$

for any $c_1, c_2 \in R$.

EXERCISE 2.7.2. Prove Corollary 2.7.1.

Hint: Use the fact that a homogeneous expression is bounded by $A(x^2 + y^2)^{\frac{1}{2}}$ and make a change of variables by translating the point of intersection of the two lines in (2.7.16) to the origin. Then apply Theorem 2.7.1 (Theorem 2.7.2).

EXAMPLE 2.7.1. The equation

$$x''(t) + \alpha \sin x + h(t) = 0$$

has a solution which satisfies the condition $x(a) = x(b) = 0$. To show this consider the solutions of

$$x'' = 0, \qquad x(a) = x(b) = 0.$$

It is clear that the only solution of this BVP is the trivial solution $x(t) \equiv 0$. Moreover $|\alpha \sin x + h(t)|/(|x| + |x'|) \to 0$ as $|x| + |x'| \to \infty$. An immediate application of Theorem 2.7.2 yields the desired result.

EXERCISE 2.7.3. Show that the boundary value problem

$$x'' + g(t,x,x') = 0, \qquad x(a) = x(b) = 0,$$

has a solution, if

$$|g(t,x,x')| \leq A|x| + B, \qquad \text{where} \quad A(b-a)^2 < \pi^2.$$

2.8 UNIQUENESS

In order to prove the uniqueness of solutions of BVP (2.5.1) - (2.5.3), we need the following inequalities:

$$\begin{aligned}
L_1(t,x_1-x_2) &\leq f(t,x_1,y) - f(t,x_2,y) \leq L_2(t,x_1-x_2) \\
&\qquad \text{if} \quad x_1 \geq x_2, \\
K_1(t,y_1-y_2) &\leq f(t,x,y_1) - f(t,x,y_2) \leq K_2(t,y_1-y_2) \\
&\qquad \text{if} \quad y_1 \geq y_2, \\
M_1(t,x_1-x_2) &\leq g(t,x_1,y) - g(t,x_2,y) \leq M_2(t,x_1-x_2) \\
&\qquad \text{if} \quad x_1 \geq x_2, \\
N_1(t,y_1-y_2) &\leq g(t,x,y_1) - g(t,x,y_2) \leq N_2(t,y_1-y_2) \\
&\qquad \text{if} \quad y_1 \geq y_2.
\end{aligned} \tag{2.8.1}$$

Assume that L_i, M_i, K_i, $N_i \in C[[a,b] \times R^+, R]$ and are positively homogeneous in their second variable. From (2.8.1), we may conclude that

$$\begin{aligned}
f(t,x_1,y_1) - f(t,x_2,y_2) &= f(t,x_1,y_1) - f(t,x_2,y_1) \\
&\quad + f(t,x_2,y_1) - f(t,x_2,y_2) \\
&\leq G_2(t,x_1-x_2,y_1-y_2);
\end{aligned}$$

and

$$f(t,x_1,y_1) - f(t,x_2,y_2) \geq G_1(t,x_1 - x_2,y_1 - y_2).$$

Similarly

$$g(t,x_1,y_1) - g(t,x_2,y_2) \leq H_2(t,x_1 - x_2,y_1 - y_2),$$

$$g(t,x_1,y_1) - g(t,x_2,y_2) \geq H_1(t,x_1 - x_2,y_1 - y_2), \quad \text{where}$$

$$G_2(t,u,v) = \begin{cases} L_2(t,u) + K_2(t,v), & u \geq 0, \quad v \geq 0, \\ L_2(t,u) + K_1(t,v), & u \geq 0, \quad v \leq 0, \\ L_1(t,u) + K_1(t,v), & u \leq 0, \quad v \leq 0, \\ L_1(t,u) + K_2(t,v), & u \leq 0, \quad v \geq 0, \end{cases}$$

$$G_1(t,u,v) = \begin{cases} L_1(t,u) + K_1(t,v), & u \geq 0, \quad v \geq 0, \\ L_1(t,u) + K_2(t,v), & u \geq 0, \quad v \leq 0, \\ L_2(t,u) + K_2(t,v), & u \leq 0, \quad v \leq 0, \\ L_2(t,u) + K_1(t,v), & u \leq 0, \quad v \geq 0, \end{cases}$$

$$H_2(t,u,v) = \begin{cases} M_2(t,u) + N_2(t,v), & u \geq 0, \quad v \geq 0, \\ M_2(t,u) + N_1(t,v), & u \geq 0, \quad v \leq 0, \\ M_1(t,u) + N_1(t,v), & u \leq 0, \quad v \leq 0, \\ M_1(t,u) + N_2(t,v), & u \leq 0, \quad v \geq 0, \end{cases}$$

and

$$H_1(t,u,v) = \begin{cases} M_1(t,u) + N_1(t,v), & u \geq 0, \quad v \geq 0, \\ M_1(t,u) + N_2(t,v), & u \geq 0, \quad v \leq 0, \\ M_2(t,u) + N_2(t,v), & u \leq 0, \quad v \leq 0, \\ M_2(t,u) + N_1(t,v), & u \leq 0, \quad v \geq 0. \end{cases}$$

Define now the functions

$$P_1(t,u,v) = \begin{cases} L_1(t,u) + K_1(t,v), & uv \geq 0, \\ L_2(t,u) + K_1(t,v), & uv < 0, \end{cases}$$

$$P_2(t,u,v) = \begin{cases} M_2(t,u) + N_2(t,v), & uv \geq 0, \\ M_2(t,u) + N_1(t,v), & uv < 0, \end{cases}$$

$$Q_1(t,u,v) = \begin{cases} L_2(t,u) + K_2(t,v), & uv \geq 0, \\ L_1(t,u) + K_2(t,v), & uv < 0, \end{cases}$$

$$Q_2(t,u,v) = \begin{cases} M_1(t,u) + N_1(t,v), & uv \geq 0, \\ M_1(t,u) + N_2(t,v), & uv < 0. \end{cases}$$

Consider the comparison systems

$$(2.8.2) \qquad u' = P_1(t,u,v), \qquad v' = P_2(t,u,v),$$

$$(2.8.3) \qquad u' = Q_1(t,u,v), \qquad v' = Q_2(t,u,v).$$

Observe that P_i, Q_i, $i = 1,2$ are positively homogeneous in u and v. Letting $\psi^+(t)$, $\psi^-(t)$ be the angular functions associated with (2.8.2) and $\theta^+(t)$, $\theta^-(t)$ be the angular functions associated with (2.8.3), we have

$$(2.8.4) \qquad \psi'(t) = P_1(t,\cos\psi,\sin\psi)\cos\psi - P_2(t,\cos\psi,\sin\psi)\sin\psi \equiv F_1(t,\psi),$$

$$(2.8.5) \qquad \theta'(t) = Q_1(t,\cos\theta,\sin\theta)\cos\theta - Q_2(t,\cos\theta,\sin\theta)\sin\theta \equiv F_2(t,\theta).$$

Since P_i, Q_i are continuous everywhere except along the coordinate axes, we agree that solutions of (2.8.2), (2.8.3) are differentiable almost everywhere. A similar statement holds for the angular functions which are not differentiable along the angular rays $n\pi/2$, $n = 0, \pm 1, \pm 2, \ldots$.

We are now able to state and prove a uniqueness result.

THEOREM 2.8.1. Assume f and g satisfy (2.8.1) and that conditions (H_1) and (H_2) hold relative to (2.8.4) and (2.8.5), respectively. Then if solutions of (2.5.1) are uniquely defined by the initial conditions, there exists one solution of the BVP (2.5.1) - (2.5.3).

Proof: Assume that there exist two solutions $z_1(t) = (x_1(t), y_1(t))$ and $z_2(t) = (x_2(t), y_2(t))$ of the BVP. Let

$$\bar{x}(t) = x_1(t) - x_2(t) \quad \text{and} \quad \bar{y}(t) = y_1(t) - y_2(t);$$

then $\bar{z}(t) = (\bar{x}(t), \bar{y}(t))$ satisfy the boundary conditions (2.5.2), (2.5.3). Also let $\bar{\varphi}(t)$ be the angular function of $\bar{z}(t)$. For convenience, we shall let $\bar{x} \equiv \bar{x}(t)$, $\bar{y} = \bar{y}(t)$, $\bar{z} = \bar{z}(t)$, and $\bar{\varphi} = \bar{\varphi}(t)$. We claim that

$$\begin{aligned} \bar{x}Q_2(t,\bar{x},\bar{y}) - \bar{y}Q_1(t,\bar{x},\bar{y}) &\leq \bar{y}'\bar{x} - \bar{x}'\bar{y} \\ &\leq \bar{x}P_2(t,\bar{x},\bar{y}) - \bar{y}P_1(t,\bar{x},\bar{y}). \end{aligned} \tag{2.8.6}$$

To see this, observe

$$\begin{aligned} \bar{y}'\bar{x} - \bar{x}'\bar{y} = \bar{x}\big(g(t,x_1,y_1) - g(t,x_2,y_2)\big) - \bar{y}\big(&f(t,x_1,y_1) \\ &- f(t,x_2,y_2)\big). \end{aligned}$$

From our previous estimates

$$\begin{aligned} \bar{x}\big(g(t,x_1,y_1) - g(t,x_2,y_2)\big) &\leq \begin{cases} \bar{x}H_2(t,\bar{x},\bar{y}) & \text{if } \bar{x} \geq 0, \\ \bar{x}H_1(t,\bar{x},\bar{y}) & \text{if } \bar{x} < 0, \end{cases} \\ &= \begin{cases} \bar{x}\big(M_2(t,\bar{x}) + N_2(t,\bar{y})\big) & \text{if } \bar{x}\,\bar{y} \geq 0, \\ \bar{x}\big(M_2(t,\bar{x}) + N_2(t,\bar{y})\big) & \text{if } \bar{x}\,\bar{y} < 0, \end{cases} \\ &= \bar{x}P_2(t,\bar{x},\bar{y}). \end{aligned}$$

Similarly,

$$-\bar{y}\big(f(t,x_1,y_1) - f(t,x_2,y_2)\big) \leq -\bar{y}P_2(t,\bar{x},\bar{y}).$$

The lower estimates in (2.8.6) are obtained in the same manner.

Hence we obtain from (2.8.6),

$$(2.8.7) \qquad F_2(t,\bar{\varphi}) \leq \bar{\varphi}' \equiv (\bar{x}\,\bar{y}' - \bar{y}\,\bar{x}')/(\bar{x}^2 + \bar{y}^2) \leq F_1(t,\bar{\varphi}),$$

where F_1, F_2 correspond to (2.8.4) and (2.8.5), respectively.

Notice that $\big(\bar{x}(a),\bar{y}(a)\big) \neq (0,0)$ by the uniqueness of initial value problems and that there exists no $t_0 \in [a,b]$ such that $\bar{z}(t_0) = \big(\bar{x}(t_0),\bar{y}(t_0)\big) = (0,0)$. This implies $\bar{\varphi}(t)$ is well defined.

From (H_1), (H_2), and (2.8.6) and the theory of differential inequalities, which is applicable since $F_1(t,\bar{\varphi}), F_2(t,\bar{\varphi})$ are continuous everywhere except at $\theta = n\pi/2$, $n = 0, \pm 1, \pm 2, \ldots$, we obtain, using exactly the same methods as in the proof of Theorem 2.7.1, that either

$$\beta + k\pi < \bar{\varphi}(b) < \beta + (k+1)\pi \qquad \text{if} \quad \varphi(a) = \alpha$$

or

$$\beta + (k+1)\pi < \bar{\varphi}(b) < \beta + (k+2)\pi \qquad \text{if} \quad \varphi(a) = \alpha + \pi.$$

In either case $\bar{x}(b),\bar{y}(b)$ cannot satisfy

$$\bar{x}(b) \sin \beta - \bar{y}(b) \cos \beta = 0,$$

unless $\big(\bar{x}(b),\bar{y}(b)\big) \equiv 0$, a contradiction. The existence of a solution follows from the next corollary.

COROLLARY 2.8.1. Assume that the hypotheses of Theorem 2.8.1 hold with the exception of the uniqueness of solutions of initial value problems. Then there exists a solution of the BVP (2.5.1) - (2.5.3).

<u>Proof</u>: In view of Theorem 2.7.1, it suffices to show (H_3) holds. Let $x_2 = y_2 \equiv 0$ and $x_1 = x$, $y_1 = y$. Then we

obtain from (2.8.6)

$$\begin{aligned} xQ_2(t,x,y) - yQ_1(t,x,y) &+ xg(t,0,0) - yf(t,0,0) \\ &\leq xg(t,x,y) - yf(t,x,y) \\ &\leq xP_2(t,x,y) - yP_1(t,x,y) + xg(t,0,0) - yf(t,0,0). \end{aligned}$$

Letting $\delta(t,x,y) = xg(t,0,0) - yf(t,0,0)$, we see that

$$\lim_{x^2+y^2\to\infty} \delta(t,x,y)/(x^2+y^2) = 0$$

uniformly for $t \in [a,b]$. Thus (H_3) holds and the proof of Corollary 2.8.1 is done.

As before, our results immediately apply to the more general boundary conditions (2.7.16).

THEOREM 2.8.2. Under the conditions of Theorem 2.8.1 (Corollary 2.8.1) the BVP (2.5.1), (2.7.16) has exactly one solution (at least one solution).

EXERCISE 2.8.1. Prove Theorem 2.8.2.

We now apply our results to the case in which $f,g \in C^{(1)}\big[[a,b]\times R^2,R\big]$. In order to do so we must consider first the case of linear systems; namely, the system

$$(2.8.8) \qquad x' = a_{11}(t)x + a_{12}(t)y, \qquad y' = a_{21}(t)x + a_{22}(t)y,$$

where $a_{ij} \in C\big[[a,b],R\big]$. Let $\theta = \theta(t)$ be the polar angle associated with (2.8.8). Then

$$(2.8.9) \quad \begin{aligned} \theta'(t) &= a_{21}\cos^2\theta + (a_{22} - a_{11})\cos\theta\sin\theta - a_{12}\sin^2\theta \\ &\equiv F_3(t,\theta). \end{aligned}$$

Assume now there exist two linear systems

$$(2.8.10) \qquad z' = \underline{A}(t)z, \qquad z' = \overline{A}(t)z,$$

where

$$\underline{A}(t) = \begin{pmatrix} \underline{a}_{11}(t) & \underline{a}_{12}(t) \\ \underline{a}_{21}(t) & \underline{a}_{22}(t) \end{pmatrix},$$

and

$$\overline{A}(t) = \begin{pmatrix} \overline{a}_{11}(t) & \overline{a}_{12}(t) \\ \overline{a}_{21}(t) & \overline{a}_{22}(t) \end{pmatrix}.$$

Letting φ, ψ be the angular functions associated with $z' = \overline{A}(t)z$ and $z' = \underline{A}(t)z$, respectively, we obtain, as in (2.8.9),

$$\text{(2.8.11)} \qquad \varphi'(t) \equiv F_3(t,\varphi) \qquad \text{and} \qquad \psi'(t) = F_3(t,\psi).$$

Since f and g are differentiable we have, for (x_1,y_1) and (x_2,y_2),

$$\text{(2.8.12)} \quad f(t,x_1,y_1) - f(t,x_2,y_2) = a_{11}(t,x_1,y_1,x_2,y_2)(x_1 - x_2) + a_{12}(t,x_1,y_1,x_2,y_2)(y_1 - y_2);$$

$$\text{(2.8.13)} \quad g(t,x_1,y_1) - g(t,x_2,y_2) = a_{21}(t,x_1,y_1,x_2,y_2)(x_1 - y_2) + a_{22}(t,x_1,y_1,x_2,y_2)(y_1 - y_2);$$

where

$$a_{11} = \int_0^1 f_x\big(t,x_2 + s(x_1 - x_2),y_2 + s(y_1 - y_2)\big)\, ds,$$

$$a_{12} = \int_0^1 f_y\big(t,x_2 + s(x_1 - x_2),y_2 + s(y_1 - y_2)\big)\, ds,$$

$$a_{21} = \int_0^1 g_x\big(t,x_2 + s(x_1 - x_2),y_2 + s(y_1 - y_2)\big)\, ds,$$

$$a_{22} = \int_0^1 g_y\big(t,x_2 + s(x_1 - x_2),y_2 + s(y_1 - y_2)\big)\, ds.$$

Assume now the following relationship between $\overline{A}(t)$, $\underline{A}(t)$, and (2.8.8):

$$\begin{aligned}(2.8.14)\quad &\underline{a}_{21}(t)u^2 + [\underline{a}_{22}(t) - \underline{a}_{11}(t)]uv - \underline{a}_{12}(t)v^2 \\ &\leq g_x(t,x,y)u^2 + [g_y(t,x,y) - f_x(t,x,y)]uv - f_y(t,x,u)v^2 \\ &\leq \overline{a}_{21}(t)u^2 + [\overline{a}_{22}(t) - \overline{a}_{11}(t)]uv - \overline{a}_{12}(t)v^2\end{aligned}$$

for all $t \in [a,b]$ and $-\infty < x,y,u,v < +\infty$. Then the following uniqueness statement is true.

COROLLARY 2.8.2. Assume in (2.5.1) that $f,g \in C^{(1)}\big[[a,b] \times R^2, R\big]$ and that there exist systems (2.8.10) satisfying (2.8.14). Assume also that the solutions $\varphi(t)$, $\psi(t)$ of (2.8.11) satisfy (H_1) and (H_2), respectively. Then the BVP (2.5.1), (2.7.16) has exactly one solution.

Proof: Replacing x by $x_2 + s(x_1 - x_2)$, y by $y_2 + s(y_1 - y_2)$, and integrating (2.8.14) from 0 to 1 with respect to s, we obtain

$$\begin{aligned}(2.8.15)\quad &\underline{a}_{21}(t)u^2 + [\underline{a}_{22}(t) - \underline{a}_{11}(t)]uv - \underline{a}_{12}(t)v^2 \\ &\leq a_{21}(t)u^2 + [a_{22}(t) - a_{11}(t)]uv - a_{12}(t)v^2 \\ &\leq \overline{a}_{21}(t)u^2 + [\overline{a}_{22}(t) - \overline{a}_{11}(t)]uv - \overline{a}_{12}(t)v^2.\end{aligned}$$

Hypotheses (H_1), (H_2), and (2.8.15) imply that there exists no nontrivial solution of the BVP (2.8.8) and (2.7.16). In (2.8.12) and (2.8.13) let $x_2 = y_2 = 0$ and $x_1 = x$, $y_1 = y$; then (2.5.1) reduces to

$$(2.8.16)\quad \begin{aligned} x' &= a_{11}(t)x + a_{12}(t)y + \delta_3(t), \\ y' &= a_{21}(t)x + a_{22}(t)y + \delta_4(t), \end{aligned}$$

where $\delta_3(t) = f(t,0,0)$ and $\delta_4(t) = g(t,0,0)$. Hence

$$\delta_3(t)/(x^2+y^2) \to 0, \qquad \delta_4(t)/(x^2+y^2) \to 0 \qquad \text{as } x^2 + y^2 \to \infty.$$

Because of Theorem 2.7.1, this assures us that (2.8.16), (2.7.16) or the equivalent problem (2.5.1), (2.7.16) has a solution.

Assume now that there are two solutions $z_1(t) = (x_1(t), y_1(t))$, $z_2(t) = (x_2(t), y_2(t))$ of (2.5.1), (2.7.16). Then $\bar{x}(t) = x_1(t) - x_2(t)$, $\bar{y}(t) = y_1(t) - y_2(t)$ is a solution of (2.8.8), (2.5.2), (2.5.3). As just mentioned, there exists no nontrivial solution of (2.8.8), (2.5.2), (2.5.3). Hence $\bar{x}(t) \equiv \bar{y}(t) \equiv 0$. Thus solutions are unique and the proof is complete.

EXAMPLE 2.8.1. We show that the equation

$$x''(t) + \delta \sin x + h(t) = 0$$

has a unique solution satisfying the boundary conditions $x(a) = x(b) = 0$ if

$$|\delta| < \pi^2/(b-a)^2.$$

In Example 2.7.1 we have shown there exists at least one solution. We now apply Corollary 2.8.2 to prove there exists at most one solution. Observe that the system

$$x''(t) \pm |\delta| x = 0$$

satisfies (2.8.14) with $x(a) = x(b) = 0$. We know that solutions of this problem are unique, if $|\delta| < \pi^2/(b-a)^2$. Hence, according to Corollary 2.8.2, the equation with conditions $x(a) = x(b) = 0$ has a unique solution.

EXERCISE 2.8.2. Show that the boundary value problem

$$x'' + g(t,x) = 0, \qquad x(a) = x(b) = 0,$$

has a unique solution if

$$\varepsilon \le g_x(t,x) \le \left(\pi^2/(b-a)^2\right) - \varepsilon, \qquad \varepsilon > 0.$$

2.9 ESTIMATION OF NUMBER OF SOLUTIONS

In some cases, it is of interest to determine if a BVP has several solutions and to estimate the number of these solutions.

Before stating our main result we need some preliminary developments. Assume that there exist two systems

$$(2.9.1) \qquad x' = r(t,x,y), \qquad y' = s(t,x,y),$$

$$(2.9.2) \qquad x' = v(t,x,y), \qquad y' = w(t,x,y),$$

where $r,s,v,w \in C[(a,b]\times R^2,R]$ and positively homogeneous. Let θ and ψ be the polar angles associated with (2.9.1) and (2.9.2), respectively. Then

$$(2.9.3) \quad \theta' = s(t,\cos\theta,\sin\theta)\cos\theta - r(t,\cos\theta,\sin\theta)\sin\theta$$

and

$$(2.9.4) \quad \psi' = w(t,\cos\psi,\sin\psi)\cos\psi - v(t,\cos\psi,\sin\psi)\sin\psi.$$

We now assume conditions similar to (H_1) - (H_3) of Theorem 2.7.1; that is, assume that:

(H_1') the maximal solutions $\psi_M^+(t)$ and $\psi_M^-(t)$ of (2.9.4) satisfying $\psi_M^+(a) = \alpha$ and $\psi_M^-(a) = \alpha + \pi$ have the property

$$\psi_M^+(b) < \beta + (\ell+1)\pi \qquad \text{and} \qquad \psi_M^-(b) < \beta + (\ell+2)\pi;$$

(H_2') the minimal solutions $\theta_m^+(t)$ and $\theta_m^-(t)$ of (2.9.3) satisfying $\theta_m^+(a) = \alpha$ and $\theta_m^-(a) = \alpha + \pi$ have the property

$$\theta_m^+(b) > \beta + \ell\pi \qquad \text{and} \qquad \theta_m^-(b) > \beta + (\ell+1)\pi;$$

(H_3') the following inequality holds

$$\begin{aligned} s(t,x,y)x - r(t,x,y)y &+ \delta_1(t,x,y) \\ &\le g(t,x,y)x - f(t,x,y)y \\ &\le v(t,x,y)x - w(t,x,y)y + \delta_2(t,x,y), \end{aligned}$$

where δ_1, δ_2 are continuous and

$$\lim_{x^2+y^2\to 0} \frac{\delta_1(t,x,y)}{x^2 + y^2} = 0 \quad \text{and} \quad \lim_{x^2+y^2\to 0} \frac{\delta_2(t,x,y)}{x^2 + y^2} = 0.$$

We are now able to estimate the number of solutions of the BVP (2.5.1) - (2.5.3).

THEOREM 2.9.1. Assume that conditions (H_1) - (H_3), (H_1') - (H_3') are satisfied. Then the BVP (2.5.1) - (2.5.3) has at least $2|k - \ell|$ nonzero solutions, whenever $k \neq \ell$.

Proof: Let $z(t,c) = \big(x(t,c),y(t,c)\big)$ be any solution of (2.5.1) satisfying the initial conditions

$$x(a) = c \cos \alpha, \qquad y(a) = c \sin \alpha$$

and let $\varphi(t,c)$ be the angular function of this solution such that

$$\varphi(a,c) = \begin{cases} \alpha & \text{if} \quad c > 0, \\ \alpha + \pi & \text{if} \quad c < 0. \end{cases}$$

Assume now $c > 0$. Using the same proof as that of Theorem 2.7.1, we conclude from (H_1) - (H_3) that there exists an $r_1 > 0$ such that for $c \geq r_1$

$$\beta + k\pi < \varphi(b,c) < \beta + (k+1)\pi.$$

Similarly, we obtain from (H_1') - (H_3') the existence of an r_0 such that for $0 < c \leq r_0$

$$\beta + \ell\pi < \varphi(b,c) < \beta + (\ell+1)\pi.$$

Since we may assume $r_0 < r_1$, we can apply Lemma 2.1.1, and conclude that $\varphi(b,[r_0,r_1])$ is connected. The previous two inequalities on $\varphi(b,c)$ then imply the existence of positive numbers $c_j^+ \in [r_0, r_1]$ such that

$$\varphi(b, c_j^+) = \beta + j\pi \qquad (j = m+1, m+2, \ldots, m + |k - \ell|),$$

where $m = \min(k, \ell)$.

Similarly the existence of negative numbers c_j^- can be found for which

$$\varphi(b, c_j^-) = \beta + \pi + j\pi \qquad (j = m+1, m+2, \ldots, m + |k - \ell|).$$

Hence there exist solutions $z(t, c_j^+)$ and $z(t, c_j^-)$ of the BVP implying that there are at least $2|k - \ell|$ solutions, completing the proof of Theorem 2.9.1.

We now present conditions on system (2.5.1) so that (H_1') - (H_3') are satisfied. Assume that $f, g \in C^{(1)}[[a,b] \times R^2, R]$ and satisfy $f(t,0,0) \equiv g(t,0,0) \equiv 0$. Let

$$A(t) = \begin{pmatrix} f_x(t,0,0) & f_y(t,0,0) \\ g_x(t,0,0) & g_y(t,0,0) \end{pmatrix}$$

and consider the linear system

$$\text{(2.9.5)} \qquad \begin{pmatrix} x \\ y \end{pmatrix}' = A(t) \begin{pmatrix} x \\ y \end{pmatrix}.$$

The angular function $\mu(t)$ then satisfies

$$\text{(2.9.6)} \qquad \mu'(t) = g_x(t,0,0) \cos^2 \mu + [g_y(t,0,0) - f_x(t,0,0)] \times \cos \mu \sin \mu - f_y(t,0,0) \sin^2 \mu.$$

The solutions of (2.9.6) are unique and thus $\mu_M^+(t) \equiv \mu_m^+(t) \equiv \mu^+(t)$ and $\mu_M^-(t) \equiv \mu_m^-(t) \equiv \mu^-(t)$ where $\mu_M^{\pm}(t)$, $\mu_m^{\pm}(t)$ are, respectively, maximal and minimal solutions of (2.9.6). Let us assume that

$$\begin{aligned} &\beta + \ell\pi < \mu^{+}(t) < \beta + (\ell+1)\pi, \\ &\beta + (\ell+1)\pi < \mu^{-}(t) < \beta + (\ell+2)\pi, \end{aligned} \tag{2.9.7}$$

for some integer ℓ.

Now let $\varphi(t,c)$ be the angular function of any solution $z(t,c) = \big(x(t,c),y(t,c)\big)$ of (2.5.1). Consequently, letting $x(t) \equiv x(t,c), y(t) \equiv y(t,c)$, we obtain

$$\begin{aligned} \varphi'(t) = g_x(t,0,0)\cos^2\varphi + [g_y(t,0,0) - f_x(t,0,0)] \\ \times \cos\varphi \sin\varphi - f_y(t,0,0)\sin^2\varphi + \delta(t), \end{aligned}$$

where

$$\begin{aligned} \delta(t) \equiv [g(t,x,y) - g_x(t,0,0)x(t) - g_y(t,0,0)y(t) - f(t,x,y) \\ - f_x(t,0,0)x(t) - f_y(t,0,0)y(t)]/\big(x^2(t) + y^2(t)\big). \end{aligned}$$

Since g is differentiable, $\delta(t) \to 0$ as $x^2 + y^2 \to 0$.

Hence (H_3') is satisfied, if we choose $\delta_1(t,x,y) \equiv \delta(t)(x^2 + y^2)^3$ and $\delta_2(t,x,y) \equiv \delta_1(t,x,y)$. Our comparison systems (2.9.1) and (2.9.2) are precisely system (2.9.5) and the conditions (H_1') and (H_2') follows immediately from (2.9.7).

We thus deduce the following result.

COROLLARY 2.9.1. Assume that (H_1) - (H_3) hold. Let $f,g \in C^{(1)}\big[[a,b]\times S_\rho, R\big]$, where $S_\rho \equiv \{(x,y)\,|\,x^2 + y^2 \le \rho^2\}$, and satisfy $f(t,0,0) \equiv g(t,0,0) \equiv 0$. If the solutions of (2.9.6) satisfy (2.9.7), then there exist at least $2|k - \ell|$ solutions of the BVP (2.5.1) - (2.5.3).

It is interesting to find conditions such that the boundary value problem has exactly $2|k - \ell|$ nonzero solutions. Clearly, this will always occur when $\varphi(t,c)$ is strictly monotonic in c, both for $c > 0$ and $c < 0$. We now present a sufficient condition for $\varphi(t,c)$ to be strictly monotonic.

LEMMA 2.9.1. Assume system (2.5.1) has the following properties:

(a) $xg(t,x,y) - yf(t,x,y) \geq 0$, $(t,x,y) \in [a,b] \times R^2$,

(b) the paths of two different solutions $z_1(t)$ and $z_2(t)$ which satisfy the conditions

$$\varphi_1(a) = \varphi_2(a) = \tan^{-1} \alpha, \qquad \varphi_1(b) = \varphi_2(b) = \tan^{-1} \beta,$$

$$\tan^{-1} \alpha \leq \varphi_1(t),\ \varphi_2(t) \leq \tan^{-1} \beta \quad \text{for} \quad t \in [a,b],$$

where $\varphi_1(t)$ and $\varphi_2(t)$ are the angular functions of $z_1(t)$ and $z_2(t)$, respectively, do not intersect;

(c) $F(t,\varphi,r) \equiv (1/r)\ [g(t,r\cos\varphi,r\sin\varphi)\cos\varphi - f(t,r\cos\varphi,r\sin\varphi)\sin\varphi]$ is strictly increasing in r.

Then for each solution $z(t,c)$ the angular function $\varphi(t,c)$ is strictly increasing function in c.

If the inequality in (a) is reversed and if $F(t,\varphi,r)$ is strictly decreasing, then $\varphi(t,c)$ is strictly decreasing in c.

EXERCISE 2.9.1. Prove Lemma 2.9.1.

Hint: Let $\varphi_1(t,c_1)$ and $\varphi_2(t,c_2)$ be angular functions for two solutions $z_1(t,c_1)$ and $z_2(t,c_2)$. Assume $c_1 > c_2$ and show $\varphi_1'(a,c_1) > \varphi_1'(a,c_2)$. This inequality holds for $t \in (a,a+\delta)$ if δ is sufficiently small. Show that the inequality holds on $(a,b]$ by using (b) and (c) and the theory of differential inequalities.

We have thus proved the following result.

COROLLARY 2.9.2. Assume that the conditions of Theorem 2.9.1 and Lemma 2.9.1 are satisfied. Then there exist exactly $2|k - \ell|$ nonzero solutions of the BVP (2.5.1) - (2.5.3).

Observe that Corollary 2.9.2 remains valid when $\varphi(t,c)$ is strictly decreasing in c.

EXAMPLE 2.9.1. We obtain an estimate, depending on δ, of the number of solutions of the equation

$$x'' + \delta \sin x = 0,$$

satisfying

$$x(0) = 0, \qquad x(1) = 0.$$

We apply Corollary 2.9.2 to obtain the result. Assume $|\delta| < \pi^2$. Then system (2.9.5) becomes

$$x' = y, \qquad y' = -\delta x \tag{2.9.8}$$

with the boundary conditions

$$x(0) = 0, \qquad x(1) = 0.$$

Since there exist no nonzero solutions, it follows easily that $\ell = -1$.

In order to find k, we observe that there exists no solution of

$$x'' = 0, \qquad x(0) = 0, \quad x(1) = 0.$$

It follows easily that $k = -1$. Thus for $|\delta| < \pi^2$, we see (even though $k = \ell$) from Corollary 2.9.2 that there exist no nontrivial solution; hence there exists only the trivial solution. If $n^2\pi^2 < |\delta| < (n+1)^2\pi^2$, $n = +1, +2, \ldots$, then an easy calculation yields $\ell = n - 1$. Hence there exists $2[(n-1)+1] \equiv 2n$ solutions.

We leave the case $|\delta| = n^2\pi^2$ for the following exercise.

EXERCISE 2.9.2. Determine the number of solutions of

$$x'' + \delta \sin x = 0, \qquad x(0) = 0 = x(1),$$

satisfying

$$x(0) = x(1) = 0$$

when $|\delta| = n^2\pi^2$, $n = 1,2,\ldots$.

2.10 EXISTENCE OF INFINITE NUMBER OF SOLUTIONS

In Section 2.7, we discussed the existence of solutions of the BVP by showing $z(b,c)$ lies in two sectors located on different sides of the straight line $x \sin \beta - y \cos \beta = 0$ when $|c|$ is sufficiently large.

It is geometrically clear that the boundary value problem also has a solution if $z(b,c)$ has a spiral character when c tends to infinity.

THEOREM 2.10.1. Consider the BVP (2.5.1) - (2.5.3). Let the inequality

$$\begin{aligned} F(t,\varphi,r) \equiv (1/r)\ [&g(t,r \cos \varphi, r \sin \varphi] \cos \varphi \\ &- f(t,r \cos \varphi, r \sin \varphi) \sin \varphi] \\ &\geq \varepsilon_0 > 0 \end{aligned}$$

hold for $r \geq \rho_0$. Suppose that the xy plane can be divided into a finite number of sectors $S_1,\ldots,S_m$;

$$S_j = [(\varphi,r);\ \varphi_j \leq \varphi \leq \varphi_{j+1},\ r \geq 0]$$
$$(\alpha \leq \varphi_1 < \varphi_2 < \cdots < \varphi_{n+1} = \varphi_1 + 2\pi)$$

such that

$$\lim_{r \to \infty} F(t,\varphi,r) = +\infty$$

uniformly in (t,φ), $\varphi_j + \varepsilon \leq \varphi \leq \varphi_{j+1} - \varepsilon$ $(j = 1,\ldots,n)$, $t \in [a,b]$, where ε is an arbitrarily small constant. Then the BVP has an infinite number of solutions.

Proof: By Lemma 2.6.3, there exists $c_0 > 0$ such that

$z(t,c)$ does not vanish for $|c| > c_0$ and therefore the angular function $\varphi(t,c)$ of this solution is well defined.

Pick any $\varepsilon > 0$ and choose $M > 0$. There exists $\{\varphi_j\}_{j=1}^n$ and a $\rho_1 \geq \rho_0$ so that

$$F(t,\varphi,r) \geq M, \qquad a \leq t \leq b, \tag{2.10.1}$$

holds for all $r \geq \rho_1$, $|\varphi - \varphi_j| \leq \varepsilon$.

Define the discontinuous function

$$H(\varphi) = \begin{cases} M & \text{where } \varphi = \varphi_j + \delta, \quad |\delta| \leq \varepsilon, \\ \varepsilon_0 & \text{for all other values of } \varphi. \end{cases}$$

We see immediately

$$F(t,\varphi,r) > H(\varphi) \tag{2.10.2}$$

for $t \in [a,b]$ and $r \geq \rho_1$.

Consider the solution $\psi(t)$ of

$$\psi'(t) = H(\psi), \qquad \psi(a) = \alpha.$$

Observe that the time τ which is required for $\psi(t)$ to assume the value $\alpha + 2\pi$ is

$$\tau = (2\varepsilon/\varepsilon_0)\, n + (1/M)\,(2\pi - 2\varepsilon n).$$

Hence $\psi(b) - \psi(a) \geq 2\pi k$, where k is the largest number of intervals of length τ contained in $[a,b]$. As $\varepsilon \to 0$ and $M \to +\infty$, $\tau \to 0$; and thus $\psi(b) \to +\infty$ as $\varepsilon \to 0$ and $M \to +\infty$.

By Lemma 2.6.3, there exists $\rho_2 > 0$ such that when $|c| \geq \rho_2$ then $\|z(t,c)\| \geq \rho_1$. Hence, the angular function $\varphi(t,c)$ $[\varphi(a,c) = \alpha,\ c \geq \rho_2]$ of $z(t,c)$ satisfies the differential inequality

$$\varphi'(t,c) = F\big(t,\varphi(t,c)\big), \qquad \|z(t,c)\| \geq H\big(\varphi(t,c)\big).$$

By the theory of differential inequalities $\varphi(t,c) \geq \psi(t)$ for $t \in [a,b]$. Hence $\varphi(b,c) \to +\infty$ as $c \to +\infty$. Also, in a similar manner, we conclude $\varphi(b,c) \to +\infty$ as $c \to -\infty$.

By Lemma 2.6.2, there exists an infinite number of values of c, both positive and negative for which

$$\varphi(b,c) = \beta \pmod{\pi}.$$

Hence our BVP has an infinite number of solutions. The proof is complete.

EXAMPLE 2.10.1. Consider the BVP

$$x' = -h(y), \qquad y' = g(x), \qquad x(a) = x(b) = 0$$

where g, h are continuous and odd. Assume $xg(x) \geq \varepsilon_0 x^2$, $yh(y) \geq \varepsilon_0 y^2$ and $\lim_{x\to\infty} g(x)/x = +\infty$, $\lim_{y\to\infty} h(y)/y = +\infty$. To show that this problem has an infinite number of solutions, we merely check the hypotheses of Theorem 2.10.1. We notice immediately

$$\begin{aligned} F(t,\varphi,r) &= \frac{1}{x^2 + y^2} [xg(x) + yh(y)] \\ &\geq \frac{1}{x^2 + y^2} [\varepsilon_0 x^2 + \varepsilon_0 y^2] \geq \varepsilon_0 > 0. \end{aligned}$$

Consider the four segments S_i, $i = 1,2,3,4$, in the xy plane defined by

$$S_i = \left[(\varphi,r) \colon \varphi \in \left(\frac{(i-1)\pi}{2} + \varepsilon, \ \frac{i\pi}{2} - \varepsilon\right), \quad 0 \leq r < \infty\right],$$

where ε is any small constant less than $\pi/4$. For $(t,\varphi,r) \in [a,b] \times S_i$, we have

$$\begin{aligned} F(t,\varphi,r) &= \frac{1}{x^2 + y^2} \left[x^2 \frac{g(x)}{x} + y^2 \frac{h(y)}{y}\right] \\ &\geq \frac{1}{x^2 + y^2} [x^2 M(r) + y^2 M(r)] = M(r), \end{aligned}$$

where $M(r)$ has the property that $g(x)/x > M(r)$ for $|x| > r$

and $h(y)/y > M(r)$ for $|y| > r$. Since $M(r)$ can be chosen so that $M(r) \to +\infty$ as $r \to +\infty$, we see the hypotheses of Theorem 2.10.1 hold.

EXERCISE 2.10.1. Show the BVP

$$x'' + x(1+x'^2)^{1/3} + h(t) = 0, \qquad x(a) = x(b) = 0$$

has an infinite number of solutions.

2.11 NONLINEAR BOUNDARY CONDITIONS

Using techniques similar to those used in Section 2.7, we investigate the existence of solutions of a two-dimensional differential system satisfying nonlinear boundary conditions. Results on the uniqueness and number of solutions may also be deduced under conditions similar to those in the previous sections, but we shall omit them here.

Before stating our existence theorem certain preliminaries need some mentioning. In particular, consider system (2.5.1). Assume there exist two functions, $W(t,x,y), V(t,x,y) \in C^{(1)}[[a,b]\times R^2, R]$ such that for each t the range of V and W is all of R and the set

$$W^{-1}(c_1) \cap V^{-1}(c_2) \neq \emptyset, \tag{2.11.1}$$

for all real numbers c_1 and c_2;

$$|V(t,x,y)| + |W(t,x,y)| \to \infty \tag{2.11.2}$$

as $\|x\| + \|y\| \to \infty$ uniformly for $t \in [a,b]$.

The derivative of V along solutions of (2.5.1) is

$$V'(t,x,y) = \partial V/\partial t\ (t,x,y) + \partial V/:x\ (t,x,y)\cdot f(t,x,y) + \partial V/\partial y\ (t,x,y)\cdot g(t,x,y).$$

A similar expression holds for $W'(t,x,y)$.

Let $x(t)$ and $y(t)$ be any solution of (2.5.1). Now

consider the angular function $\varphi(t)$ and polar radius $\rho(t)$ defined in the VW plane. Thus

$$\rho(t) = \left[W^2\big(t,x(t),y(t)\big) + V^2\big(t,x(t),y(t)\big)\right]^{\frac{1}{2}},$$

and

$$\tan \varphi(t) = W\big(t,x(t),y(t)\big)/V\big(t,x(t),y(t)\big).$$

Hence,

$$\varphi'(t) = \Big[V\big(t,x(t),y(t)\big)W'\big(t,x(t),y(t)\big) - W\big(t,x(t),y(t)\big)V'\big(t,x(t),y(t)\big)\Big]/(V^2 + W^2).$$

The following inequalities on V' and W' will play a major role in applying the comparison theorems:

$$(2.11.3) \qquad \begin{aligned} &\delta_3(t,V,W) + F_2(t,V,W) \le V' \le F_1(t,V,W) + \delta_1(t,V,W), \\ &\delta_4(t,V,W) + G_2(t,V,W) \le W' \le G_1(t,V,W) + \delta_2(t,V,W), \end{aligned}$$

where $F_i, G_i \in C\big[[a,b]\times R^2, R\big]$ are positively homogeneous in V and W; and $\delta_i(t,V,W) \in C\big[[a,b]\times R^2, R\big]$ such that

$$(2.11.4) \qquad \delta_i(t,V,W)/(|V| + |W|) \to 0 \quad \text{as} \quad |V| + |W| \to \infty$$

uniformly for $t \in [a,b]$.

Define

$$P_2(t,V,W) = \begin{cases} G_1(t,V,W) & \text{if } V \ge 0, \\ G_2(t,V,W) & \text{if } V < 0, \end{cases}$$

$$P_1(t,V,W) = \begin{cases} F_2(t,V,W) & \text{if } W \ge 0, \\ F_1(t,V,W) & \text{if } W < 0, \end{cases}$$

$$Q_2(t,V,W) = \begin{cases} G_2(t,V,W) & \text{if } V \ge 0, \\ G_1(t,V,W) & \text{if } V < 0, \end{cases}$$

$$Q_1(t,V,W) = \begin{cases} F_1(t,V,W) & \text{if } W \geq 0, \\ F_2(t,V,W) & \text{if } W < 0. \end{cases}$$

Using (2.11.3) we obtain

$$(2.11.5) \qquad VQ_2(t,V,W) - WQ_1(t,V,W) + \delta^0(t,V,W) \leq VW' - WV'$$
$$\leq VP_2(t,V,W) - WP_1(t,V,W) + \delta^1(t,V,W),$$

where

$$\delta^1(t,V,W) = \begin{cases} V\delta_2(t,V,W) - W\delta_3(t,V,W), & V \geq 0, \quad W \geq 0, \\ V\delta_2(t,V,W) - W\delta_1(t,V,W), & V \geq 0, \quad W < 0, \\ V\delta_4(t,V,W) - W\delta_3(t,V,W), & V < 0, \quad W \geq 0, \\ V\delta_4(t,V,W) - W\delta_1(t,V,W), & V < 0, \quad W < 0; \end{cases}$$

and

$$\delta^0(t,V,W) = \begin{cases} V\delta_4(t,V,W) - W\delta_1(t,V,W), & V \geq 0, \quad W \geq 0, \\ V\delta_4(t,V,W) - W\delta_3(t,V,W), & V \geq 0, \quad W < 0, \\ V\delta_2(t,V,W) - W\delta_1(t,V,W), & V < 0, \quad W \geq 0, \\ V\delta_2(t,V,W) - W\delta_3(t,V,W), & V < 0, \quad W < 0. \end{cases}$$

In view of (2.11.5), the comparison systems are

$$(2.11.6) \qquad A' = Q_1(t,A,B), \qquad B' = Q_2(t,A,B),$$

$$(2.11.7) \qquad S' = P_1(t,S,T), \qquad T' = P_2(t,S,T).$$

As before we see that systems (2.11.6) and (2.11.7) may have discontinuities along the lines $A = 0$ and $B = 0$ as well as $S = 0$, $T = 0$. Assume that all solutions of (2.11.6) and (2.11.7) exist on $[a,b]$ and are continuously differentiable everywhere except at the previously mentioned lines. Observe that P_1, P_2, Q_1, Q_2 are positively homogeneous in their last two variables.

Let $\theta(t)$ and $\psi(t)$ be the angular functions corresponding to (2.11.6) and (2.11.7), respectively. Since $\tan\theta(t) = B(t)/A(t)$, where $B(t)$ and $A(t)$ are solutions of (2.11.6), we have

$$\theta'(t) = Q_2(t,\cos\theta,\sin\theta)\cos\theta - Q_1(t,\cos\theta,\sin\theta)\sin\theta. \tag{2.11.8}$$

Similarly,

$$\psi'(t) = P_2(t,\cos\psi,\sin\psi)\cos\psi - P_1(t,\cos\psi,\sin\psi)\sin\psi. \tag{2.11.9}$$

Consider the BVP

$$x' = f(t,x,y), \qquad y' = g(t,x,y), \tag{2.11.10}$$

with the boundary conditions

$$\begin{aligned} V\big(a,x(a),y(a)\big)\sin\alpha - W\big(a,x(a),y(a)\big)\cos\alpha &= 0,\\ V\big(b,x(b),y(b)\big)\sin\beta - W\big(b,x(b),y(b)\big)\cos\beta &= 0, \end{aligned} \tag{2.11.11}$$

where $0 \le \alpha < \pi$ and $0 < \beta \le \pi$.

We now assume essentially the same hypotheses (H_1) - (H_3) as in Section 2.7.1. Since the notation is the same we will not repeat the definitions. Therefore, assume the following:

(G_1) The maximal solutions $\psi_M^+(t)$ and $\psi_M^-(t)$ of (2.11.9) such that $\psi_M^+(a) = \alpha$, $\psi_M^-(a) = \alpha + \pi$ satisfy

$$\psi_M^+(b) < \beta + (k+1)\pi \qquad \text{and} \qquad \psi_M^-(b) < \beta + (k+2)\pi$$

for some integer k;

(G_2) The minimal solutions $\theta_m^-(t)$ and $\theta_m^+(t)$ of (2.11.8) such that $\theta_m^+(a) = \alpha$, $\theta_m^-(a) = \alpha + \pi$ satisfy

$$\theta_m^-(b) > \beta + (k+1)\pi \quad \text{and} \quad \theta_m^+(b) > \beta + k\pi$$

for the same integer k as in (G_1).

We are now able to state our main existence result. Since the proof is similar to that of Theorem 2.7.1 we leave it as an exercise.

THEOREM 2.11.1. For any $V(t,x,y)$ and $W(t,x,y)$ satisfying (2.11.1) - (2.11.3) there exists a solution of the BVP (2.11.10), (2.11.11) provided (G_1) and (G_2) hold.

EXERCISE 2.11.1. Prove Theorem 2.11.1.

Hint: For any c let $\big(x(a,c),y(a,c)\big)$ be a point in R^2 satisfying

$$V\big(a,x(a,c),y(a,c)\big) = c \cos \alpha,$$

$$W\big(a,x(a,c),y(a,c)\big) = c \sin \alpha.$$

Letting $\big(x(t,c),y(t,c)\big)$ be any solution of (2.11.10) through $\big(x(a,c),y(a,c)\big)$, we obtain the angular function $\varphi(t,c)$ where

$$\tan \varphi(t,c) = W\big(t,x(t,c),y(t,c)\big)/V\big(t,x(t,c),y(t,c)\big).$$

Proceed now as in the proof of Theorem 2.7.1 and obtain for sufficiently large N the two points

$$V\left(b,x\left(b, \frac{+N}{\cos \alpha}\right), y\left(b, \frac{+N}{\cos \alpha}\right)\right), \quad W\left(b,x\left(b, \frac{+N}{\cos \alpha}\right), y\left(b, \frac{+N}{\cos \alpha}\right)\right)$$

and

$$V\left(b,x\left(b, \frac{-N}{\cos \alpha}\right), y\left(b, \frac{-N}{\cos \alpha}\right)\right), \quad W\left(b,x\left(b, \frac{-N}{\cos \alpha}\right), y\left(b, \frac{-N}{\cos \alpha}\right)\right)$$

lie on different sides of the straight line

$$V \sin \beta - W \cos \beta = 0$$

in the VW plane.

Show then that there exists a set $S \subset R^2$ such that $V(a,S) = [-N,N]$. Thus the set

$$\bigcup_{(x_0,y_0)\in S} \left[V\big(b,x(b,a,x_0),y(b,a,y_0)\big),\ W\big(b,x(b,a,x_0),y(b,a,y_0)\big)\right]$$

is connected in R^2. The result then follows.

COROLLARY 2.11.1. Let the hypotheses of Theorem 2.11.1 hold. Then there exists a solution of (2.11.10) satisfying the boundary conditions

$$R_1V\big(a,x(a),y(a)\big) + R_2W\big(a,x(a),y(a)\big) = c_1,$$
$$R_3V\big(b,x(b),y(b)\big) + R_4W\big(b,x(b),y(b)\big) = c_2,$$

for any real numbers R_i, $i = 1,2,3,4$ and c_i, $i = 1,2$.

Under the conditions of Theorem 2.11.1 and Corollary 2.11.1, we notice that when solutions of systems (2.11.8) and (2.11.9) are uniquely determined by initial conditions, then (G_1) and (G_2) together are equivalent to the fact that there exist no solutions of (2.11.6) and (2.11.7) satisfying the boundary conditions (2.11.11).

EXAMPLE 2.11.1. Consider the boundary value problem

$$\text{(2.11.12)} \qquad x' = x + f(y), \qquad y' = 2y + g(x);$$
$$x^2(a) - y(a) = c_1, \qquad x(b) = c_2,$$

where $f,g \in C\big[[a,b]\times R^2,R\big]$ satisfy

$$|f(y)| \to 0 \quad \text{as} \quad |y| \to \infty, \qquad |g(x)|/|x| \to 0 \quad \text{as} \quad |x| \to \infty.$$

Define

$$V = x, \qquad W = x^2 - y.$$

We are thus interested in finding solutions of (2.11.12) satisfying

$$W(a) = c_1, \qquad V(b) = c_2.$$

Observe that V and W satisfy (2.11.1) since the level curves $V = k_1$ and $W = k_2$ intersect in R^2 for every k_1 and k_2. It is not difficult to verify (2.11.2). Now

$$V' = x' = x + f(y) = V + f(y),$$

$$\begin{aligned} W' = 2xx' - y' &= 2x^2 + 2xf(y) - 2y - g(x) \\ &= 2y - 2W + 2xf(y) - 2y - g(x) \\ &= -2W + 2xf(y) - g(x). \end{aligned}$$

Let $F_1 \equiv F_2 \equiv V$, $\delta_1 = \delta_3 = f(y)$, $G_1 = G_2 = -2W$, and $\delta_4 = \delta_2 = 2xf(y) - g(x)$. Since $|2xf(y) - g(x)|/(|x^2 - y| + |x|) \to 0$ as $|x^2 - y| + |x| \to \infty$, we see that (2.11.4) holds. Systems (2.11.6) and (2.11.7) are then the same, with $P_1 \equiv Q_1 \equiv V$ and $P_2 \equiv Q_2 \equiv -2W$. Because the solutions of (2.11.8) and (2.11.9) are uniquely determined by initial conditions due to the linearity of P_i and Q_i, $i = 1,2$, it is sufficient to show there exist no solutions of

$$V' = V, \qquad W' = -2W,$$

satisfying

$$W(a) = 0, \qquad V(b) = 0;$$

that this is true is clear for any a and b.

EXERCISE 2.11.2. Show there exists a solution of the BVP

$$x' = x + 2y \sin^2 x,$$

$$y' = (y/2) + \sin^2 x;$$

$$y(a) = c_1, \qquad x(b) - y^2(b) = c_2.$$

EXERCISE 2.11.3. Keeping in mind the results of Sections 2.8 and 2.11, state and prove a uniqueness theorem for a BVP with nonlinear boundary conditions.

2.12 NOTES AND COMMENTS

Theorem 2.1.1 is due to Hartman [4]. Example 2.1.1 is taken from Lasota and Opial [8]. For Corollary 2.1.1, see Lasota and Opial [7] and Jackson [2]. Exercise 2.1.1 is a result of Lees [1] while Exercise 2.1.2 is of Levin [1]. Exercise 2.1.3 may be found in Jackson [2]. Theorem 2.2.1 is due to Hartman [4]. Exercise 2.2.1 is a classical result of De La Vallée Poussin [1]. Theorem 2.2.2 is due to Tutaj [1]. Corollary 2.2.2 is taken from Keller [1] and Exercise 2.2.4 is based on Heidel [1]. Section 2.3 is adapted from Schrader and Waltman [4], while Section 2.4 is the work of Waltman [3]. For related discussion on uniqueness questions, see Bailey et al. [3], Shampine [1] and Sherman [1].

The contents of Sections 2.5 - 2.10 are based on the work of Perov [1] and Krasnoselskii et al. [2]. For related results on existence and uniqueness, see Waltman [3]. Section 2.11 is due to Bernfeld and Lakshmikanthan [2].

Chapter 3
TOPOLOGICAL METHODS

3.0 INTRODUCTION

In this chapter, various topological principles are utilized in solving boundary value problems. For example, the Wazewski's topological method together with the connectedness properties of solution funnels is used to prove the existence of solutions. In this set up, the boundary conditions consist of sets which, in particular cases, can be constructed from the knowledge of upper and lower solutions and Nagumo's condition. We also develop the Wazewski-like method for boundary value problems associated with contingent equations and set valued differential equations. Finally, the continuous dependence of solutions on boundary data is discussed.

3.1 SOLUTION FUNNELS

We shall consider the existence of solutions of boundary value problems associated with the two-dimensional system

$$(3.1.1) \qquad x' = f(t,x,y), \qquad y' = g(t,x,y).$$

The following preliminary notation will be needed. Let $\psi(t,y)$ and $\varphi(t,y)$ be continuous functions such that $\psi(t,y) \geq \varphi(t,y)$ for $t \geq 0$, $|y| < \infty$. Define the following sets:

$$Q(t_0) = [(t,x,y)\colon 0 \leq t \leq t_0,\ |x| + |y| < \infty],$$

$$Q(\infty) = [(t,x,y)\colon 0 \leq t < \infty,\ |x| + |y| < \infty],$$

$$R = [(t,x,y)\colon 0 \leq t < \infty,\ |y| < \infty,\ \varphi(t,y) \leq x \leq \psi(t,y)],$$

$$C(t) = [(x,y)\colon (t,x,y) \in R],$$

$$S_\psi(t) = [(x,y)\colon (x,y) \in C(t),\ x = \psi(t,y)],$$

$$S_\varphi(t) = [(x,y)\colon (x,y) \in C(t),\ x = \varphi(t,y)].$$

Let $f(t,x,y)$ and $g(t,x,y)$ be continuous in a set E which is open relative to $Q(t_0)$ where $Q(t_0) \cap R \subset E$.

Let $t_1, t_2 \geq 0$ and let S be a subset of $C(t_1)$. Denote by $I_E(S,t_1,t_2)$, the set of all points (x,y) such that there is a solution $\big(x(t),y(t)\big)$ of (3.1.1) on $[t_1,t_2]$ in which $\big(x(t_1),y(t_1)\big) \in S$, $\big(x(t_2),y(t_2)\big) = (x,y)$ and $\big(t,x(t),y(t)\big) \in E$ for all $t \in [t_1,t_2]$. This set, as we have seen is called the <u>solution funnel cross section</u> at $t = t_2$.

Before stating our main results, we shall need the following hypotheses which restrict the behavior of solutions as they cross the φ and ψ surfaces. We shall always assume:

(H_1) for all $t_1 \in [0,t_0]$, $(x_1,y_1) \in I_E(S_1,0,t_1) \cap S_\psi(t_1)$ implies that there exists a solution of (3.1.1) emanating from (t_1,x_1,y_1) with a trajectory which is on or above the ψ-surface on some right neighborhood of t_1.

(H_2) for all $t_1 \in [0,t_0]$, $(x_1,y_1) \in I_E(S_1,0,t_1) \cap S_\varphi(t_1)$ implies that there exists a solution of (3.1.1) which emanates from (t_1,x_1,y_1) with a trajectory which is on or below the φ-surface on some right neighborhood of t_1.

Our first result describes qualitatively the behavior of solutions of (3.1.1) assuming (H_1) and (H_2) hold.

THEOREM 3.1.1. Let S_1 be a compact connected set in $C(0)$ which intersects $S_\psi(0)$ and $S_\varphi(0)$. Then either

(i) $I_E(S_1,0,t)$ contains a compact connected component in $C(t)$ which intersects both $S_\psi(t)$ and $S_\varphi(t)$ for all $t \in [0,t_0]$, or

(ii) there is a solution of (3.1.1) with $\big(x(0),y(0)\big) \in S_1$ having a maximal right interval of $[0,t^+] \subset [0,t_0]$ such that

$|y(t)| \to \infty$ as $t \to t^+$.

Proof: Let $E' \equiv [(t,x,y) \in E: \varphi(t,y) - 1 < x < \psi(t,y) + 1]$. Assume there is no number K such that for all $t \in [0,t_0]$, $(x,y) \in I_{E'}(S_1,0,t)$ implies $|x| \leq K$ and $|y| \leq K$. Then there is some interval $[0,t^+] \subset [0,t_0]$ such that for any $K > 0$ there exists an $\varepsilon(K) > 0$ such that either $|x| > K$ or $|y| > K$ for $t \in [t^+ - \varepsilon, t^+)$. Pick any sequence $t_n \to t^+$ and let $(x_n,y_n) \in I_{E'}(S_1,0,t_n)$. Then, either $|x_n| \to \infty$ or $|y_n| \to \infty$. By the continuity of $\psi(t,y)$ and $\varphi(t,y)$, it follows that if $|x_n| \to \infty$, then $|y_n| \to \infty$. Hence in either case we have $|y_n| \to \infty$.

By a standard diagonalization process, using the solutions associated with $\{(x_n,y_n)\}$, we may construct a solution $\big(x(t),y(t)\big)$ of (3.1.1) with $\big(x(0),y(0)\big) \in S_1$ which exists on $[0,t] \subset [0,t^+)$ and such that $|y(t)| \to \infty$ as $t \to t^+$. Moreover, $\big(t,x(t),y(t)\big) \in E' \subset E$. This is case (ii).

Hence, to complete the proof we may assume there exists a number $K > 0$ such that for all $t \in (0,t_0]$, $(x,y) \in I_{E'}(S_1,0,t)$ implies $|x| \leq K$ and $|y| \leq K$.

Let T be the set of all points $\overline{t} \in [0,t_0]$ such that, for all $t \in [0,\overline{t}]$, the set $I_{E'}(S_1,0,t)$ contains a component in $C(t)$ which intersects both $S_\psi(t)$ and $S_\varphi(t)$. Then T is nonempty since $0 \in T$ and is bounded above by t_0. We shall show T is closed and thus if we let $s = \sup T$, then $s = t_0$ implies the conclusion of Theorem 3.1.1.

Let $\{s_i\}$ be a sequence of points in T converging to s with $C_i \subset C(s_i)$ a component of $I_{E'}(S_1,0,s_i)$ which intersects both $S_\psi(s_i)$ and $S_\varphi(s_i)$. Let L denote the limit set of $\{C_i\}$ and let $(a,b) \in L$ since L is nonempty. Then there exists a sequence $\{(x_i',y_i')\}$ converging to (a,b) where $(x_i',y_i') \in C_i'$ and $\{C_i'\}$ is a subsequence of $\{C_i\}$.

Let L' be the set of limit points of $\{C_i'\}$; hence L' is compact and since C_i' intersects both $S_\varphi(s_i')$ and $S_\psi(s_i')$ for all i then L' intersects both $S_\varphi(s)$ and $S_\psi(s)$.

If L' contains no component which intersects both $S_\psi(s)$ and $S_\varphi(s)$, then since L' is compact and intersects both $S_\varphi(s)$ and $S_\psi(s)$, L' is the union of two nonempty compact sets M and N which are separated by an arc A such that $A \cap L' = \varphi$.

Assume $(a,b) \in M$. Let (c,d) be any point in N and let $(\bar{x}_i'', \bar{y}_i'') \to (c,d)$, where $(\bar{x}_i'', \bar{y}_i'') \in C_i''$, a subsequence of $\{C_i'\}$. Choose $\{(x_i'', y_i'')\}$ to be the subsequence of $\{(x_i', y_i')\}$ contained in C_i''. For i sufficiently large the connected set $I_{E'}(C_i'', s_i'', s)$ intersects A and let (p_i, q_i) be the points of intersection. Thus $(p_i, q_i) \in I_{E'}(C_i'', s_i'', s) \cap S$. There exists a point (p,q) which is a limit point of $\{(p_i, q_i)\}$ and since $s_i'' \to s$, (p,q) is a limit point of $\{C_i''\}$. However, then $(p,q) \in L'$ and $(p,q) \in A \cap L'$. This is a contradiction and we conclude L' and thus $I_{E'}(S_1, 0, s)$ contains a component C in $C(s)$ intersecting both $S_\varphi(s)$ and $S_\psi(s)$. Thus T is closed.

By assumptions (H_1) and (H_2) there exists a $\delta > 0$ such that

$$I_{E'}(C,s,t) \subset [(x,y)\colon (t,x,y) \in E']$$

and $I_{E'}(C,s,t)$ contains a component which intersects both $S_\psi(t)$ and $S_\varphi(t)$ for all $t \in (s, s+\delta)$. However, this contradicts the fact that $s = \sup T$. Thus $s = t_0$ and (i) is proven. This concludes the proof of Theorem 3.1.1.

Theorem 3.1.1 with conditions that restrict the possibility of (ii) occurring can be used to deduce existence theorems. We have seen in Chapter 1 that one such condition is Nagumo's condition. The following condition will thus be imposed:

(H_3) Given any $n > 0$ and $t_0 > 0$ there exists a number $N(t_0,n)$ such that for any solution $(x(t),y(t))$ of (3.1.1) with $|y(0)| < n$ and $(t,x(t),y(t)) \in E$ for $t \in [0,t_0)$ we have $|y(t)| < N(t_0,n)$ for all $t \in [0,t_0)$.

We now present an existence theorem.

THEOREM 3.1.2. Assume conditions (H_1) - (H_3) hold. Let S_1 be a compact connected set in $C(0)$ which intersects both $S_\psi(0)$ and $S_\varphi(0)$. Let S_2 be a closed connected subset of $C(t_0)$ such that $S_2 \cap [(x,y)\colon y$ arbitrary$] \neq \emptyset$, then there exists a solution of $(x(t),y(t))$ of (3.1.1) on $[0,t_0]$ such that $(x(0),y(0)) \in S_1$, $(x(t_0),y(t_0)) \in S_2$ with $(t,x(t),y(t)) \in E$ for all $t \in [0,t_0]$.

Proof: Since S_1 is a compact set, let $n = \sup |y|$ for $(x,y) \in S_1$, and let $N = N(t_0,n)$ be as in (H_3). Then $|y(t)| < N(t_0,n)$ for any $(x(t),y(t))$ of (3.1.1) with $(x(0),y(0)) \in S_1$ and all $t \in [0,t_0]$. By Theorem 3.1.1, $I_E(S_1,0,t_0)$ contains a compact connected component C in $C(t_0)$ which intersects both $S_\psi(t_0)$ and $S_\varphi(t_0)$. The conditions imposed on S_2 insure that $S_2 \cap I_E(S_1,0,t_0) \neq \emptyset$. This concludes the proof of the theorem.

Another approach to this problem is the application of the Wazewski's method. We shall introduce this method here. Consider the differential system

$$(3.1.2) \qquad x' = f(t,x), \qquad x(t_0) = x_0, \qquad t_0 \geq 0,$$

where $f \in C[\Omega,R^n]$, Ω being any open set in R^{n+1}. Let Ω_0 be an open set of Ω, $\partial\Omega_0$, the boundary, and $\overline{\Omega}_0$ the closure of Ω_0.

DEFINITION 3.1.1. A point $(t_0,x_0) \in \Omega \cap \partial\Omega_0$ is said to be an egress point of Ω_0 with respect to the system (3.1.2) if, for every solution $x(t)$ of (3.1.2), there is an $\varepsilon > 0$ such that $(t,x(t)) \in \Omega_0$ for $t_0 - \varepsilon \leq t < t_0$. An egress point (t_0,x_0) of Ω_0 is called a strict egress point of Ω_0, if $(t,x(t)) \notin \overline{\Omega}_0$ for $t_0 < t \leq t_0 + \varepsilon$ for a small $\varepsilon > 0$. Denote the set of all points of egress (strict egress) as S (S^*).

DEFINITION 3.1.2. If $A \subset B$ are any two sets of a topological space and $\pi: B \to A$ is a continuous mapping from B into A such that $\pi(p) = p$ for every $p \in A$, then π is said to be a retraction of B onto A. When there exists a retraction of B onto A, A is called a retract of B.

The following theorem of Wazewski is quite useful.

THEOREM 3.1.3. Let $f \in C[\Omega,R^n)$, Ω open in R^{n+1}. Assume that through every point of Ω there passes a unique solution of (3.1.2). Let Ω_0 be an open subset of Ω. Suppose that all egress points of Ω_0 are strict egress points. Let Z be a nonempty subset of $\Omega_0 \cup S$ such that $Z \cap S$ is a retract of S, but not a retract of Z. Then there exists at least one point $(t_0,x_0) \in Z \cap \Omega_0$ such that the solution $(t,x(t))$ of (3.1.2) remains in Ω_0 on its maximal interval of existence to the right of t_0.

We now apply Theorem 3.1.3 to the BVP prescribed in Theorem 3.1.2. Although we assume uniqueness of solutions of (3.1.1) this is not essential as a Wazewski-like theorem for nonuniqueness has been developed. Let $\Omega = [(t,x,y): t \geq 0] - S_2$; then Ω is a relatively open subset of the half space $[(t,x,y): t \geq 0]$. Let $\Omega_0 = [(t,x,y): 0 \leq t < t_1,\ \varphi(t) < x < \psi(t),\ |y| < \infty]$ where now $\psi(t,y)$, $\varphi(t,y)$ are independent of y. Let $Z \equiv S_1$. From hypotheses (H_1) and (H_2), it is not difficult to see that $S = S^*$, and S consists of the union of the sets

$$\{(0,x,y) \in S_\varphi(0) \colon y \le \varphi'(0)\}, \qquad \{(0,x,y) \in S_\psi(0) \colon y \ge \psi'(0)\},$$

$$\{(t,x,y) \colon 0 < t \le t_1,\ x = \psi(t),\ y > \psi'(t)\},$$

$$\{(t,x,y) \colon 0 < t \le t_1,\ x = \varphi(t),\ y < \varphi'(t)\}, \quad \text{and} \quad C(t_2) - S_2.$$

We see from the properties of S and the set S_2 that $S_1 \cap S$ is a retract of S. Since $S_1 \cap S$ is not connected $S_1 \cap S$ is not a retract of S_1. Hence from Theorem 3.1.2 there is a solution $(x(t),y(t))$ of (3.1.1) such that $(t_1,x(t_1),y(t_1)) \in S_1 \cap C(t_1)$ and such that $(t,x(t),y(t))$ remains in Ω_0 on its right maximal interval of existence. This implies $(t_2,x(t_2), y(t_2)) \in S_2$ due to (H_3). Hence there exists a solution satisfying the BVP prescribed in Theorem 3.1.2.

We now can obtain under the same hypotheses the existence of a solution remaining in $R \cap Q(t_0)$.

THEOREM 3.1.4. Under the same hypotheses as in Theorem 3.1.2, there exists a solution $(x(t),y(t))$ of (3.1.1) such that $(x(0),y(0)) \in S_1$, $(x(t_0),y(t_0)) \in S_2$, and $(t,x(t), y(t)) \in R \cap Q(t_0)$.

EXERCISE 3.1.1. Prove Theorem 3.1.4.

Hint: Choose a sequence of open sets $\{E_n\}_{n=0}^{\infty}$, where $E_n \supset E_{n+1} \supset R \cap Q(t_0)$, $E_0 = E$, and $\bigcap_{n=0}^{\infty} E_n = R \cap Q(t_0)$. Show there exist solutions which lie in E_n and satisfy the boundary conditions by applying Theorem 3.1.2. Then apply Ascoli's theorem to obtain the desired solution.

We can now prove a result for a semiinfinite BVP.

THEOREM 3.1.5. Let $f,g \in C\big[[0,\infty) \times R^2, R\big]$ and assume

(H_1) - (H_3) hold. Let S_1 be a compact connected set in $C(0)$ intersecting both $S_\psi(0)$ and $S_\varphi(0)$. Then (3.1.1) has a solution $(x(t),y(t))$ existing on $[0,\infty)$ with $(x(0),y(0)) \in S_1$ and $(t,x(t),y(t)) \in R \cap Q(t_0)$ for all $t \geq 0$.

EXERCISE 3.1.2. Prove Theorem 3.1.5.

Hint: Use Theorem 3.1.4 and a standard diagonalization argument.

3.2 APPLICATION TO SECOND-ORDER EQUATIONS

In this section we apply the results of Section 3.1 to the case in which $f(t,x,y) = y$ in (3.1.1). That is, we consider the equivalent second-order scalar equation

$$x'' = g(t,x,x'). \tag{3.2.1}$$

Recall that a function $\alpha(t) \in C^{(2)}[0,t_0]$ is a lower solution of (3.2.1) if

$$\alpha''(t) \geq g(t,\alpha(t),\alpha'(t)), \qquad t \in [0,t_0]. \tag{3.2.2}$$

Moreover, α becomes a strict lower solution if, in (3.2.2), $\geq$ is replaced by $>$. Similarly a function $\beta(t) \in C^{(2)}[0,t_0]$ is an upper solution of (3.2.1) if

$$\beta''(t) \leq g(t,\beta(t),\beta'(t)), \qquad t \in [0,t_0]. \tag{3.2.3}$$

If in (3.2.3) $\leq$ is replaced by $<$, then β is a strict upper solution.

We now present a general existence result. Notice that in Chapter 1 other techniques have been used to prove similar types of theorems.

THEOREM 3.2.1. Assume that:

(i) there exists an upper solution $\beta(t)$ and a lower

solution $\alpha(t)$ of (3.2.1) with $\alpha(t) \le \beta(t)$ on $[0,t_0]$. Let E be any open set containing the set $[(t,x,x'): 0 \le t \le t_0, \alpha(t) \le x \le \beta(t), |x'| < \infty]$;

(ii) (H_3) holds.

Let S_1 be a compact connected set in $C(0)$ such that $S_1 \cap S_\alpha(0) \ne \emptyset$ with $x' \le \alpha'(0)$ for some $(x,x') \in S_1 \cap S_\alpha(0)$ and such that $S_1 \cap S_\beta(0) \ne \emptyset$ with $y' \ge \beta'(0)$ for some $(y,y') \in S_1 \cap S_\beta(0)$. Let S_2 be a closed connected subset of $C(t_0)$ such that $S_2 \cap [(x,x'): x' \text{ arbitrary}] \ne \emptyset$. Then there exists a solution $x(t)$ of (3.2.1) such that $\big(x(0),x'(0)\big) \in S_1$, $\big(x(t_0),x'(t_0)\big) \in S_2$ with $\alpha(t) \le x(t) \le \beta(t)$ on $[0,t_0]$.

Proof: We first prove the result with $\alpha(t)$ and $\beta(t)$ assumed to be strict lower and upper solutions, respectively.

Since (H_3) is satisfied the possibility of case (ii) in Theorem 3.1.1 is eliminated. Let $C_t \subset C(t)$ be a component of $I_E(S_1,0,t)$ which intersects both $S_\alpha(t)$ and $S_\beta(t)$. Define P by

$$P \equiv \big[\zeta\text{: there exists } (x,x') \in S_\alpha(t) \cap C_t \text{ such that } x' \le \alpha'(t) \text{ and there exists } (y,y') \in C_t \cap S_\beta(t) \text{ such that } y' \ge \beta'(t) \text{ for all } t \in [0,\zeta]\big].$$

Notice P is nonempty since $0 \in P$. Let $n = \sup P$. It is sufficient to show $n = t_0$. Exactly as in the proof of Theorem 3.1.1 it follows that $n \in P$. We omit the details.

There either exists $(x,x') \in C_n \cap S_\alpha(n)$ such that $x = \alpha(n)$, $x' = \alpha'(n)$ or there exists $(y,y') \in C_n \cap S_\beta(n)$ such that $y = \beta(n)$, $y' = \beta'(n)$ for if not, then there exists $n_1 > n$ such that $n_1 \in P$. This is impossible.

Consider then the case where $(y,y') \in C_n \cap S_\beta(n)$ such that $y = \beta(n)$, $y' = \beta'(n)$. The other case is similar. If

$y(t)$ is any solution of (3.2.1) with $y(n) = \beta(n)$, $y'(n) = \beta'(n)$, then there exists an $\varepsilon_1 > 0$ such that $y(t) > \beta(t), y'(t) > \beta'(t)$ on $(n, n+\varepsilon_1)$. This follows since $\beta(t)$ is a strict upper solution on $[0, t_0]$. Moreover, for any $\bar{t} > n$, any solution $y(t)$ of (3.2.1) with $y(\bar{t}) = \beta(\bar{t})$, $y'(\bar{t}) < \beta'(\bar{t})$ satisfies $y(t) < \beta(t)$ for $\bar{t} < t < \bar{t} + \delta$ and $y(t) > \beta(t)$ on $\bar{t} - \delta < t < \bar{t}$ for some $\delta > 0$. By the continuity of g, it follows that there exists an $\varepsilon_2 > 0$ and $\rho_1 > 0$ such that $\beta''(t) < g(t, y, y')$ for all $t \in [n, n+\varepsilon_2]$ and all (y, y') such that $d\big((y, y'), (\beta(t), \beta'(t))\big) < \rho_1$, where d denotes the Euclidean metric. By standard arguments there exist an $\varepsilon_3 > 0$ and $\rho_2 > 0$ such that any solution $y(t)$ of (3.2.1) with $y(n) = y_0$, $y'(n) = y_0'$ with $d\big((y_0, y_0'), (\beta(n), \beta'(n))\big) < \rho_2$ satisfies $d\big(y(t), y'(t)), (\beta(t), \beta'(t))\big) < \rho_1$ for all $t \in [n, n+\varepsilon_3]$. Moreover, $\beta''(t) < y''(t)$ for all $t \in [n, n+\min[\varepsilon_2, \varepsilon_3]$. Finally, there exists an $\varepsilon_4 > 0$ such that for any solution $y_1(t)$ of (3.2.1) with $y_1(n) = y_1$, $y_1'(n) = y_1'$ where $(y_1, y_1') \in C_n$ and $d\big((y_1, y_1'), (\beta(n), \beta'(n))\big) < \rho_1$ both $y_1(t) \geq \beta(t)$ and $y_1'(t) \leq \beta'(t)$ are not possible for any $t \in [n, n+\varepsilon_4]$.

If we let $\varepsilon = \min[\varepsilon_1, \varepsilon_2, \varepsilon_3, \varepsilon_4]$, then it follows from the preceding observations that $C_t \cap S_\beta(t) \neq \emptyset$ and $y' \geq \beta'(t)$ for some $(y, y') \in C_t \cap S_\beta(t)$ for all $t \in [0, n+\varepsilon]$. A similar argument leads to the conclusion that $C_t \cap S_2(t) \neq \emptyset$ and $x' \leq \alpha'(t)$ for some $(x, x') \in C_t \cap S_\alpha(t)$ for $t \in [0, n+\bar{\varepsilon}]$. Then $n_1 = \min[n+\varepsilon, n+\bar{\varepsilon}] \in P$ and $n_1 > n$, a contradiction. Thus $n = t_0$ and we conclude that $I_E(S_1, 0, t_0)$ contains a compact connected component intersecting both $S_\alpha(t_0)$ and $S_\beta(t_0)$. Hence $S_2 \cap I_E(S_1, 0, t_0) \neq \emptyset$ and hence there exists a solution $x(t)$ of (3.2.1) with $\big(x(0), x'(0)\big) \in S_1$, $\big(x(t_0), x'(t_0)\big) \in S_2$, and $\big(t, x(t), x'(t)\big) \in E$. Using the same type of proof as in Theorem 3.1.4, we may conclude that $\alpha(t) \leq x(t) \leq \beta(t)$ for all $t \in [0, t_0]$.

We now prove Theorem 3.2.1, assuming that $\alpha(t)$ and $\beta(t)$ are lower and upper solutions, respectively. For $0 < \gamma < 1$, let $A_\gamma(t) = \alpha(t) - \gamma$ and $B_\gamma(t) = \beta(t) + \gamma$. Define

$$G(t,x,x') = \begin{cases} g\big(t,\beta(t),x'\big) + x - \beta(t), & x > \beta(t), \\ g(t,x,x'), & \alpha(t) \leq x \leq \beta(t), \\ g\big(t,\alpha(t),x'\big) + x - \alpha(t), & x < \alpha(t). \end{cases}$$

Let $M = \max_{(x,x')\in S_1} |x'| + 1$ and let N be the number associated with M in (H_3). Choose $K = \max\big(N, \max_{t\in[0,t_0T]} (|\alpha'(t)|, |\beta'(t)|)\big)$ and define

$$H(t,x,x') = \begin{cases} 0, & |x'| \geq K + 2, \\ G(t,x,x'), & |x'| \leq K + 1, \\ (K+2-x')G(t,x,K+1), & K + 1 < x' < K + 2, \\ (K+2+x')G(t,x,-K-1), & -K-2 < x' < -K-1. \end{cases}$$

Moreover

$$\begin{aligned} A''_\gamma(t) = \alpha''(t) \geq g\big(t, \alpha(t), \alpha'(t)\big) &> g\big(t, \alpha(t), A'_\gamma(t)\big) - \gamma \\ &= H\big(t, A_\gamma(t), A'_\gamma(t)\big), \\ B''_\gamma(t) = \beta''(t) \leq g\big(t, \beta(t), \beta'(t)\big) &< g\big(t, \beta(t), B'_\gamma(t)\big) + \gamma \\ &= H\big(t, B_\gamma(t), B'_\gamma(t)\big). \end{aligned}$$

Hence A_γ and B_γ are strict lower and upper solutions of

$$x'' = H(t,x,x'). \tag{3.2.4}$$

Let $(p,q) \in S_1 \cap [(x,x'): x = \alpha(0),\ x' \leq \alpha'(0)]$, and let $(u,v) \in S_1 \cap [(y,y'): y = \beta(0),\ y' \geq \beta(0)]$. Let X_γ be the union of S_1, the line segment from (p,q) to $(p-\gamma,q)$ and the line segment from (u,v) to $(u+\gamma,v)$.

Consider now the BVP (3.2.4) with the boundary conditions

$$\big(x(0),x'(0)\big) \in X_\gamma, \qquad \big(x(t_0),x'(t_0)\big) \in S_2. \tag{3.2.5}$$

Let $E = [(t,x,x'): 0 \le t \le t_0, \alpha(t) - 1 < x < \beta(t) + 1, |x'| < \infty]$. Since $H(t,x,x') = 0$ for $|x'| \ge K + 2$, we may apply the proof of the first half of this theorem, recalling that $A_\gamma(t)$, $B_\gamma(t)$ are strict lower and upper solutions respectively, to conclude the existence of a solution $X_\gamma(t)$ of the BVP (3.2.4), (3.2.5) with the property that $A_\gamma(t) \le X_\gamma(t) \le B_\gamma(t)$ for $t \in [0,t_0]$. Also, since $H(t,x,x') = 0$ for $|x'| \ge K + 2$ it follows that $|X'_\gamma(t)| \le K + 2$ on $[0,t_0]$.

By Kamke's convergence theorem a subsequence of $\{X_{1/n}\}_{n=1}^{\infty}$ converges to a solution $X_0(t)$ of (3.2.4) on $[0,t_0]$ such that

$$\left(X_0(0),X_0'(0)\right) \in S_1, \qquad \left(X_0(1),X_0'(1)\right) \in S_2.$$

Also $\alpha(t) \le X_0(t) \le \beta(t)$, since

$$A_{1/n}(t) = \alpha(t) - (1/n) \le X_{1/n}(t) \le \beta(t) + (1/n) = B_{1/n}(t).$$

For $\alpha(t) \le x \le \beta(t)$, $|x'| \le K + 1$ we have $H(t,x,x') = g(t,x,x')$. It follows, from (H_3), that $|X_0'(t)| < N \le K$. Hence X_0 satisfies the BVP and this concludes the proof of Theorem 3.2.1.

It is not necessary to assume S_1 intersects the upper and lower surface at $t = 0$ provided there exist solutions which eventually hook S_1 onto the lower and upper surfaces. More precisely, we have the following result.

THEOREM 3.2.2. Assume (H_3) holds relative to E. Let α, β be lower and upper solutions for (3.2.1) with $\alpha(t) \le \beta(t)$ on $[0,t_0]$. Let S_1 be a compact connected set in $C(0)$ and let S_2 be a closed connected subset of $C(t_0)$ such that $S_2 \cap [(x,x'): x' \text{ arbitrary}] \ne \emptyset$. If $\left(x_1(t),x_1'(t)\right)$ and $\left(x_2(t),x_2'(t)\right)$ are solutions of (3.2.1) with $\left(x_1(0),x_1'(0)\right)$, $\left(x_2(0),x_2'(0)\right) \in S_1$ such that $\left[\left(x_1(u),x_1'(u)\right)\right] \cap S_\beta(u) \ne \emptyset$

for some $u \in (0,t_0]$ and $[(x_2(v),x_2'(v))] \cap S_2(v) \neq \emptyset$ for some $v \in (0,t_0]$, then there exists a solution $x(t)$ of (3.2.1) with $(x(0),x'(0)) \in S_1$, $(x(t_0),x'(t_0)) \in S_2$, and $\alpha(t) \leq x(t) \leq \beta(t)$ for all $t \in [0,t_0]$.

Proof: The proof is similar to that of Theorem 3.2.1 since $I_E(S_1,0,t)$ has a component intersecting $S_\alpha(t)$ and $S_\beta(t)$ for $t \geq \max[u,v]$.

In Chapter 1, we have provided sufficient conditions for (H_3) to hold. One such condition was Nagumo's condition. We now obtain a result to show how Nagumo's condition can be utilized to determine S_1 and S_2. Let $\alpha(t)$ and $\beta(t)$ be lower and upper solutions of (3.2.1) with $\alpha(t) \leq \beta(t)$ for $t \in [0,t_0]$. Define

$$\lambda = \max\left[\frac{|\alpha(0)-\beta(t_0)|}{t_0}, \frac{|\alpha(t_0)-\beta(0)|}{t_0}, \max_{[0,t_0]} |\alpha'(t)|, \max_{[0,t_0]} |\beta'(t)|\right]$$

and assume $g(t,x,y)$ satisfies Nagumo's condition

$$|g(t,x,y)| \leq \varphi(|y|)$$

for all $t \in [0,t_0]$, $\alpha(t) \leq x \leq \beta(t)$, $|y| < \infty$ such that

$$\int^{\infty} \frac{s\,ds}{\varphi(s)} = +\infty.$$

Let $N(t)$, $t \in [0,t_0]$, be given by

$$\int_{\lambda}^{N(t)} \frac{s\,ds}{\varphi(s)} = \max_{u\in[0,t]} \beta(u) - \min_{u\in[0,t]} \alpha(u).$$

Letting

$$F(x) = \int_{\lambda}^{x} \frac{s\ ds}{\varphi(s)} ,$$

we observe that

$$N(t) = F^{-1}\left(\max_{[0,t]} \beta(u) - \min_{[0,t]} \alpha(u) \right)$$

is a continuous function on $[0,t_0]$. Let $N = \min_{[0,t_0]} N(t)$. Define

$$S_3 = \{(0,x,N): \alpha(0) \leq x \leq \beta(0)\}$$
$$\cup \{0,\beta(0),x'): \beta'(0) \leq x' \leq N\},$$

$$S_4 = \{(0,x,-N): \alpha(0) \leq x \leq \beta(0)\}$$
$$\cup \{(0,\alpha(0),x'): -N \leq x' \leq \alpha'(0)\},$$

$$S_5 = \{(t_0,x,N(t_0)): \alpha(t_0) \leq x \leq \beta(t_0)\}$$
$$\cup \{(t_0,\alpha(t_0),x'): \alpha'(t_0) \leq x' \leq N(t_0)\},$$

$$S_6 = \{(t_0,x,-N(t_0)): \alpha(t_0) \leq x \leq \beta(t_0)\}$$
$$\cup \{(t_0,\beta(t_0),x'): -N(t_0) \leq x' \leq \beta'(t_0)\}.$$

THEOREM 3.2.3. Assume $g(t,x,y)$ satisfies Nagumo's condition with respect to $\alpha(t),\beta(t) \in C^{(2)}[0,t_0]$ with $\alpha(t) \leq \beta(t)$. If S_1 is any closed connected subset of $C(0)$ such that $S_1 \cap S_3 \neq 0$, $S_1 \cap S_4 \neq 0$ and if S_2 is any closed connected subset of $C(t_0)$ such that $S_2 \cap S_5 \neq \emptyset$ and $S_2 \cap S_6 \neq \emptyset$, then there exists a solution $x(t)$ of (3.2.1) such that $\left(x(0),x'(0)\right) \in S_1$, $\left(x(t_0),x'(t_0)\right) \in S_2$ with $\alpha(t) \leq x(t) \leq \beta(t)$.

Proof: Choose $(0,x_1,x_1') \in S_1 \cap S_3$ and $(0,x_2,x_2') \in S_1 \cap S_4$ such that both points belong to the same component $\overline{S}_1$ of $S_1 \cap [(0,x,y): |y| \leq N]$. Choose $\left(0,\beta(0),x_3'\right) \in S_\beta(0)$ as follows. If $x_1 = \beta(0)$, let $x_3' = x_1'$; if $x_1 < \beta(0)$, choose $x_3' > N$. Let L_1 be the line segment joining $(0,x_1,x_1')$

to $(0,\beta(0),x_3')$, where, in the case when $x_1 = \beta(0)$, $L_1 = [(0,x_1,x_1')] \subset \overline{S}_1$. In a similar manner, choose $(0,\alpha(0),x_4') \in S_\alpha(0)$ by letting $x_4' = x_2'$ if $x_2 = \alpha(0)$ or $x_4' < -N$ if $x_2 > \alpha(0)$ and take L_2 to be the line segment joining $(0,x_2,x_2')$ to $(0,\alpha(0),x_4')$. Let $S_1^* = L_1 \cup \overline{S}_1 \cup L_2$ and observe S_1^* is compact, connected, and intersects both $[(0,\alpha(0),x'): x' \le \alpha'(0)]$ and $\{(0,\beta(0),x'): x' \ge \beta'(0)\}$.

Pick $(t_0,x_5,x_5') \in S_2 \cap S_5$ and $(t_0,x_6,x_6') \in S_2 \cap S_6$ such that both points belong to the same component $\overline{S}_2$ of $S_2 \cap [(t_0,x,x'): |x'| \le N(t_0)]$. Let L_5 and L_6 be the half lines given by $[(t_0,x_5,y): y \ge x_5']$ and $[(t_0,x_6,y): y \le x_6']$ and let $S_2^* = L_5 \cup \overline{S}_2 \cup L_6$.

Consider the BVP (3.2.1) and

$$(3.2.6) \qquad (0,x(0),x'(0)) \in S_1^*, \qquad (t_0,x(t_0),x'(t_0)) \in S_2^*.$$

By the construction of S_1^* and the assumptions on $g(t,x,y)$, there exists a compact, connected subset $C \subset C(t_0)$ of the funnel cross section $F(t_0,0,S_1^*) = \bigcup_{s \in S_1^*} F(t_0,0,s)$ which intersects both $[(t_0,x,y) \in S_\beta(t_0): y \ge \beta'(t_0)]$ and $[(t_0,x,y) \in S_2(t_0): y \le \alpha'(t_0)]$. Hence, by Theorem 3.2.1 the BVP (3.2.1), (3.2.6) has a solution $x(t)$ with $\alpha(t) \le x(t) \le \beta(t)$, $|x'(t_0)| \le N(t_0)$ and $(t_0,x(t_0),x'(t_0)) \in S_2$. From Nagumo's condition $|x'(0)| \le N$ and by the construction of S_1^*, $(0,x(0),x'(0)) \in S_1$. This concludes the proof of Theorem 3.2.3.

EXAMPLE 3.2.1. As an application of this result consider the following BVP arising in the dynamics of certain chemical reactions.

$$(3.2.7) \qquad x'' = -L_1 x' - L_2 k(x),$$

$$(3.2.8) \qquad x(0) = 0, \qquad x'(1) = -L_1 x(1).$$

The positive constants L_1 and L_2 depend on various parameters of the physical problem with $k(x)$ continuous and nonincreasing on $[0,c]$ such that $k(x) > 0$ for $0 \leq x < c$ and $k(c) = 0$. Theorem 3.2.3 gives the existence of a solution $x(t)$ of the BVP (3.2.7), (3.2.8) with $0 \leq x(t) \leq c$ since $\alpha(t) = 0$ and $\beta(t) = c$ are lower and upper solutions, respectively, and Nagumo's condition is clearly satisfied.

3.3 WAZEWSKI RETRACT METHOD

This section is devoted to an application of the Wazewski retract method to (3.2.1) with boundary conditions

$$(3.3.1) \qquad L_1(x) = r_1, \qquad L_1(x) \equiv a_1x(0) - b_1x'(0),$$

$$(3.3.2) \qquad L_2(x) = r_2, \qquad L_2(x) \equiv a_2x(t_0) + b_2x'(t_0).$$

Without loss of generality, we may assume $b_1, b_2 \geq 0$. Let $\alpha(t)$ and $\beta(t)$ be lower and upper solutions of (3.2.1) such that

$$(3.3.3) \qquad L_i(\alpha) \leq r_i, \qquad i = 1,2,$$

$$(3.3.4) \qquad L_i(\beta) \geq r_i, \qquad i = 1,2.$$

We shall assume that g satisfies Nagumo's condition, that is,

$$(3.3.5) \qquad |g(t,x,x')| \leq h\,|x'|\,,$$

for all $(t,x,x') \in D = \big[(t,x,x');\ t \in [0,t_0],\ \alpha(t) \leq x \leq \beta(t),$ and $x' \in R\big]$ such that

$$(3.3.6) \qquad \int_0^\infty \frac{s\,ds}{h(s)} > \max_{[0,t_0]} \beta(t) - \min_{[0,t_0]} \alpha(t) \equiv K.$$

As an application of Theorem 1.4.1, we obtain the following lemmas.

LEMMA 3.3.1. Assume that g satisfies Nagumo's condition. Then, for every $c > 0$, there exists a constant $M(c)$ such that each solution $x(t)$ satisfying $\alpha(t) \leq x(t) \leq \beta(t)$ for $t \in I \subset [0,t_0]$ and which satisfies

$$|x'(s_0| \leq c \quad \text{for some} \quad s_0 \in I$$

also has the property

$$|x'(t)| \leq M(c) \quad \text{for all} \quad t \in I.$$

LEMMA 3.3.2. Assume that there exists a strict lower solution $\alpha(t)$ and strict upper solution $\beta(t)$ with respect to (3.2.1), satisfying (3.3.3), (3.3.4) with $\alpha(t) < \beta(t)$ for all $t \in [0,t_0]$. Then, there exists a solution $x_0(t)$ of (3.2.1) defined on $[0,h+\varepsilon] \subset [0,t_0]$ for some $h \geq 0$, $\varepsilon > 0$, satisfying (3.3.1) and such that $\alpha(t) \leq x_0(t) \leq \beta(t)$ for all $t \in [0,h]$, $x_0(h) = \beta(h)$ and $x_0(t) > \beta(t)$ for $t \in (h,h+\varepsilon]$. Similarly there exists a solution $z_0(t)$ satisfying (3.3.1), existing on an interval $[0,k+\delta] \subset [0,t_0]$ for some $k \geq 0$, $\delta > 0$, such that $\alpha(t) \leq z_0(t) \leq \beta(t)$ for $t \in [0,k]$, $z_0(h) = \alpha(h)$ and $z_0(t) < \alpha(t)$ for all $t \in [k,k+\delta]$.

Proof: We show the existence of $x_0(t)$ [a similar proof yields the existence of $z_0(t)$]. First assume $b_1 \neq 0$; then, letting $x_0(0) = \beta(0)$, we obtain from (3.3.4) that $x_0'(0) > \beta'(0)$. Hence the result holds for $h = 0$.

Let $b_1 \neq 0$ and consider a solution $x_n(t)$ of (3.2.1) satisfying, for each integer n,

$$x_n(0) = r_1 a_1^{-1}, \qquad x_n'(0) = n.$$

Define

$$B = \left\{ \max_{t\in[0,t_0]} \beta(t) - \min_{t\in[0,t_0]} \alpha(t), \max_{t\in[0,t_0]} \alpha'(t) \right\}.$$

For $N > M(B)$, there exists $s_0 \in [0,t_0)$ satisfying $x_N(s_0) = \beta(s_0)$ and $\alpha(t) < x_N(t) < \beta(t)$, $t \in (0,s_0)$. Assume not; then either $\alpha(t) < x_N(t) < \beta(t)$ for all $t \in (0,t_0)$ or there exists a $T_n < t_0$ such that $x_n(T_n) = \alpha(T_n)$, $\alpha(t) < x_n(t) < \beta(t)$, for $t \in (0,T_N)$. In either case, there exists a $t_N \in [0,t_0]$ such that $|x_N'(t_N)| \leq B$. [In the second case, we make use of the fact $x_N'(T_N) < \alpha'(t_N)$, since $\alpha(t)$ is a strict lower solution.]

By Lemma 3.3.1, $|x_N'(t)| \leq M(B)$ for $t \in [0,t_N]$ and since $x_N'(0) = N$, we have a contradiction for $N > M(B)$. Thus for some $h \in [0,t_0], \alpha(t) \leq \beta(t)$ for all $t \in [0,h]$ and $x_N(h) = \beta(h)$. Since $\beta(t)$ is a strict upper solution, we have $x_N'(h) > \beta'(h)$. Thus, there exists an $\varepsilon > 0$ such that $x_N(t) > \beta(t)$, $t \in (h,h+\varepsilon]$. The proof is complete.

LEMMA 3.3.3. Let $x(t,\mu)$ be a family of functions indexed by a parameter μ. Let $x(t,\mu)$ be continuous in t and μ simultaneously. Let C be a simple closed curve in the (t,x) plane and denote by Int C and Ext C the open interior and open exterior of C. Suppose that for each μ, $\big(0,x(0,\mu)\big) \in C \cup \text{Int } C$ and $x(t,\mu)$ is defined on an interval $[0,T_\mu]$ where $\big(T_\mu,x(T_\mu,\mu)\big) \in \text{Ext } C$. Define $\gamma(\mu) = t_\mu$ where $t_\mu = \sup[t \geq 0: \big(t,x(t,\mu)\big) \in \text{Int } C]$. Then γ is lower semi-continuous. If, in addition, each function $x(t,\mu)$ has the additional property that $\big(t,x(t,\mu)\big) \in \text{Ext } C$ for $t \in (t_\mu,t_\mu+h_\mu)$, then γ is continuous. (Lemma 3.3.3 is another version of the Wazewski method.)

THEOREM 3.3.1. Assume that there exists an upper solution $\beta(t)$ and a lower solution $\alpha(t)$ of (3.2.1) with $\alpha(t) \leq \beta(t)$ for all $t \in [0,t_0]$, satisfying (3.3.3) and (3.3.4). Let f satisfy Nagumo's condition with respect to α and β. Then there is a solution $x(t)$ of the BVP (3.2.1), (3.3.1), (3.3.2)

$\alpha(t) \leq x(t) \leq \beta(t)$ for all $t \in [0, t_0]$.

Proof: We may assume that (3.3.5) holds for all $x \in R$ since we may redefine g outside of D so that the Nagumo's condition will hold.

We first prove Theorem 3.3.1 under the stronger hypotheses that $\alpha(t)$ and $\beta(t)$ are strict lower and upper solutions and g is locally Lipschitzian. We will then show how these hypotheses may be weakened.

Consider the set of solutions of (3.2.1) and (3.3.1) such that $\alpha(0) \leq x(0) \leq \beta(0)$. These solutions form a one-parameter family of functions $\{x(t,\mu)\}$, where the set of values of μ are connected. Define a map

$$\varphi:\ \mu \to \big(t_\mu, x(t_\mu, \mu)\big),$$

where t_μ is defined in Lemma 3.3.3 in which C is the Jordan curve forming the boundary of the region D. Now $x(t,\mu)$ is continuous in both t and μ since g is locally Lipschitz and thus applying Lemma 3.3.3 we have $\gamma:\ \mu \to t_\mu$ is continuous. Thus φ is continuous and $\{\varphi(\mu)\}$ is connected. Combining Lemma 3.3.2 with the connectedness of $\{\varphi(\mu)\}$, we have the existence of μ_1 and μ_2, such that

$$t_{\mu_1} = t_{\mu_2} = t_0, \qquad x(t_0, \mu_1) = \alpha(t_0), \qquad x(t_0, \mu_2) = \beta(t_0).$$

Since α and β are strict solutions

$$x'(t_0, \mu_1) < \alpha'(t_0), \qquad x'(t_0, \mu_2) > \beta'(t_0).$$

Thus $L_2\big(x(\cdot, \mu_1)\big) \leq r_2$ and $L_2\big(x(\cdot, \mu_2)\big) \geq r_2$. From the continuity of L_2, there exists a $\mu_0 \in [\mu_1, \mu_2]$ such that

$$L_2\big(x(\cdot, \mu_0)\big) = r_2.$$

The solution $x(t,\mu_0)$ is a solution of BVP (3.2.1), (3.3.1), (3.3.2).

We no longer assume g satisfies a Lipschitz condition. From (3.3.6), there exists an $\varepsilon_0 > 0$ such that if $0 < \varepsilon < \varepsilon_0$, then

$$\int_0^\infty \frac{s\,ds}{h(s)+\varepsilon} > K.$$

There exists a sequence of locally Lipschitzian functions $\{g_n(t,x,y)\}_{n=1}^\infty$ which converge to $g(t,x,y)$ uniformly on $[0,t_0]\times R^2$. Assume that

$$|g_n(t,x,y) - g(t,x,y)| \le 1/n \qquad \text{for} \quad (t,x,y) \in [0,t_0]\times R^2.$$

Pick N so large such that $1/N \le \varepsilon_0$ and

$$1/N < \inf\left[|\alpha'' - g(t,\alpha,\alpha')|,\ |\beta'' - g(t,\beta,\beta')|\right],$$

Thus, for $n > N$, α and β will be strict lower and strict upper solutions for

$$x'' = g_n(t,x,x'), \tag{3.3.7}$$

where

$$|g_n(t,x,y)| \le h(|y|) + (1/n).$$

By our previous analysis, there exists a function $x_n(t)$ satisfying (3.3.7), (3.3.1), and (3.3.2).

From the mean value theorem, there exists a $T \in [0,t_0]$ such that $|x'(T)| \le K/t_0 \equiv K_1$.

Choose $M = M(K)$ so large that

$$\int_{K_1}^M \frac{s\,ds}{h(s)} > K.$$

Then, there exists N_1, such that for $n > N_1$

$$\int_{K_1}^{M} \frac{s\,ds}{h(s) + (1/n)} > K.$$

As in the proof of Theorem 1.4.1, we have for $n > N_1$,

$$|x_n'(t)| \leq M(K) \qquad \text{for all} \quad t \in [0,t_0].$$

Since

$$|x_n(t)| \leq K \qquad \text{for all} \quad t \in [0,t_0],$$

it follows that

$$|x_n''(t)| \leq \max\Big[|g(t,x,y)| + 1 \colon t \in [0,t_0], \\ |x| \leq K, \ |y| \leq M\Big].$$

Therefore the sequence $\{x_n(t)\}$, for $n \geq \max(N,N_1)$ has a subsequence which converges in the $C^{(1)}$ norm. The limit function is then a solution of (3.2.1) and satisfies (3.3.1) and (3.3.2).

We now remove the restriction that $\alpha(t)$ and $\beta(t)$ are strict lower and upper solutions. Let $p\colon [0,t_0] \to (0,t_0]$ satisfy $p''(t) \leq 0$ for all $t \in [0,t_0]$, $L_i(p) \geq 0$ for $i = 1,2$, and

$$\sup_{t\in[0,t_0]} \big[\{|p(t)| + |p'(t)|\} \leq 1\big].$$

Define for $\varepsilon > 0$

$$\beta_\varepsilon(t) = \beta(t) + \varepsilon p(t), \qquad \alpha_\varepsilon(t) = \alpha(t) - \varepsilon p(t).$$

Then $\alpha_\varepsilon(t) < \beta_\varepsilon(t)$,

$$L_i(\alpha\varepsilon) < r_1 \qquad \text{for} \quad i = 1,2,$$

$$L_i(\beta\varepsilon) > r_2 \qquad \text{for} \quad i = 1,2.$$

Define

$$\gamma_\varepsilon(t,x,y) = \frac{x - \frac{1}{2}(\beta_\varepsilon(t) + \alpha_\varepsilon(t))}{\beta_\varepsilon(t) - \alpha_\varepsilon(t)},$$

and let

$$g_\varepsilon(t,x,y) = g(t,x,y) + \frac{4\gamma_\varepsilon(t,x,y)}{1 + 2|\gamma_\varepsilon(t,x,y)|}\,\delta(\varepsilon),$$

where $\delta(\varepsilon)$ is the modulus of continuity of $g(t,x,y)$ on the domain $D' = [(t,x,y)\colon t \in [0,t_0],\ |x| \le K,\ |y| \le S]$, where K is defined as before and $S > \max\,(M(K),\ \sup_{t\in[0,t_0]} |\alpha'(t)|,\ \sup_{t\in[0,t_0]} |\beta'(t)|)$. Since $|\alpha_\varepsilon(t) - \alpha(t)| + |\alpha'_\varepsilon(t) - \alpha'(t)| < \varepsilon$ and $|g(t,\alpha,\alpha') - g(t,\alpha_\varepsilon,\alpha'_\varepsilon)| < \delta(\varepsilon)$ we have

$$g(t,\alpha,\alpha') > g(t,\alpha_\varepsilon,\alpha'_\varepsilon) - \delta(\varepsilon).$$

Moreover, since $\gamma_\varepsilon(t,\alpha_\varepsilon,\alpha'_\varepsilon) = -2$

$$g_\varepsilon(t,\alpha_\varepsilon,\alpha'_\varepsilon) = g(t,\alpha_\varepsilon,\alpha'_\varepsilon) - \delta(\varepsilon).$$

Therefore

$$g(t,\alpha,\alpha') > g_\varepsilon(t,\alpha_\varepsilon,\alpha'_\varepsilon).$$

Hence

$$\alpha''_\varepsilon \ge \alpha'' \ge g(t,\alpha,\alpha') > g_\varepsilon(t,\alpha_\varepsilon,\alpha'_\varepsilon).$$

Thus α_ε is a strict lower solution of

$$x'' = g_\varepsilon(t,x,x'). \tag{3.3.8}$$

Similarly β_ε is a strict upper solution of (3.3.8).

Furthermore, $g_\varepsilon(t,x,y)$ converges to $g(t,x,y)$ uniformly as $\varepsilon \to 0$. Therefore $g_\varepsilon(t,x,y)$ satisfies Nagumo's condition and as before there exists a solution $x_\varepsilon(t,x,y)$ of (3.3.8), (3.3.1), and (3.3.2). Moreover, $\{x_\varepsilon\}$ converges in the $C^{(1)}$

norm as $\varepsilon \to 0$. The limiting function is then a solution of the BVP (3.2.1), (3.3.1), and (3.3.2). This concludes the proof of Theorem 3.3.1.

REMARK 3.3.1. The significance of this result is that we can show existence of solutions of the Neumann problem, that is, when $a_1 \equiv a_2 \equiv 0$ in (3.3.1) and (3.3.2). Observe that except for this very important case Theorem 3.3.1 is a consequence of Lemma 3.3.2 and Theorem 3.2.2.

The following example illustrates Theorem 3.3.1.

EXAMPLE 3.3.1. Assume $r_1 = r_2 = 0$ and $g\colon [0,1] \times R^2 \to R$ and satisfies Nagumo's condition. Let g_x exist and satisfy

$$g_x \geq \varepsilon > 0. \tag{3.3.9}$$

Then the boundary value problem (3.2.1), (3.3.1), (3.3.2) has a solution. To see this let

$$C > 1/\varepsilon \max_{t \in [0,1]} |g(t,0,0)|; \tag{3.3.10}$$

then, from (3.3.9) we have

$$g(t,0,0) - g(t,-C,0) \geq C\varepsilon.$$

Using (3.3.10), we then obtain

$$g(t,-C,0) < 0.$$

Thus $\alpha(t) = -C$ is a lower solution and similarly $\beta(t) = C$ is an upper solution for (3.2.1), (3.3.1), and (3.3.2). By Theorem 3.3.1, there exists a solution of the BVP (3.2.1), (3.3.1), (3.3.2).

3.4 GENERALIZED DIFFERENTIAL EQUATIONS

In this section we develop a Wazewski theorem for contingent differential equations and utilize it to prove existence theorems for BVP's.

The following preliminaries will prove useful in the discussion. Let $c(R^n)$ $(cc(R^n))$ be the collection of all nonempty compact (compact and convex) subsets of R^n. For $x \in R^n$ and $A,B \in c(R^n)$, let $q(x,B) = \inf[\|x-b\|: b \in B]$ and $q(A,B) = \sup[q(x,B): x \in A]$. Then $d(A,B) \equiv \max[q(B,A),q(A,B)]$ is the Hausdorff metric on $c(R^n)$ and $c(R^n)$ is a complete metric space.

Let V be a subset of $R \times R^n$ and denote the points of V by $p = (t,x)$. A mapping $F: V \to c(R^n)$ is upper semi-continuous (USC) at $p \in V$ if, for each $\varepsilon > 0$, there exists a $\delta > 0$ such that $\|Q - p\| < \delta$ implies $q(F(Q),F(p)) < \varepsilon$. If we replace $q(F(Q),F(p))$ by $d(F(Q),F(p))$, then F is continuous.

Let X be an open subset of $R \times R^n$ and let $F: X \to cc(R^n)$ be USC. A solution of the generalized differential equation

$$x' \in F(t,x) \tag{3.4.1}$$

is a function $\varphi: I \to R^n$ which is absolutely continuous on each compact subinterval of I and $\varphi'(t) \in F(t,\varphi(t))$ almost everywhere on I.

For $E \subset X$ and $p_0 = (t_0,x_0)$, an accumulation point of E, the positive contingent of E at p_0 is $D^+(E,p_0) = [y \in R^n$: there exists $\{(t_n,x_n)\} \subset E$, $(t_n,x_n) \to (t_0,x_0)$, $t_n > t_0$, $(x_n - x_0)/(t_n - t_0) \to y$ as $n \to \infty]$. The negative contingent and contingent of E at p_0 are defined in a similar manner. If E is the graph of a function $\varphi: I \to R^n$, then write $D^+\varphi(t_0)$ instead of $D^+(E,p_0)$. A solution of the contingent equation

$$Dx \subset F(t,x) \tag{3.4.2}$$

is a continuous function $\varphi: I \to R^n$ such that $D\varphi(t) \subset F(t,\varphi(t))$ for all $t \in I$. A function $\varphi(t)$ is a solution of (3.4.1) if and only if it is a solution of (3.4.2).

The basic theory for generalized equations hold; namely, the Peano's existence theorem, extendability of solutions, and the Kamke convergence theorem are true for generalized equations, and we shall not expound on this further.

Denote the maximal interval of existence of a solution $\varphi(t)$ of (3.4.1) by D_φ. Before stating a Wazewski theorem for generalized equations, we will need the following definition.

DEFINITION 3.4.1. A set $A \subset X$ is (positively) weakly invariant with respect to (3.4.1) if for each $p_0 = (t_0, x_0) \in A$, there is a solution $\varphi(t)$ of (3.4.1) with $\varphi(t_0) = x_0$, $(t, \varphi(t)) \in A$ on $D_\varphi \cap [t_0, \infty)$.

REMARK 3.4.1. A closed set E is positively weakly invariant if and only if $D^+(E, p_0) \cap F(p_0) \neq \emptyset$ for all $p_0 \in E$.

LEMMA 3.4.1. Assume E_1 and E_2 are relatively closed, positively weakly invariant subsets of X with $X = E_1 \cup E_2$. Then $H = E_1 \cap E_2$ is positively weakly invariant.

Proof: Let $p_0 = (t_0, x_0) \in H$. By hypotheses, there exist two solutions $\varphi_i(t)$ of (3.4.1) with $\varphi_i(t_0) = x_0$, $i = 1,2$, and an $a > 0$ such that $(t, \varphi_i(t)) \in E_i$ for $t_0 \le t < t_0 + a$. Let $L(t)$ be the segment joining $(t, \varphi_1(t))$ to $(t, \varphi_2(t))$ and let $(t, x(t)) \in L(t) \cap (E_1 \cap E_2)$. Then $x(t) = \alpha(t)\varphi_1(t) + (1 - \alpha(t))\varphi_2(t)$, $0 \le \alpha(t) \le 1$. Choose a sequence $\{t_n\}$; $t_n \to t_0^+$, and $\alpha(t_n) \to \alpha_0$. Then

$$\frac{x(t_n) - x_0}{t_n - t_0} = \frac{\alpha(t_n)(\varphi_1(t_n) - x_0) + (1 - \alpha(t_n))(\varphi_2(t_n) - x_0)}{t_n - t_0} .$$

We now can choose a subsequence $\{t_k\}$ of $\{t_n\}$ such that in the limit the left side belongs to $D^+(H, p_0)$ and the right side equals $\alpha_0 v_1 + (1 - \alpha_0)v_2$ where $v_i \in D^+\varphi_i(t_0)$.

By convexity, $\alpha_0 v_1 + (1-\alpha_0)v_2$ is in $F(p_0)$ and thus $D^+(H,p_0) \cap F(p_0) \neq \emptyset$ and as a result, by our previous remark, H is positively weakly invariant. This concludes the proof of Lemma 3.4.1.

The following ideas will be needed for the Wazewski theorem. For $p_0 = (t_0,x_0) \in V \subset X$, the zone of emission relative to V is $E_V(p_0) = [(\tau,y)\colon y = \varphi(\tau),\ \varphi(t)$ is a solution of (3.4.1) with $\varphi(t_0) = x_0$ and $(t,\varphi(t)) \in V$ for all t between t_0 and $\tau]$, and the right zone of emission relative to V is $E_V^+(p_0) = [(\tau,y)\colon (\tau,y) \in E_V(p_0),\ \tau \geq t_0]$. If $A \subset X$, $E_V(A) = \bigcup[E_V(p)\colon p \in A]$ and $E_V^+(A)$ is similarly defined.

Let W be a relatively closed subset of X. For $p_0 \in W$, the trace of emission relative to W is defined to be $T_W^+(p_0) = E_W^+(p_0) \cap (\partial W \cap W)$ and, for $A \subset W$, $T_W^+(A) = E_W^+(a) \cap (\partial W \cap W)$.

DEFINITION 3.4.2. A point $Q \in \partial W \cap W$ is a <u>strict egress point</u> relative to (3.4.1) if for every solution $\varphi(t,Q)$, $c_\varphi = \sup[t\colon (s,\varphi(s,Q)) \in \partial W \cap W,\ t_Q \leq s \leq t] < \infty$, and there exists a sequence $\{t_n\}$, $t_n \to c_\varphi$ with $(t_n,\varphi(t_n,Q)) \in X - W$.

We will denote the set of strict egress points by S. Finally, a solution $\varphi(t,p)$, $p \in W$, of (3.4.1) "leaves W," if there exists some $t_1 \in D_\varphi \cap [t_p,\infty)$ such that $(t_1,\varphi(t_1,p)) \in X - W$.

Before stating the Wazewski theorem we will need the following lemmas.

LEMMA 3.4.2. Let $z \subset \operatorname{int} W \cup S$ and assume that all points of $T^+(z) \equiv T_W^+(z)$ are strict egress points. If all solutions through $p \subset W \cup S$ leave W, then $T^+(p)$ is compact.

<u>Proof</u>: We shall show any sequence $\{Q_n\} \subset T^+(p)$ contains

a subsequence which converges to a point of $T^+(p)$. Let $\{\varphi_n(t)\}$ be solutions of (3.4.1) through p such that $(t_n, \varphi_n(t_n)) = Q_n$. By an extension of Kamke's convergence theorem to contingent equations, there is a subsequence $\{\varphi_k(t)\}$ of $\{\varphi_n(t)\}$ converging to a solution $\varphi(t)$ of (3.4.1) with $\varphi(t_p) = x_p$. By assumption $\varphi(t)$ leaves W. Thus there exists $t_a > t_p$ such that $(t_a, \varphi(t_a)) \in X - \overline{W}$. For all k sufficiently large, $(t_a, \varphi_k(t_a)) \in X - \overline{W}$, hence $t_p < t_k < t_a$. Choose a subsequence $\{t_m\}$ such that $t_m \to t_b < t_a$. Then $Q_m = (t_m, \varphi_k(t_m)) \to (t_b, \varphi(t_b))$.

Now $(t_b, \varphi(t_b)) \notin T^+(p)$ implies for m sufficiently large, that Q_m is not a strict egress point. This leads to a contradiction and thus $T^+(p)$ is compact.

LEMMA 3.4.3. Let Z be a nonempty subset of $W \cup S$. If all solutions through p leave W, for each $p \in Z$, and if all points $T^+(Z)$ are strict egress points, then T^+ is an USC map on Z.

<u>Proof</u>: By Lemma 3.4.2, $T^+(p)$ is compact. If T^+ is not USC at some $p_0 \in Z$, then there is an $\varepsilon > 0$ and a subsequence $\{p_n\} \in Z$ such that $p_n \to p_0$ and $q(T^+(p_n), T^+(p_0)) \geq \varepsilon$ as $n \to \infty$. Hence for each n, there is a $Q_n \in T^+(p_n)$ such that $q(Q_n, T^+(p_0)) > \varepsilon$. Let $\varphi_n(t)$ be a solution of (3.4.1) through p_n and Q_n. By the convergence theorem there is a subsequence $\{\varphi_k(t)\}$ of $\{\varphi_n(t)\}$ converging to a solution $\varphi(t)$ of (3.4.1) with $\varphi(t_0) = x_0$. By assumption, $\varphi(t)$ leaves W. As in the proof of Lemma 3.4.2, we can find a subsequence $\{t_m\}$ of $\{t_k\}$ such that $(t_m, \varphi_m(t_m)) = Q_m \to (t_b, \varphi(t_b)) \in T^+(p_0)$ which is a contradiction. This completes the proof of Lemma 3.4.2.

We now state and prove the Wazewski theorem which will then be used to obtain the existence of solutions of BVP's associated with (3.4.1).

THEOREM 3.4.1. Let $Z \subset \text{int } W \cup S$ be connected. If all points of $T^+(Z)$ are strict egress points and if $T^+(Z)$ is not connected, then there exists $p_0 \in Z$ and a solution $\varphi(t,p_0)$ of (3.4.1) such that $(t,\varphi(t,p_0)) \in W$ on $D_\varphi \cap [t_0,\infty)$.

<u>Proof</u>: Assume not; then for any $p \in Z$, every solution $\varphi(t,p)$ leaves W and there exists $Q \in T^+(p) \subset \partial W \cap W$ (Q depends on φ) such that $(t_Q,\varphi(t_Q,p)) \in \partial W \cap W$ and $(t,\varphi(t,p)) \in W$ on $[t_p,t_Q]$. By Lemma 3.4.2, $T^+(p)$ is compact for each $p \in Z$.

For each $p \in Z$, we will show $T^+(p)$ is connected. If $p \in Z \cap S$, then $p \in T^+(p)$ and p is a strict egress point. Clearly, $T^+(p)$ is then connected. Let $V = \text{int } W$, $p_1 \in V \cap Z$, and assume that $T^+(p_1) = C_1 \cup C_2$ is a separation where C_1, C_2 are nonempty disjoint compact sets.

For a solution $\varphi(t,p)$ of (3.4.1) with $p \in V$, let $I_\varphi = D_\varphi \cap [t_p,\infty) = [t_p,w)$ be the right maximal interval of existence of $\varphi(t,p)$ relative to X and let $J_\varphi = [t_p,\gamma)$ be the right maximal interval of existence of $\varphi(t,p)$ relative to V. Let $\Phi(p)$ denote the right trajectory of $\varphi(t,p)$ relative to V, and let $\sigma(A,B) = \inf[\|a-b\|: a \in A, b \in B]$, where A,B are arbitrary sets.

Define $E_1 = [p \in V$: there exists a solution $\varphi(t,p)$ with $I_\varphi = J_\varphi$ or there exists a solution $\varphi(t,p)$ such that $I_\varphi \neq J_\varphi$ with $\sigma(\Phi(p),C_1) \leq \sigma(\Phi(p),C_2)]$. Define E_2 similarly with the inequality reversed.

It follows that E_1, E_2 are closed relative to V, are positively weakly invariant, such that $V = E_1 \cup E_2$ and $p_1 \in E_1 \cap E_2$. By Lemma 3.4.1, there is a solution $\varphi(t,p_1)$ of (3.4.1) such that $(t,\varphi(t,p_1)) \in E_1 \cap E_2$ on J_φ.

However, $\varphi(t,p_1)$ must leave W. Hence $(\gamma,\varphi(\gamma,p_1)) \in$

$T^+(p_1) = C_1 \cup C_2$. Assume that $(\gamma, \varphi(\gamma, p_1)) \in C_1$. Let $\{Q_n\}$ be a sequence of points on $\Phi(p_1)$ such that $Q_n \to (\gamma, \varphi(\gamma, p_1)) \in C_1$. Since $Q_n \in E_2$, there exists a solution $\varphi(t, Q_n)$ of (3.4.1) such that $\sigma(\Phi(Q_n), C_1) \geq \sigma(\Phi(Q_n), C_2)$ for each n. Since

$$\psi_n(t, P_1) = \begin{cases} \varphi(t, p_1), & t_{p_1} \leq t \leq t_{Q_n}, \\ \varphi(t, Q_n), & t_{Q_n} < t < W_n, \end{cases}$$

is a solution of (3.4.1) through $p_1 \in Z$, $\psi_n(t, p_1)$ must leave W through C_2 and hence $(\gamma_n, \varphi(\gamma_n, Q_n)) \in C_2$.

This implies that there exists a solution $\psi(t, p_1)$ of (3.4.1) with $\psi(\gamma, p_1) = \varphi(\gamma, p_1)$ and a subsequence $\{\psi_k(t)\}$ of $\{\psi_n(t)\}$ such that for any compact interval $I \subset D_\psi$ and all k sufficiently large, $\psi_k(t, p_1)$ is defined on I and $\psi_k(t, p_1) \to \psi(t, p_1)$ uniformly on I as $k \to \infty$. However, then $(\gamma, \varphi(\gamma, p_1)) \in C_1 \subset T^+(p_1)$ is not a strict egress point, which is a contradiction. Thus $T^+(p_1)$ is connected. Finally, from Lemma 3.4.3, T^+ is an USC map on Z from which it follows $T^+(Z)$ is connected. This is a contradiction to the assumption $T^+(Z)$ is not connected. Hence we conclude that there exists $p_0 \in Z$ and a solution $\varphi(t, p_0)$ of (3.4.1) such that $(t, \varphi(t, p_0)) \in W$ on $I_\varphi = D_\varphi \cap [t_0, \infty)$. This concludes the proof of Theorem 3.4.1.

An alternate way of stating Theorem 3.4.1 is in terms of retracts. Let A, B be subsets of R^{n+1} with $B \subset A$. If there exists an USC mapping $G: A \to c(R^{n+1})$ such that $G(x) \subset B$ is connected for all $x \in A$ and $x \in G(x)$ for all $x \in B$, then B is a retract of A.

THEOREM 3.4.2. Let Z be a subset of $\text{int}\, W \cup S$ such that all points of $T^+(Z)$ are strict egress points. If

$Z \cap T^+(Z)$ is a retract of $T^+(Z)$, but not of Z, then there exists $p_0 \in Z$, such that $\bigl(t,\varphi(t,p_0)\bigr) \in W$ for all $t \in D_\varphi \cap [t_0,\infty)$.

Proof: If the conclusion does not hold, then for all $p \in Z$, every solution $\varphi(t,p)$ of (3.4.1) leaves W, and hence T^+ is USC on Z. Let $H\colon T^+(Z) \to T^+(Z) \cap Z$ be a retraction of $T^+(Z)$ onto $T^+(Z) \cap Z$ which is assumed to exist. Then $H\colon T^+$ is a retraction of Z onto $T^+(Z) \cap Z$. This is a contradiction and concludes the proof of Theorem 3.4.2.

The preceding results remain valid if X is a relatively open subset of $[a,\infty) \times R^n$, W is a relatively closed subset of X, and Z is a connected subset of $\operatorname{int} W \cup S$. We now apply Theorems 3.4.1 and 3.4.2 to BVP's associated with second-order contingent equations.

Consider the generalized second-order differential equation

$$x'' \in G(t,x,x'), \tag{3.4.3}$$

where $G\colon [a,b] \times R^2 \to cc(R)$ is upper semicontinuous (USC). By extension, we may assume G is USC on $X = [a,\infty) \times R^2$. Letting $x = y_1$, $x' = y_2$, $H(t,y) = \bigl(y_2, G(t,y_1,y_2)\bigr)$, where $y = (y_1,y_2)$, we see $H\colon X \to cc(R^2)$ is USC and

$$y' \in H(t,y) \tag{3.4.4}$$

is equivalent to (3.4.3). A function $x(t) \in C^{(1)}(I)$ is a solution of (3.4.3) on $I \subset [a,\infty)$ if and only if

$$Dx'(t) \subset G\bigl(t,x(t),x'(t)\bigr) \tag{3.4.5}$$

for all $t \in I$ and $Dx'(t) \neq \emptyset$ for all $t \in I$.

DEFINITION 3.4.3. The functions $\psi(t)$, $\varphi(t)$ are called <u>strict upper, lower solutions</u> for (3.4.3), respectively, if, $\psi(t)$, $\varphi(t) \in C^{(1)}(I)$, $I \subset [a,\infty)$ and if

$$\max D^+\psi'(t) < \min G(t,\psi(t),\psi'(t)) \tag{3.4.6}$$

and

$$\min D^+\varphi'(t) > \max G(t,\varphi(t),\varphi'(t)) \tag{3.4.7}$$

for each t, with $D^+\psi'(t)$ and $D^+\varphi'(t) \neq \emptyset$ for all $t \in I$.

We will assume that $\Phi(t) < \psi(t)$ for all $t \in I$. Define

$$W = [(t,x,x'): t \in [a,b],\ \varphi(t) \leq x \leq \psi(t),\ x' \in R],$$

$$A_1 = [(t,x,x'): t \in [a,b],\ x = \psi(t),\ x' \geq \psi'(t)],$$

and

$$A_2 = [(t,x,x'): t \in [a,b],\ x = \varphi(t),\ x' \leq \varphi'(t)].$$

Note that W is a relatively closed subset of X. The upper surface of W is the set $S_1 = [(t,x,x'): t \in [a,b],\ x = \psi(t),\ x' \in R]$ and the lower surface of W is the set $S_2 = [(t,x,x'): t \in [a,b],\ x = \varphi(t),\ x' \in R]$. Let Z_1 be a connected subset of $W \cap [t = a]$ such that Z_1 intersects S_1 at a single point of A_1 and Z_1 intersects S_2 at a single point of A_2. We now present a theorem which is in the same spirit as Theorem 3.2.1.

THEOREM 3.4.3. If $\psi(t)$ and $\varphi(t)$ are strict upper and lower solutions of (3.4.3), then either

(i) $T^+(Z_1) \cap [t = b] \neq \emptyset$, or

(ii) there is a solution $x(t)$ of (3.4.3) such that $(a,x(a),x'(a)) \in Z_1$, $\varphi(t) \leq x(t) \leq \psi(t)$ on $[a,w^+)$, and $|x'(t)| \to \infty$ as $t \to w^+$ where $[a,w^+)$ $(w^+ \leq b)$ is the right maximal interval of existence of $x(t)$.

Proof: Assume $T^+(Z_1) \cap [t = b] = \emptyset$. Let $x(t)$ be a solution of (3.4.3) such that $(a,x(a),x'(a)) \in Z_1$, $(t,x(t),x'(t)) \in W$ for $t \in [a,t_0]$ and $(t_0,x(t_0),x'(t_0)) \in S_1$.

Then

$$x(t_0) = \psi(t_0) \quad \text{and} \quad x(t) \leq \psi(t) \quad \text{for} \quad t \in [a,t_0],$$

so $x'(t_0) \geq \psi'(t_0)$. Hence $\big(t_0,x(t_0),x'(t_0)\big) \in A_1$. Similarly a solution from Z_1 can intersect the lower surface S_2 only at points of A_2. Thus $T^+(Z_1) \subset A_1 \cup A_2$. Observe also

$$T^+(Z_1) \cap A_1 \neq \emptyset \quad \text{and} \quad T^+(Z_1) \cap A_2 \neq \emptyset$$

since $A_1 \cap Z_1 \neq \emptyset$ and $A_2 \cap Z_1 \neq \emptyset$.

We now show any point of $T^+(Z_1)$ is a strict egress point. Let $Q = (t_0,x_0,x_0') \in A_1 \cap T^+(Z_1)$. Assume Q is not a strict egress point; then there exists an interval $[t_0,\tau]$, $t_0 < \tau$, and a solution $x(t)$ emanating from Q such that $x(t) \leq \psi(t)$ on $[t_0,\tau]$. Clearly, $x'(t_0) = \psi'(t_0)$. Let $t_n \to t_0$ and consider the sequence $\{[x'(t_n) - x'(t_0)/(t_n - t_0)\}$. Since $Dx'(t_0) \subset G\big(t_0,x(t_0),x'(t_0)\big)$ and since $G\big(t_0,x(t_0),x'(t_0)\big)$ is compact, the sequence is bounded. Thus there exists a subsequence $\{t_k\}$ such that

$$\begin{aligned}\lim_{k\to\infty} \frac{x'(t_k) - x'(t_0)}{t_k - t_0} &= y \in G\big(t_0,x(t_0),x'(t_0)\big) \\ &= G\big(t_0,\psi(t_0),\psi'(t_0)\big).\end{aligned}$$

Consider now the sequence $\{[\psi'(t_k) - \psi'(t_0)]/(t_k - t_0)\}$. Because $\psi(t)$ is a strict upper solution, this sequence contains a subsequence converging to a point in $(-\infty,y)$. Therefore, there exists a subsequence $\{t_j\}$ of $\{t_k\}$ such that $x'(t_j) > \psi'(t_j)$ for j sufficiently large. Hence there exists $\theta \in (t_0,\tau]$ such that $x'(t) > \psi'(t)$ for $t \in (t_0,\theta)$. This implies $x(\theta) > \psi(\theta)$, a contradiction. Similarly, it can be shown that any point of $A_2 \cap T^+(Z_1)$ is a strict egress point.

We can now apply Theorem 3.4.1 because all points of

$T^+(Z_1)$ are strict egress points and $T^+(Z_1)$ is not connected. Hence there is a solution $x(t)$ of (3.4.3) such that $(a,x(a),x'(a)) \in Z_1$ and $\varphi(t) \leq x(t) \leq \psi(t)$ on $[a,w^+)$ with $w^+ \leq b$. Furthermore $|x'(t)| \to \infty$ as $t \to w^+$, for if not, $T^+(Z_1) \cap [t=b] \neq \emptyset$. Hence if (i) does not hold, then (ii) holds. This concludes the proof of Theorem 3.4.3.

REMARK 3.4.2. The proof of Theorem 3.4.3 is different than that of Theorem 3.2.1. We could have, however, proved Theorem 3.4.3 in a manner similar to that of Theorem 3.2.1. This follows from the fact that it can be shown that there exists a connected subset of $T^+(Z_1)$ in $W \cap [t=b]$ which intersects A_1 and A_2.

We now extend the concept of Nagumo's condition to second-order contingent equations.

DEFINITION 3.4.4. The function $G: [a,b] \times R^2 \to cc(R)$ satisfies Nagumo's condition relative to W if there exists a positive function $h(s) \in C([0,\infty))$ such that $\max[|z| : z \in G(t,x,x')] \leq h(|x'|)$ for $(t,x,x') \in W$ and $\int^\infty s\, ds/h(s) = +\infty$.

The following lemma can be proved similarly to that of Lemma 1.4.1.

LEMMA 3.4.4. If $\varphi(t), \psi(t) \in C^{(1)}[a,b]$ with $\varphi(t) \leq \psi(t)$ and if $G(t,x,x')$ satisfies Nagumo's condition on $[a,c]$, then, given an interval $[a,c]$, $a < c \leq b$, there exists an $N > 0$ such that for any solution $x(t)$ of (3.4.3) with $\varphi(t) \leq x(t) \leq \psi(t)$ on $[a,c]$, then $|x'(t)| \leq N$ on $[a,c]$.

EXERCISE 3.4.1. Prove Lemma 3.4.4.

Let Z_2 be a subset of $W \cap [t=b]$ so that there exists a separation $W \cap [t=b] - Z_2 = E_1 \cup E_2$ with $A_1 \cap [t=b] \subset E_1$ and $A_2 \cap [t=b] \subset E_2$. We now state an existence theorem for a BVP.

THEOREM 3.4.4. If there exist strict upper and lower solutions for (3.4.3) on $[a,b]$, and if $G(t,x,x')$ satisfies Nagumo's condition with respect to W, then there is a solution $x(t)$ of (3.4.3) such that $\varphi(t) \leq x(t) \leq \psi(t)$ on $[a,b]$ and satisfies the boundary conditions $\big(a,x(a),x'(a)\big) \in Z_1$ and $\big(b,x(b),x'(b)\big) \in Z_2$.

Proof: If $T^+(Z_1) \cap Z_2 \neq \emptyset$, then the conclusion is immediate. Assume then that $T^+(Z_1) \cap Z_2 = \emptyset$. Then

$$T^+(Z_1) \subset S_1 \cup S_2 \cup (W \cap [t=b] - Z_2),$$

and as in the proof of Theorem 3.4.3, $T^+(Z_1) \cap S_i \subset A_i$ for $i = 1,2$. All points of $E_1 \cup E_2 \subset W \cap [t=b]$ are obviously strict egress points. This together with the fact, that $Q \in T^+(Z_1) \cap A_1$, $i = 1,2$, with $t_Q < b$, is a strict egress point implies that all points of $T^+(Z_1)$ are strict egress points (see the proof of Theorem 3.4.3).

Since $T^+(Z_1) \subset (A_1 \cup E_1) \cup (A_2 \cup E_2)$ and since $E_1 \cup E_2$ is a separation of $W \cap [t=b] - Z_2$, it follows that $T^+(Z_1) \cap Z_1$ is a retract of $T^+(Z_1)$. However, $T^+(Z_1) \cap Z_1$ is not a retract of Z_1. By Theorem 3.4.2, there exists a solution $x(t)$ of (3.4.3) with $\big(a,x(a),x'(a)\big) \in Z_1$ such that $\big(t,x(t),x'(t)\big) \in W$ on its maximal interval of existence $[t_0,w)$ relative to X. This means $w \leq b$ and $|x'(t)| \to \infty$ as $t \to w^-$. By Lemma 3.4.4 this is impossible. Therefore $T^+(Z_1) \cap Z_2 \neq \emptyset$ and this concludes the proof of Theorem 3.4.4.

3.5 DEPENDENCE OF SOLUTIONS ON BOUNDARY DATA

This section utilizes the topological properties of

solution funnels to deduce results on the continuous dependence on boundary values of solutions of BVP's for systems of differential equations with generalized boundary conditions.

Let I be an open interval in R, $x = (x_1, x_2)$, and $f = (f_1, f_2)$. We will consider the two point BVP

$$x' = f(t,x), \tag{3.5.1}$$

$$g\big(x(a)\big) = r, \qquad h\big(x(b)\big) = q \qquad (a,b \in I), \tag{3.5.2}$$

where $f \in C[I \times R^2, R^2]$, $g \in C[R^2, R]$, $h \in C[R^2, R]$.

We shall use some of the notions dealing with compact sets that were developed in the preceding section. For completeness we introduce similar notation here. If A is a set, let the $\operatorname{int} A$ (∂A) denote the interior (boundary) of A and let $\{a\}$ be the one-point set containing a.

Let $c(R^2)$ denote the family of all compact, nonempty subsets of R^2. Let $q(x,B) = \inf\big[\|x - y\|: y \in B\big]$, $q(A,B) = \sup[q(x,B): x \in A]$, $d(A,B) = \max\big(q(A,B), q(B,A)\big)$. When $A, B \in R^2$ for $A \subset R^2$ and $\varepsilon > 0$ let $N(A,\varepsilon) = [x: q(x,A) \leq \varepsilon)$ will be denoted by $N(x,\varepsilon)$.

For $X \subset R^2$, the mapping $F: X \to c(R^2)$ is said to be <u>upper semicontinuous</u> (continuous) at $x \in X$ if $y \to x$ implies $q\big(F(y), F(x)\big) \to 0$.

As we have observed, (3.5.1) generates a mapping F of $R^2 \times I \times I$ into the family of all closed subsets of R^2 defined by

$$F(x, t_0, t_1) = [y: y = x(t_1),\ x = x(t) \text{ is a solution of (3.5.1)},\ x(t_0) = x].$$

This is precisely the cross section at t_1 of the integral funnel through (t_0, x).

The mapping F has the following well-known properties:

(I) $F(A,t_0,t_1) = \bigcup[F(x,t_0,t_1): x \in A]$ is a continuum (compact and connected) provided A is a continuum and all solutions of (3.5.1), $x(t_0) = y$, $y \in A$ exist on $[t_0,t_1]$.

(II) If every solution of (3.5.1) through $(t_0,x_0) \in Q \subset I \times R^2$ exist on $[t_0 - h, t_0 + h]$, then given a compact set $A \times \{t_0\} \subset Q$ and $\varepsilon > 0$ there are $\delta > 0$ and $\eta > 0$ such that $q(B,A) < \delta$, $|s - t_0| < \delta$ implies

$$q\big(F(B,s,t_1), F(A,t_0,t_1)\big) < \varepsilon,$$

and $|t_1 - t_2| < \eta$ implies

$$d\big(F(A,t_0,t_1), F(A,t_0,t_2)\big) < \varepsilon,$$

provided $B \times \{s\} \subset Q$, $s, t_1, t_2 \in I$, $|s - t_1| \le h$, $|t_0 - t_1| \le h$, $|t_2 - t_0| \le h$.

Let $x_0(t)$ be a solution of (3.5.1) defined on $[a,b] \subset I$ and let 0 be a subset of $[a,b] \times R^2$. The BVP (3.5.1), (3.5.2) is said to be unique relative to $x_0(t)$, $[a,b]$, and 0, if for any solution $y(t)$ of (3.5.1), the conditions $\big(t,y(t)\big) \in Q$ for $t \in [c,d] \subset [a,b]$, $g\big(y(c)\big) = g\big(x_0(c)\big)$, $h\big(y(d)\big) = h\big(x_0(d)\big)$ imply $y(t) = x_0(t)$ on $[c,d]$.

We are now able to state a continuous dependence result. The proof of this will follow after a series of lemmas.

THEOREM 3.5.1. Let f, g, h satisfy the conditions:

(i) For all $r \in R$, the curves $K(r) = [x: g(x) = r]$, $H(r) = [x: h(x) = r]$ are simple arcs separating R^2;

(ii) For all r, $q \in R$, $K(r) \cap H(q)$ is a one-point set.

Let $x \equiv x_0(t)$ be a solution of the BVP (3.5.1), and

$$g\big(x(0)\big) = r_0, \qquad h\big(x(1)\big) = q_1. \tag{3.5.3}$$

Let the BVP (3.5.1), (3.5.2) be unique relative to $x_0(t)$, $[0,1]$, and some open set $Q \in [0,1] \times R^2$ containing the arc $[(t,x): x = x_0(t),\ t \in [0,1]]$.

Then there is a neighborhood U of $(0,1,r_0,q_1)$ such that for all $(a,b,r,q) \in U$, the BVP (3.5.1), (3.5.2) has a solution $x \equiv x(t,a,b,r,q)$ satisfying $x(t,a,b,r,q) \to x_0(t)$ uniformly in $t \in [0,1]$ as $(a,b,r,q) \to (0,1,r_0,q_1)$.

If uniqueness of the BVP in the "usual sense" is assumed, we obtain immediately the following corollary.

COROLLARY 3.5.1. Assume (i) and (ii) hold and that the solutions of the BVP (3.5.1), (3.5.2), whenever they exist, are unique. If the solution of the BVP (3.5.1), (3.5.3) exists, then the BVP (3.5.1), (3.5.2) has a solution $x = x(t,a,b,r,q)$ for all (a,b,r,q) in a certain neighborhood U of $(0,1,r_0,q_1)$ and $x(t,a,b,r,q)$ is continuous on $[0,1] \times U$.

In order to prove Theorem 3.5.1, we will need some preliminary results. Choose $\varepsilon > 0$ such that the set $\big[(t,x): \|x - x_0(t)\| \le 2\varepsilon,\ t \in [0,1]\big]$ is in Q. Let $h > 0$ be the number with the property that if $s \in [0,1]$, $\|y - x_0(s)\| \le \varepsilon$, then every solution $x = x(t)$ of (3.5.1) satisfying $x(s) = y$ exists on $[s-h,\ s+h]$ and satisfies $\|x(t) - x_0(t)\| \le 2\varepsilon$ on this interval.

Choose a sequence of numbers $0 = t_0 < t_1 < \cdots < t_m = 1$ such that $t_{i+1} - t_i < h$ if $i = 0,\ldots,m-1$, and let $x_i = x_0(t_i)$, $N_i = N(x_i,\varepsilon)$, $r_i = g(x_i)$, $q_i^* = h(x_i)$, $U_i = [x: h(x) < q_i^*]$, $V_i = [x: h(x) > q_i^*]$.

By the choice of h, if $A \subset N\big(x_0(s),\varepsilon\big)$, $0 < t - s \le h$, then F satisfies (I) and (II) and $F(A,s,t) \subset N\big(x_0(t),2\varepsilon\big)$.

For $A \subset N_{i-1}$, write $F_i(A) \equiv F(A,t_{i-1},t_i)$. Define the sets S_i, $i = 0,\ldots,m-1$, by S_0 as the component of $K(r_0) \cap N_0$ containing x_0, S_{i+1} as the component of

$F_{i+1}(S_i) \cap N_{i+1}$ containing x_{i+1}. Since S_0 and N_i are compact, it follows from (I) that S_i are compact. The uniqueness condition on the BVP (3.5.1), (3.5.2) implies that

$$F_{i+1}(S_i) \cap H(q^*_{i+1}) = \{x_{i+1}\}, \qquad i = 0,1,\ldots,m-1. \tag{3.5.4}$$

LEMMA 3.5.1. Let A_0, B_0 satisfy $A_0 \cup B_0 = K(r_0) \cap \partial N_0$, $F_1(A_0) \subset U_1$, $F_1(B_0) \subset V_1$. Let $A_i = F_i(S_{i-1}) \cap U_i \cap \partial N_i$, $B_i = F_i(S_{i-1}) \cap V_i \cap \partial N_i$. Then given $T \in (t_1, 1]$ there exists a $\delta > 0$ and $\eta > 0$ such that for every solution $x = x(t)$ of (3.5.1) defined on $[0,T] \cup [p,T]$, the conditions $|x(t) - x_0(t)| \leq 2\varepsilon$ for $t \in [0,T] \cup [p,T]$; $|p - t_i| < \eta$, $x(p) \in N(A_i,\eta)$ [or $x(p) \in N(B_i,\eta)$] for some $t_i \in [0,T)$ imply

$$h(x(t)) < h(x_0(t)) - \delta \quad \text{or} \quad h(x(t) \quad > h(x_0(t)) + \delta \tag{3.5.5}$$

on $[t,T] \cap [p,T]$.

<u>Proof</u>: Assume the result is not true. Let $y_n(t)$, $n = 1,2,\ldots$, be a sequence of solutions of (3.5.1) such that $y_n(p_n) \in N(A_i, 1/n)$, $h(y_n(s_n)) \geq h(x_0(s_n)) - (1/n)$, where $|p_n - t_i| < 1/n$, $s_n \in [p_n,T] \cap [t,T]$. We may assume $s_n \to s_0$ and $y_n(t) \to y_0(t)$ uniformly on $[0,T]$.

Hence $y_0(t_i) \in A_i \subset U_i$, $h(y_0(s_0)) \geq h(x_0(s_0))$, $s_0 > t_i$; this implies that there is a solution $z(t)$ satisfying $z(0) \in K(r_0)$, $z(t_i) = y_0(t_i)$. Let $w(t) = z(t)$ on $[0,t_i]$, $w(t) = y_0(t)$ on $[t_i, s_0]$.

We have $g(w(0)) = g(x_0(0))$, $h(w(t_i)) < h(x_0(t_i))$, $h(w(s_0)) \geq h(x_0(s_0))$, contradicting the uniqueness. This completes the proof of Lemma 3.5.1.

LEMMA 3.5.2. For a given $\varepsilon_{i+1} > 0$ there is a $\delta_i = \delta_i(\varepsilon_{i+1})$ such that if $|s - t_i| < \delta_i$, $|s - t_{i+1}| \leq h$,

$D \subset N(S_i, \delta_i) \cap N(x_0(s), \varepsilon)$, then all components of $F(D,s,t_{i+1}) \cap N_{i+1}$ which intersect $H(q^*_{i+1})$ are in $N(S_{i+1}, \varepsilon_{i+1})$.

Proof: Since $S = F_{i+1}(S_i) \cap N_{i+1}$ is compact for any $\varepsilon_{i+1} > 0$, there is a decomposition of S into compact sets $X_1, X_2, S = X_1 \cup X_2$ such that

$$(3.5.6) \qquad X_1 \cap X_2 = \emptyset, \qquad S_{i+1} \subset X_1 \subset N(S_{i+1}, \varepsilon_{i+1}).$$

From (3.5.4) and the definition of S_i,

$$(3.5.7) \qquad X_1 \cap H(q^*_{i+1}) \neq \emptyset, \qquad \left(F_{i+1}(S_i) - X_1\right) \cap H(q^*_{i+1}) = \emptyset.$$

Let $D_1(\delta)$, $\delta > 0$ be the union of all components of $N\left(F_{i+1}(S_i) - S, \delta\right) \cap N_{i+1}$ which intersect $N(X_1, \delta)$, and let $D_2(\delta) = \left(N(F_{i+1}(S_i) - S, \delta) \cap N_{i+1}\right) - D_1(\delta)$. Then by (3.5.6) and (3.5.7), the sets $C_k(\delta) = N(X_k, \delta) \cup D_k(\delta)$ $(k = 1,2)$ satisfy

$$(3.5.8) \qquad \begin{array}{ll} C_1(\delta) \cap C_2(\delta) = \emptyset, & C_1(\delta) \cap H(q^*_{i+1}) \neq \emptyset, \\ C_2(\delta) \cap H(q^*_{i+1}) = \emptyset, & C_1(\delta) \subset N(S_{i+1}, \varepsilon_{i+1}), \end{array}$$

provided δ is sufficiently small.

Applying the semicontinuity of F, choose $\delta_i = \delta_i(\varepsilon_{i+1})$ such that if $D \subset N(S_i, \delta) \cap N(x_0(s), \varepsilon)$, $|t_i - s| < \delta_i$, $|s - t_{i+1}| \leq h$, then

$$F(t_{i+1}, s, D) \subset N\left(F_{i+1}(S_i), \delta\right).$$

The assertion follows immediately with the use of (3.5.8) and the formula $N\left(F_{i+1}(S_i), \delta\right) \cap N_{i+1} = C_1(\delta) \cup C_2(\delta)$.

We are now in a position to prove Theorem 3.5.1.

Proof of Theorem 3.5.1: Let $S_0(r,a)$ be the component of the set $K(r) \cap N\left(x_0(a), \varepsilon\right)$ which intersects the sets

$\{x\colon h(x) < h(x_0(a))\}$, $\{x\colon h(x) > h(x_0(a))\}$.

Let $S_i(r,a)$, $i = 1,\ldots,m-1$, denote the component of $F(S_i(r,a), s_{i-1}, t_i) \cap N_i$, where $s_0 = a$, $s_j = t_j$ for $j \geq 1$, satisfying $S_i(r,a) \cap U_i \neq \emptyset$, $S_i(r,a) \cap V_i \neq \emptyset$.

Set $C(r,a,b) = F(S_{m-1}(r,a), t_{m-1}, b)$. To prove the theorem, it is enough to show that

$$(3.5.9) \qquad C(r,a,b) \cap H(q) \neq \emptyset \qquad \text{for } (a,b,r,q) \in U,$$

where U is a certain neighborhood of $(0,1,r_0,q_1)$.

In fact, (3.5.9) implies the existence of a solution $x(t,a,b,r,q)$ of the BVP (3.5.1), (3.5.2) for $(a,b,r,q) \in U$. Since $x(t_i,a,b,r,q) \in N_i$ for $i = 1,\ldots,m-1$, by the choice of t_i, $|x(t,a,b,r,q) - x_0(t)| < 2\varepsilon$ on some interval $[c,d]$ containing $[0,1]$. Thus, the family $\{x(t,a,b,r,q)\}$, $(a,b,r,q) \in U$ is equicontinuous on $[c,d]$. By Ascoli's theorem and the uniqueness of $x_0(t)$, we deduce the uniform convergence on $[0,1]$ of $x(t,a,b,r,q)$ as $(a,b,r,q) \to (0,1,r_0,q_1)$.

In order to show that (3.5.9) is true, we first prove that all sets $S_i(r,a)$ exist if $|r - r_0|$ and $|a|$ are small enough. For this, define constants $\alpha_i > 0$, $\varepsilon_i > 0$ by

$$(3.5.10) \qquad N(K(r_0), \alpha_0) \cap \partial N_0 \subset N(A_0 \cup B_0, \eta),$$

$$N(F_i(S_{i-1}), \alpha_i) \cap \partial N_i \subset N(A_i \cup B_i, \eta) \qquad \text{for } i = 1,\ldots,m-1,$$

$$(3.5.11) \qquad \varepsilon_{m-1} = \alpha_{m-1}, \qquad \varepsilon_i = \min(\alpha_i, \delta_i(\varepsilon_{i+1})) \qquad \text{if } i = m-2,\ldots,0,$$

where A_i, B_i, η are as in Lemma 3.5.1 and $\delta_i(\cdot)$ is defined for $F(\cdot, s_i, t_{i+1})$ as in Lemma 3.5.2. We now proceed by induction.

Let $\ell > 0$ be so chosen that $S_0(r,a)$ exists and

$$(3.5.12) \qquad S_0(r,a) \subset N(S_0, \varepsilon_0), \qquad \text{where}$$

$$(3.5.13) \qquad S_0(r,a) \cap N(A_0,\eta) \neq \emptyset, \qquad S_0(r,a) \cap N(B_0,\eta) \neq \emptyset$$

for all $|a| \leq \ell$, $|r - r_0| \leq \ell$. The possibility of such an ℓ follows from (i), (ii), and the continuity of g, h, and x_0.

Let $k = \min(\ell, \varepsilon_0, \eta)$. If $|a| \leq k$, $|r - r_0| \leq k$, then all $S_i(r,a)$ exist. To see this, observe that if $S_i(r,a)$ exists and if it satisfies the condition

$$(3.5.14)_i \qquad S_i(r,a) \subset N(S_i, \varepsilon_i),$$

and the condition that

$(3.5.15)_i$ there are solutions $u_i(t)$, $v_i(t)$ of (3.5.1) such that $u_i(s_j)$, $v_i(s_j) \in S_j(r,a)$ for $j = 0,\ldots,i$; $u_i(s_h) \in N(A_h,\eta)$, $v_i(s_k) \in N(B_k,\eta)$ for some h, k ($s_0 = a$, $s_j = t_j$ for $1 \leq j \leq i$),

then, $S_{i+1}(r,a)$ exists and satisfies $(3.5.14)_{i+1}$ and $(3.5.15)_{i+1}$.

In fact $(3.5.15)_i$ and Lemma 3.5.1 imply that $F_{i+1}(u_i(t_i))$, $F_{i+1}(v_i(t_i))$ are separated by $H(q^*_{i+1})$; hence at least one component of $F_{i+1}(S_i(r,a)) \cap N_{i+1}$ intersects U_{i+1} and V_{i+1}. By Lemma 3.5.2, $S_{i+1}(r,a) \subset N(S_{i+1}, \varepsilon_{i+1})$, that is, $(3.5.14)_{i+1}$ holds.

To show $(3.5.15)_{i+1}$ holds, choose any point u_{i+1} in $S_{i+1}(r,a) \cap U_{i+1} \cap \partial N_{i+1}$, if this set is nonempty; otherwise let $u_{i+1} \in F_{i+1}(u_i(t_i))$ (v_{i+1} is chosen similarly). If $F_{i+1}(S_i(r,a)) \cap U_{i+1} \cap \partial N_{i+1} \neq \emptyset$, then by (3.5.10), $A_{i+1} \neq \emptyset$. Hence by $(3.5.14)_{i+1}$, (3.5.11), and (3.5.10), $u_{i+1} \in N(A_{i+1},\eta)$. Thus in every case, there is a solution $u_{i+1}(t)$ through (t_{i+1}, u_{i+1}) satisfying $(3.5.15)_{i+1}$. By (3.5.12) and (3.5.13), $S_0(r,a)$ exists and satisfies $(3.5.14)_0$, $(3.5.15)_0$ for all $|a| \leq k$, and $|r - r_0| \leq k$. Hence, by induction, $S_i(r,a)$ exists for all $i > 0$.

To prove (3.5.9), observe that the existence of $S_{m-1}(r,a)$ implies the existence of $C(r,a,t)$ for all $|a| \le k$, $|r - r_0| \le k$, and $t \in [t_{m-1}, t_{m-1}+h]$. By property (II), $C(r,a,t)$ is continuous in t on $[t_{m-1}, t_{m-1} + h]$. Thus $(3.5.15)_{m-1}$ and Lemma 3.5.1 imply that

$$\begin{aligned} &\min\{h(x)\colon x \in C(r,a,b)\} < h(x_0(b)) - \delta, \\ &\max\{h(x)\colon x \in C(r,a,b)\} > h(x_0(b)) + \delta \end{aligned} \tag{3.5.16}$$

hold for $|a| \le k$, $|r - r_0| \le k$, $|b - 1| \le k_1$, where $k_1 > 0$ is sufficiently small. Since $C(r,a,b)$ is connected, then (i) and (3.5.16) imply that (3.5.9) holds, thus completing the proof of Theorem 3.5.1.

Equation (3.5.1) with the boundary condition

$$g(x(a)) = r, \qquad g(x(b)) = q \tag{3.5.17}$$

may be considered as a special case of the BVP (3.5.1), (3.5.2).

EXERCISE 3.5.1. State and prove a theorem similar to Theorem 3.5.1 for the BVP (3.5.1), (3.5.17).

We now consider a special case of Theorem 3.5.1, when $f_1(t,x_1,x_2) = x_2$ and g and h are linear.

THEOREM 3.5.2. Assume that

(i) $x_1 < x_2$ implies $f(t,x_1,y) \le f(t,x_2,y)$ for all $(t,y) \in [a,b] \times R$;

(ii) $|f(t,x,y) - f(t,x,z)| \le M|y - z|$ on $[a,b] \times R^2$;

(iii) $a_0, a_1, b_0, b_1 \ge 0$, $a_0 + b_0 > 0$, $a_0 + a_1 > 0$, $b_0 + b_1 > 0$.

Then the BVP

$$x'' = f(t,x,x'), \qquad a_0x(a) - a_1x'(a) = r, \qquad b_0x(b) + b_1x'(b) = q,$$

has a unique solution $x(t,r,q)$ for any r, q and $x(t,r,q)$, $x'(t,r,q)$ are continuous on $[a,b] \times R^2$.

EXERCISE 3.5.2. Prove Theorem 3.5.2.

<u>Hint</u>: Show that $x(t,r,q)$ exists and is unique by utilizing (i) and (ii). By (iii) $a_0 b_1 + a_1 b_0 > 0$. This implies condition (ii) of Theorem 3.5.1 holds. Apply finally Corollary 3.5.1.

In the spirit of Exercise 3.5.1, we state the following result.

THEOREM 3.5.3. Let $f: I \times R^2 \to R$ be continuous. Let $x_0(t)$ be a solution of the BVP

$$x'' = f(t,x,x'), \quad x(a) = r, \quad x(b) = q, \quad a,b \in I \tag{3.5.18}$$

corresponding to $(a,b,r,q) = (0,1,r_0,q_1)$. Assume that there is a $\delta > 0$ such that if $x(t)$ is a solution of $x'' = f(t,x,x')$ with $x(a) = x_0(a)$, $x(b) = x_0(b)$, $a,b \in [0,1]$, $|x(t) - x_0(t)| < \delta$, $|x'(t) - x_0'(t)| < \delta$ on $[a,b]$, then $x(t) \equiv x_0(t)$ on $[a,b]$.

Then the BVP (3.5.18) has a solution $x(t,a,b,r,q)$ for all (a,b,r,q) sufficiently close to $(0,1,r_0,q_1)$ and $\big(x(t,a,b,r,q),\ x'(t,a,b,r,q)\big) \to \big(x_0(t), x_0'(t)\big)$ uniformly on $[0,1]$ as $(a,b,r,q) \to (0,1,r_0,q_1)$.

EXERCISE 3.5.3. Prove Theorem 3.5.3 by applying Theorem 3.5.1 to the special case in which $g(x,y) \equiv h(x,y) \equiv x$ and $f_1(t,x,y) = y$.

3.6 NOTES AND COMMENTS

Theorems 3.1.1 and 3.1.2 of Section 3.1 are taken from Bebernes and Wilhelmsen [6], while the subsequent discussion dealing with Wazewski's method is based on Jackson and Klaasen [3] who treat a more general problem not demanding uniqueness. Theorems 3.1.4 and 3.1.5 are also taken from Berbernes and Wilhelmsen [6]. The results contained in Theorems 3.2.1 and 3.2.2 are due to Bebernes and Wilhelmsen [6] (see also Sedziwy [3]). Theorem 3.2.3 may be found in Bebernes and Fraker [9]. Exercise 3.2.1 is due to Markus and Amundsen [1]. Section 3.3 contains the results of Kaplan et al. [1]. Lemmas 3.4.1 - 3.4.3 are from Bebernes and Schuur [7] while Remark 3.4.1 is based on Bebernes and Schuur [7] and Yorke [1]. The other results of Section 3.4 are due to Bebernes and Kelley [10]. Section 3.5 consists of the results of Sedziwy [1].

For related work on continuous dependence, see Bebernes and Gaines [4], Gaines [1,2], Klaasen [1], and Ingram [1].

Chapter 4
FUNCTIONAL ANALYTIC METHODS

4.0 INTRODUCTION

Many diverse problems in the qualitative theory of differential equations are concerned with the existence of a solution which belongs to a specified subset of a given Banach space. These problems can be treated in a unified setting by techniques commonly used in functional analysis. These techniques suggest themselves in a natural way, when the considerations involving the integral equation obtained by the method of variation of parameters are translated into a suitable abstract setting. This is the underlying theme of this important chapter.

A variety of nonlinear functional analytic techniques like the Fredholm alternative, the Schauder's fixed point theorem, the method of a priori estimates, the notion of admissibility of spaces, Leray-Schauder's alternative, and the degree theory are employed to investigate existence results for boundary value problems in various ways. Some of the results are concerned with periodic boundary conditions.

We state Schauder's fixed point theorem and the contraction mapping principle in a generalized normed space and then utilizing these results, prove existence and uniqueness of solutions of a system of integral equations. As an application of the latter results, we derive existence and uniqueness theorems for various boundary value problems including a generalized Nicoletti problem.

We present an existence and uniqueness result for nonlinear functional equations in terms of the theory of set-valued mappings. As an application of this result, we consider the question of existence of solutions of general linear problems. We then proceed to develop a general theory of linear problems for set-valued differential equations utilizing the fixed point theorems for set-valued mappings. Finally, we prove an existence result for boundary value problems associated with set-valued differential equations.

4.1 LINEAR PROBLEMS FOR LINEAR SYSTEMS

For an n-dimensional vector $x = (x_1, \ldots, x_n)$ and an $n \times n$ real matrix $A = \{a_{ij}\}$, let

$$\|x\| = \sum_{i=1}^{n} |x_i|, \qquad \|A\| = \sum_{i,j=1}^{n} |a_{ij}|.$$

Consider any compact interval $\Delta = [0,h]$ and define

$$C(M) = C(M)[\Delta,R]$$

to be the space of continuous matrix functions $A(t) = \{a_{ij}(t)\}$ defined on Δ with norm

$$\|A\|_0 = \max\left[\|A(t)\|: t \in \Delta\right].$$

As before, $C = C[\Delta,R]$ will be the set of all continuous vector functions $x(t)$ with norm

$$\|x\|_0 = \max\left[\|x(t)\|: t \in \Delta\right].$$

Let L_1 be the linear space of all real n-dimensional vector functions defined and integrable in Δ with norm

$$\|x\|_1 = \int_\Delta \|x(t)\|\, dt.$$

Define $L_1(M)$ to be the linear space of all real $n \times n$ matrix functions $A(t) = \{a_{ij}(t)\}$ defined and integrable in Δ with the norm

$$\|A\|^{\circ} = \max\left[\sum_{i,j=1}^{n} \left| \int_0^t a_{ij}(s)\, ds \right| : t \in \Delta\right].$$

For any $A(t) \in L_1(M)$, define

$$\tilde{A}(t) = \int_0^t A(s)\, ds = \left\{\int_0^t a_{ij}(s)\, ds\right\};$$

thus $\|A\|^{\circ} = \|\tilde{A}\|_{\circ}$. Then the mapping $J: A(t) \to \tilde{A}(t)$ maps $L_1(M)$ isometrically onto the subspace $C_{\circ}(M)$ of all absolutely continuous $n \times n$ matrix functions that vanish at $t = 0$.

We will also be using the norm on $A(t)$

$$\|A\|_1 = \int_{\Delta} \|A(t)\|\, dt.$$

Let $L(C,R^n)$ denote the linear space of all continuous linear mappings T of C into R^n with the usual norm

$$\|T\| = \sup\left[\|Tx(t)\|: x(t) \in C,\ \|x\|_{\circ} \leq 1\right].$$

For a given T in $L(C,R^n)$, T^n will be the induced mapping of $C(M)$ into the space of real matrices which to every matrix $A(t)$ in $C(M)$ assigns the matrix obtained by the application of T to every column of $A(t)$. Hence T^n is a linear continuous mapping and

$$T[A(t)c] = [T^n A(t)]c$$

for any $c \in R^n$.

The following preliminary lemma will be important in subsequent discussions.

LEMMA 4.1.1. For any nonnegative Lebesgue integrable function $\alpha(t)$ on Δ, the set

$$K = [A(t) \in L_1(M): \|A(t)\| \leq \alpha(t) \text{ almost everywhere in } \Delta]$$

is compact.

Proof: We first show K is a closed set. Let $\{A_k(t)\}$ be any sequence in K such that $\|A_k - A_\circ\|^\circ \to 0$ as $k \to \infty$. For any interval $[t, t+s] \subset \Delta$, we have

$$\left\| \int_t^{t+s} A_k(u)\,du \right\| \leq \int_t^{t+s} \|A_k(u)\|\,du \leq \int_t^{t+s} \alpha(u)\,du.$$

Letting $k \to \infty$, we obtain

$$\left\| \frac{1}{s} \int_t^{t+s} A_\circ(u)\,du \right\| \leq \frac{1}{s} \int_t^{t+s} \alpha(u)\,du.$$

Now letting $s \to 0$, it is clear that $\|A_\circ(t)\| \leq \alpha(t)$ at every Lebesgue point t and this implies $A_\circ(t)$ belongs to K.

It suffices to show K is relatively compact in $L_1(M)$. Since the mapping J^{-1}: $C_\circ(M) \to L_1(M)$ is continuous, we need only show that $J(K)$ is compact in $C_\circ(M)$. From the definition of K, it follows that $J(K)$ is uniformly bounded and equicontinuous. By Ascoli's theorem, the set $J(K)$ is relatively compact in $C(M)$. Since a uniform limit of equicontinuous functions, which are absolutely continuous, is also absolutely continuous, it follows that $J(K)$ is relatively compact in $C_\circ(M)$. This completes the proof.

We shall consider the linear nonhomogeneous system of differential equations

$$x' = A(t)x + b(t), \tag{4.1.1}$$

where $A(t) \in L_1(M)$ and $b(t) \in L_1$. Let T: $C \to R^n$ be continuous and linear. We shall be interested in solutions of (4.1.1) satisfying

$$Tx(t) = r \tag{4.1.2}$$

for a given $r \in R^n$.

We now state a well-known result exhibiting a relationship between homogeneous and nonhomogeneous systems.

THEOREM 4.1.1. The BVP (4.1.1), (4.1.2) has a unique solution for every $r \in R^n$ and every $b(t) \in L_1$ if and only if the corresponding homogeneous linear BVP

$$x' = A(t)x, \tag{4.1.3}$$

$$Tx(t) = 0, \tag{4.1.4}$$

has only the trivial solution $x(t) = 0$.

Proof: Let $U(t)$ be the fundamental matrix solution of (4.1.3) with $U(0) = I$. The general solution of (4.1.1), then satisfies

$$x(t) = U(t)c + x_o(t),$$

where

$$x_o(t) = U(t)\int_0^t U^{-1}(s)b(s)\,ds \tag{4.1.5}$$

is a solution of (4.1.1) such that $x_o(0) = 0$ and c is an arbitrary element of R^n. Thus the boundary condition (4.1.2) takes the form

$$T[U(t)c + x_o(t)] = r$$

or

$$[T^n U(t)]c = r - Tx_o(t). \tag{4.1.6}$$

Then (4.1.6) has a unique solution for any $r \in R^n$ if and only if

$$\det[T^n U(t)] \neq 0, \tag{4.1.7}$$

that is, if and only if (4.1.3) and (4.1.4) has only the trivial solution.

Furthermore, when (4.1.6) is satisfied, the solution of (4.1.1), (4.1.2) is uniquely represented by the formula

$$x(t) = U(t)[T^n U(t)]^{-1} \; r - Tx_\circ(t) \; + x_\circ(t). \tag{4.1.8}$$

This completes the proof of Theorem 4.1.1.

We now let S be a set in $L_1(M)$ satisfying the following two properties:

(i) there exists a positive constant α such that $\|A\|_1 \leq \alpha$ for each $A(t) \in S$,

(ii) for each $A(t)$ in S, the BVP (4.1.3), (4.1.4) has only the trivial solution.

From (ii), it follows that (4.1.1), (4.1.2) has a unique solution for each $A(t) \in S$, $b(t) \in L_1$, and $r \in R^n$.

In the following discussion, we let the mapping $T \in L(C,R^n)$ and the vector r be fixed and only $A(t)$ and $b(t)$ will change in S and L_1, respectively.

Let $Q: S \times L_1 \to C$ be the mapping which assigns to each pair $(A(t), b(t))$ in $S \times L_1$ the unique solution of (4.1.1), (4.1.2). The continuity properties of Q will be inferred from the cartesian product topology induced in the space $L_1(M) \times L_1$ by the norm $\|\cdot\|^\circ$ in $L_1(M)$ and the norm $\|\cdot\|_1$ in L_1.

LEMMA 4.1.2. If the set $S \subset L_1(M)$ satisfies conditions (i) and (ii), then the map Q is continuous.

<u>Proof</u>: We can consider the mapping $Q \equiv Q_2 Q_1 Q_0$, where

$$(A(t),b(t)) \xrightarrow{Q_0} (U(t),b(t)) \xrightarrow{Q_1} (U(t),x_\circ(t)) \xrightarrow{Q_2} x(t).$$

Here $U(t)$ denotes the fundamental solution of (4.1.3), $x_o(t)$ is given by (4.1.5), and $x(t)$ by (4.1.8). Let $F: L_1(M) \to C(M)$ denote the mapping which to each $A(t) \in L_1(M)$ assigns the fundamental matrix of (4.1.3). In the commutative diagram

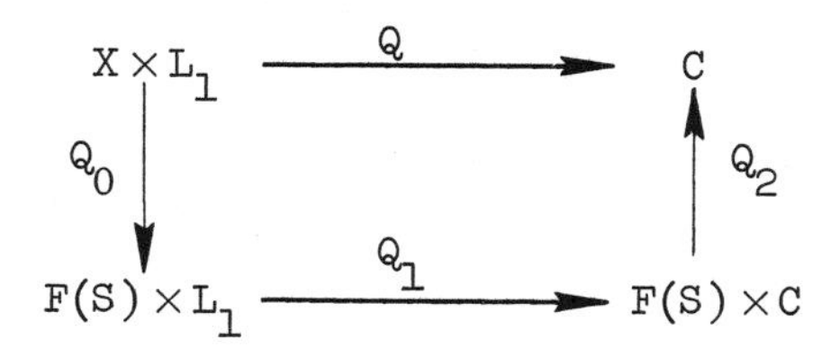

the mapping Q_0 is continuous since from (i) it follows that the restriction of F to S is continuous. The continuity of Q_1 follows immediately from (4.1.5) and from the continuity of the mapping of $F(S)$ into $C(M)$ which to every $U(t)$ in $F(S)$ assigns its inverse matrix $U^{-1}(t)$. Finally, the mapping Q_2 is also continuous since from (ii) it follows that (4.1.7) holds for any $U(t)$ in $F(S)$ so that the matrix $[T^n U(t)]^{-1}$ is a continuous function of $U(t)$. Hence, we see that Q is continuous thus completing the proof of Lemma 4.1.2.

LEMMA 4.1.3. If a compact set $S \subset L_1(M)$ satisfies (i) and (ii), then there exists positive constants β, γ depending only on S such that

$$\|QA(t)b(t)\| \leq \gamma|r| + \beta\|b\|_1 \tag{4.1.9}$$

for each $A(t)$ in S and $b(t)$ in L_1.

Proof: From the continuity of the restriction of the mapping F to the set S and from the compactness of S, it follows that the set $F(S)$ is compact in $C(M)$. Defining

$$(4.1.10) \qquad a = \max_{U \in F(S)} \|U\|, \qquad b^* = \max_{U \in F(S)} \|U^{-1}\|,$$

$$c = \max_{U \in F(S)} \|(T^n U)^{-1}\|,$$

we deduce easily from (4.1.5) and (4.1.8) that

$$\|x\| \leq ac|r| + \left(1 + ac\|T\|\right)ab^*\|b\|_1,$$

so that, in (4.1.9), $\gamma = ac$ and $\beta = \left(1 + ac\|T\|\right)ab^*$. Thus Lemma 4.1.3 is proven.

Although Q does not in general map bounded subsets of $S \times L_1$ into compact subsets of C, this property does hold for certain subsets of $S \times L_1$. In particular, we have the following lemma.

LEMMA 4.1.4. If a compact set $S \subset L_1(M)$ satisfies (i) and (ii) and if D is a set in L_1 such that

$$(4.1.11) \qquad \|b(t)\| \leq m(t)$$

for $b(t) \in D$ and $t \in \Delta$, where $m(t)$ is integrable in Δ, then $Q(S \times D)$ is a relatively compact set in C.

<u>Proof</u>: From the diagram in the proof of Lemma 4.1.2 we have

$$(4.1.12) \qquad Q(S \times D) = Q_2\left[Q_1\left(F(S) \times D\right)\right],$$

since $Q_0(S \times D) = F(S) \times D$. From (4.1.5), (4.1.11) it follows that, for $U(t)$ in $F(S)$ and $b(t)$ in D, the functions $x_\circ(t)$ satisfy

$$\|x_\circ\|_\circ \leq ab^* \int_\Delta m(t)\, dt \equiv d.$$

Since

$$x_o'(t) = A(t)x_o(t) + b(t),$$

we deduce, because of (i), that

$$\|x_o(t) - x_o(s)\| \leq \int_s^t \|x_o'(u)\|\, du \leq \alpha d(t - s) + \int_s^t m(u)\, du$$

for $0 \leq s < t \leq h$, so that the functions $x_o(t)$ are equicontinuous. Then, the set $\left(Q_1\, F(S) \times D\right)$ is relatively compact in $C(M) \times C$. Since Q_2 is continuous, we see, from (4.1.12), that $Q(S \times D)$ is also compact, thus completing the proof of Lemma 4.1.4.

4.2 LINEAR PROBLEMS FOR NONLINEAR SYSTEMS

Using the results in Section 4.1, we prove existence theorems for the BVP

$$(4.2.1) \qquad x' = A(t,x)x + b(t,x)$$

and (4.1.2). We assume that the $n \times n$ matrix functions $A(t,x)$, and the vector function $b(t,x)$ are continuous in x for almost all t in Δ and measurable in t for all x in R^n. Assume, moreover, that the functions

$$\alpha(t) = \sup_{x \in R^n} \|A(t,x)\|, \quad \beta_k(t) = \sup_{\|x\| \leq k} \|b(t,x)\| \quad (k = 1,2,\ldots)$$

are integrable in Δ. This implies, by the well-known Carathéodory theorem, that solutions of initial value problems of (4.2.1) exist locally.

For a given linear mapping $T\colon C \to R^n$ and an arbitrary vector r in R^n, we consider the BVP (4.2.1) and (4.1.2). By an application of the Schauder fixed point theorem and the results of the previous section, we prove the following existence theorem. We use the same notation and symbols as in the previous section.

THEOREM 4.2.1. Let S be a closed set in $L_1(M)$ such that for any $A(t) \in S$, the BVP (4.1.3), (4.1.4) has only the trivial solution. Assume that $A(t,y(t)) \in S$ for each $y(t)$ in C and that

$$\liminf_{k\to\infty} \frac{1}{k} \int_{\Delta} \beta_k(t)\,dt = 0. \tag{4.2.2}$$

Then there exists at least one solution of the BVP (4.2.1), (4.1.2).

Proof: The closure $\overline{N}$ of the set $N = [A(t,y(t)): y(t) \in C]$ is a closed subset of S so, without loss of generality, we may assume $S = \overline{N}$. From Lemma 4.1.1, S is a closed subset of a compact set and therefore is itself compact. We have for S that (i) is satisfied with $\alpha \equiv \int_{\Delta} \alpha(t)\,dt$.

Let $W: C \to R^n$ be the mapping which to each $y(t)$ in C assigns the unique solutions of

$$x' = A(t,y(t))x + b(t,y(t)) \tag{4.2.3}$$

satisfying (4.1.2). To prove Theorem 4.2.1., it suffices to show that W has a fixed point in C.

The mapping $P: C \to S \times L_1$ which to each $y(t)$ in C assigns the pair $(A(t,y(t)), b(t,y(t)))$ is continuous. Indeed, for a sequence $\{y_k(t)\} \subset C$ such that

$$\lim_{k\to\infty} \|y_k - y_0\|_\circ = 0$$

we have by the continuity of $A(t,x)$ and $b(t,x)$ with respect to x, that

$$\lim_{k\to\infty} A_k(t) \equiv A(t,y_k(t)) = A(t,y_0(t)) \equiv A_0(t),$$

$$\lim_{k\to\infty} b_k(t) \equiv b\big(t,y_k(t)\big) = b\big(t,y_0(t)\big) \equiv b_0(t),$$

almost everywhere in Δ so by the Lebesgue dominated convergence theorem, we see that

$$\|A_k - A_0\|^{\circ} \to 0 \qquad \text{and} \qquad \|b_k - b_0\|_1 \to 0.$$

Since $W = QP$, it follows from Lemma 4.1.2 that the mapping W is continuous. Furthermore, since P maps

$$B_k = |y(t) \in C: \|y\| \leq k|, \qquad k = 1,2,\ldots$$

into the set $S \times D_k$, where $D_k \equiv [b(t) \in L_1: \|b(t)\| \leq \beta_k(t),$ $t \in \Delta]$, we infer from Lemma 4.1.4 that W is a compact mapping. Finally, from (4.1.9) we have

$$\begin{aligned}\|Wy\| = \|Q\big(A(t,y(t)),\ b(t,y(t))\big)\| &\leq \gamma|r| + \beta\|b\big(t,y(t)\big)\|_1\\ &\leq \gamma|r| + \beta\int_{\Delta} \beta_k(t)\ dt\end{aligned}$$

for any $y(t)$ in B_k. From (4.2.2), we see that for k sufficiently large, W maps B_k into itself. Therefore W satisfies Schauder's theorem and has at least one fixed point. This completes the proof of Theorem 4.2.1.

From Theorem 4.2.1 we derive the following theorem.

THEOREM 4.2.2. If there is a positive constant δ such that

$$|\det T^n U(t)| \geq \delta \tag{4.2.4}$$

for every $A(t)$ in $N \equiv \big[A\big(t,y(t)\big): y(t) \in C\big]$ and if (4.2.2) holds, the BVP (4.2.1), (4.1.2) has a solution.

Proof: From Lemma 4.1.1, it follows immediately that the closed set $S = \overline{N}$ satisfies condition (i). From continuous dependence, we obtain that the mapping $F: A(t) \to U(t)$ is continuous in S, thus implying (4.2.4) holds for any $A(t)$ in $S = \overline{N}$ and hence the set S satisfies (ii). An application of the proof of Theorem 4.2.1 completes the proof.

4.3 INTERPOLATION PROBLEMS

We now apply Theorem 4.2.1 to the well-known Nicoletti problem for nth-order nonlinear differential equations.

For the nth-order scalar differential equation

$$x^{(n)} = f(t,x,x^{(1)},\ldots,x^{(n-1)}), \tag{4.3.1}$$

we consider the following problem: given n distinct points $t_1 < t_2 < \cdots < t_n$ in Δ and n arbitrary real numbers $r_1, r_2, \ldots, r_n$, does there exist a solution $x(t)$ of (4.3.1) such that

$$x(t_i) = r_i, \qquad i = 1,\ldots,n? \tag{4.3.2}$$

Assume that $f(t,x) \equiv f(t,x_1,x_2,\ldots,x_n)$ is defined in the set $\Delta \times R^n$, continuous in x for almost all t in Δ and measurable in t for each $x \in R^n$. Furthermore, assume that there exist n nonnegative integrable functions $p_1(t),\ldots,p_n(t)$ and a nonnegative function $p_0(t,x)$ satisfying the same regularity conditions as $f(t,x)$ such that

$$\begin{aligned} |f(t,x)| \le p_1(t)|x_1| + p_2(t)|x_2| + \cdots \\ + p_n(t)|x_n| + p_0(t,x), \end{aligned} \tag{4.3.3}$$

for $k = 1,2,\ldots,$ the functions

$$q_k(t) = \sup_{\|x\| \le k} p_0(t,x) \tag{4.3.4}$$

are integrable in Δ, and

$$\liminf_{k\to\infty} \frac{1}{k} \int_\Delta q_k(t)\, dt = 0. \tag{4.3.5}$$

Letting

$$g(t,x) \equiv p_1(t)|x_1| + p_2(t)|x_2| + \cdots + p_n(t)|x_n| + p_0(t,x)$$

we have

$$f(t,x) = \frac{f(t,x)}{g(t,x)}\left[\sum_{i=1}^{n} (p_i(t) \operatorname{sgn} x_i)x_i + p_0(t,x)\right]$$

$$= \sum_{i=1}^{n} \frac{p_i(t)f(t,x)(\operatorname{sgn} x_i)x_i}{g(t,x)} + \frac{p_0(t,x)f(t,x)}{g(t,x)} .$$

Letting

$$a_i(t,x) = \frac{p_i(t)f(t,x) \operatorname{sgn} x_i}{g(t,x)}, \qquad a_0(t,x) = \frac{p_0(t,x)f(t,x)}{g(t,x)},$$

we arrive at

$$(4.3.6) \qquad |a_i(t,x)| \le p_i(t), \qquad |a_0(t,x)| \le p_0(t,x).$$

We can now write (4.3.1) in the form

$$x^{(n)} = \sum_{i=1}^{n} a_i(t,x,\ldots,x^{(n-1)})\, x^{(i-1)} + a_0(t,x,\ldots,x^{(n-1)})$$

which is equivalent to the following system of differential equations

$$(4.3.7) \qquad \begin{aligned} x_i' &= x_{i+1} \qquad (i = 1,\ldots,n-1), \\ x_n' &= \sum_{i=1}^{n} a_i(t,x)x_i + a_0(t,x), \end{aligned}$$

and condition (4.3.2) becomes

$$(4.3.8) \qquad x_1(t_i) = r_i \qquad (i = 1,\ldots,n).$$

The linear problem (4.3.7), (4.3.8) is a special case of (4.2.1), (4.1.2) with

$$A(t,x) = \begin{pmatrix} 0 & 1 & 0 & \cdot & \cdot & \cdot & 0 \\ 0 & 0 & 1 & 0 & \cdot & \cdot & 0 \\ 0 & 0 & 0 & 1 & 0 & \cdot & 0 \\ \cdot & \cdot & \cdot & \cdot & \cdot & \cdot & \cdot \\ \cdot & \cdot & \cdot & \cdot & \cdot & \cdot & \cdot \\ \cdot & \cdot & \cdot & \cdot & \cdot & \cdot & \cdot \\ 0 & 0 & 0 & 0 & \cdot & \cdot & 1 \\ a_1(t,x) & a_2(t,x) & \cdot & \cdot & \cdot & \cdot & a_n(t,x) \end{pmatrix},$$

$$b(t,x) = \big(0,0,\ldots,a_0(t,x)\big),$$

$$T: \big(x_1(t),\ldots,x_n(t)\big) \to \big(x_1(t_1),\ldots,x_1(t_n)\big).$$

We easily see that the family S of all matrix functions in $L_1(M)$ of the form

$$\begin{pmatrix} 0 & 1 & 0 & \cdot & \cdot & \cdot & \cdot & \cdot & 0 \\ 0 & 0 & 1 & 0 & \cdot & \cdot & \cdot & \cdot & 0 \\ 0 & 0 & 0 & 1 & 0 & \cdot & \cdot & \cdot & 0 \\ \cdot & \cdot & \cdot & \cdot & \cdot & \cdot & \cdot & \cdot & \cdot \\ \cdot & \cdot & \cdot & \cdot & \cdot & \cdot & \cdot & \cdot & \cdot \\ \cdot & \cdot & \cdot & \cdot & \cdot & \cdot & \cdot & \cdot & \cdot \\ 0 & 0 & 0 & 0 & 0 & \cdot & \cdot & \cdot & 1 \\ a_1(t) & a_2(t) & \cdot & \cdot & \cdot & \cdot & \cdot & \cdot & a_n(t) \end{pmatrix}$$

with

$$(4.3.9) \qquad |a_i(t)| \le p_i(t) \qquad (t \in \Delta, \quad i = 1,\ldots,n)$$

is closed in $L_1(M)$ and satisfies condition (i) of Section 4.1. Condition (ii) is equivalent to the requirement that the function $x(t) \equiv 0$ is the unique solution of the problem

$$(4.3.10)\qquad \begin{aligned} &x^n = a_n(t)x^{(n-1)} + \cdots + a_1(t)x, \\ &x(t_i) = 0, \qquad i = 1,\ldots,n \end{aligned}$$

with the coefficients $a_i(t)$ satisfying (4.3.9).

This requirement is equivalent to the condition that $x(t) \equiv 0$ is the unique function with an absolutely continuous $(n-1)$st derivative satisfying conditions (4.3.10) almost everywhere in Δ, satisfies the inequality

$$(4.3.11)\qquad |x^{(n)}(t)| \le p_n(t)|x^{n-1}(t)| + \cdots + p_1(t)|x(t)|.$$

Finally, from the definition of $A(t,x)$ and (4.3.6) it follows that $A(t,x(t)) \in S$ for any $x(t)$ in C and that (4.3.5) implies (4.2.2).

Thus, by appealing to Theorem 4.2.1, we have proved the following result:

THEOREM 4.3.1. For (4.3.1) assume conditions (4.3.3) - (4.3.5) hold. If $x(t) \equiv 0$ is the unique solution of inequality (4.3.11) with absolutely continuous $(n-1)$th derivative satisfying the boundary condition in (4.3.10), then there exists at least one solution of the problem (4.3.1), (4.3.2).

The following uniqueness theorem follows readily.

THEOREM 4.3.2. In (4.3.1), let

$$(4.3.12)\qquad |f(t,x_1',\ldots,x_n') - f(t,x_1,x_2,\ldots,x_n)| \le \sum_{i=1}^{n} p_i(t)|x_i' - x_i|.$$

Assume $x(t) \equiv 0$ is the unique solution of

$$x^n(t) = a_n(t)x^{n-1}(t) + \cdots + a_1(t)x(t)$$

with absolutely continuous $(n-1)$th derivative satisfying (4.3.10), whenever $|a_i(t)| \le p_i(t)$. Then there exists a unique solution of the BVP (4.3.1) and (4.3.2).

EXERCISE 4.3.1. Prove Theorem 4.3.2.

Hint: Show that (4.3.12) implies that conditions (4.3.3) - (4.3.5) hold. Then apply Theorem 4.3.1 to show the existence of a solution. The uniqueness then follows readily.

We now employ these last two theorems by giving sufficient conditions on the functions $p_i(t)$ and the length of the interval $[a,b]$ to ensure that the only solution of (4.3.10) and (4.3.1) is the trivial solution.

THEOREM 4.3.3. Assume f satisfies (4.3.3) - (4.3.5), where $\{p_i\}_{i=1}^n$ satisfy

$$(4.3.13) \qquad \sum_{k=0}^{n} \frac{h^k}{2^k k\big((k-1)/2\big)!\,(k/2)!}\, p_{n-k} < 1$$

with $h = b - a$. Then the BVP (4.3.1), (4.3.2) has at least one solution in $[a,b]$.

EXERCISE 4.3.2. Prove Theorem 4.3.3.

Hint: Show (4.3.13) implies the only solution of (4.3.10) and (4.3.1) is the trivial solution. Then use Theorem 4.3.1.

We now apply our result to the BVP

$$(4.3.14) \qquad x'' = f(t,x,x'),$$

$$(4.3.15) \qquad x(t_1) = r_1, \quad x(t_2) = r_2.$$

Consider the system

$$x'' = p_2(t)x' + p_1(t)x \tag{4.3.16}$$

and assume

$$2m \equiv \max_{t\in[a,b]} |p_2(t)|, \qquad k = \max_{t\in[a,b]} |p_1(t)|.$$

Consider the following property of (4.3.16).

LEMMA 4.3.1. If

$$4m(b - a) + k(b - a)^2 < \pi^2, \tag{4.3.17}$$

then the only solution of (4.3.16) satisfying $x(t_1) = 0$, $x(t_2) = 0$ is the trivial solution.

EXERCISE 4.3.3. Prove Lemma 4.3.1.

Finally we state the following result.

THEOREM 4.3.4. Assume in (4.3.14) that

$$|f(t,x,x')| \leq M + p_2(t)|x'| + p_1(t)|x'|$$

and that (4.3.17) holds. Then (4.3.14), (4.3.15) has at least one solution.

Proof: The proof follows immediately from Lemma 4.3.1 and Theorem 4.3.1.

4.4 FURTHER NONLINEAR PROBLEMS

We now apply the results of Section 4.3 in a slightly different manner. Namely, we provide conditions other than those used in Section 4.1 to verify the hypotheses of Theorem 4.2.1. For example, instead of condition (4.2.2), we assume a relationship on a term involving the operator T and the perturbation term $b(t,x)$. Let us first consider the BVP

$$x' = A(t)x + b(t,x), \tag{4.4.1}$$

(4.4.2) $$Tx = r.$$

As before, let $U(t)$ be the fundamental matrix of

(4.4.3) $$x' = A(t)x,$$

and assume that

(a) $\det[T^n U(t)] \neq 0$;

this says that the restriction of the operator T, call it T_A, to the space of solutions of (4.4.3) is invertible. This is equivalent, as we have seen, to the fact that the only solution of (4.4.3) satisfying $Tx = 0$ is the trivial solution.

We shall also suppose that

(b) $\|A(t)\| \leq \mu(t), \quad t \in \Delta,$

where $\mu(t)$ is integrable on Δ.

Observe that conditions (a) and (b) are of the same nature as conditions (ii) and (i) in Section 4.2, respectively. We now present an existence theorem in which we use the variation of parameters formula as well as the Brouwer fixed point theorem.

THEOREM 4.4.1. Assume that conditions (a) and (b) are satisfied. In addition, suppose that

(c) $b(t,x)$ verifies the Carathéodary conditions in $R \times R^n$ and that there exists, for some sufficiently large k, an integrable function $v_k(t)$, in Δ, such that

(4.4.4) $$\|b(t,x)\| \leq v_k(t), \quad t \in \Delta, \quad \|x\| \leq k,$$

and

(4.4.5) $$k \geq \|T_A^{-1}\| + (1 + \|T_A^{-1}T\|)n \exp \int_\Delta \mu(\theta)\, d\theta \int_\Delta v_k(\theta)\, d\theta.$$

Then the BVP (4.4.1), (4.4.2) has at least one solution for each $r \in R^n$.

Proof: We first prove the result in the case that $v_k(t)$ is independent of k, that is, (4.4.4) holds with $v(t) \equiv v_k(t)$. As before, let $U(t)$ be the fundamental matrix of (4.4.3) for which $U(t_0) = I$ for some $t_0 \in R$.

For each integer $i > 1$, define for each $c \in R^n$

$$\text{(4.4.6)}\qquad \begin{aligned} x(t) &= c \quad \text{for } t \leq t_0 \\ x(t) &= U(t)c + U(t)\int_{t_0}^{t} U^{-1}(s)b\big(s, x(s - h/i)\big)\, ds, \end{aligned}$$

where h is the length of Δ.

The mapping

$$c \to U(t)\int_{t_0}^{t} U^{-1}(s)b\big(s, x(s - h/i)\big)\, ds$$

is a continuous transformation of R^n into $C[\Delta, R^n]$.

For each $r \in R^n$, the mapping

$$c \to r - TU(t)\int_{t_0}^{t} U^{-1}(s)b\big(s, x(s - h/i)\big)\, ds$$

is then a continuous transformation of R^n into itself. Moreover, from (a), T_A^{-1} exists, is continuous, and thus the mapping

$$\text{(4.4.7)}\qquad c \to U^{-1}(t)T_A^{-1}r - U^{-1}(t)T_A^{-1}TU(t) \times \int_{t_0}^{t} U^{-1}(s)b\big(s, x(s - (h/i))\big)\, ds$$

is a continuous transformation of R^n into itself. Moreover, since $U(t)$, $U^{-1}(t)$ are bounded on Δ, we have

$$\|U^{-1}(t)T_A^{-1}r - U^{-1}(t)T_A^{-1}TU(t)\int_{t_0}^{t} U^{-1}(s)b\big(s,x(s-(h/i))\big)\,ds$$
$$- U^{-1}(t)T_A^{-1}r\| \leq \|U^{-1}\|_0^2\ \|U\|_0\ \|T_A^{-1}T\|_0 \int_\Delta v(s)\,ds;$$

hence the transformation (4.4.7) maps R^n into a bounded set M in R^n.

By the Brouwer fixed point theorem there exists a $c \in R^n$ such that

$$(4.4.8)\qquad c = U^{-1}(t)T_A^{-1}r - U^{-1}(t)T_A^{-1}TU(t) \times \int_{t_0}^{t} U^{-1}(s)b\big(s,x(s-(h/i))\big)\,ds.$$

Therefore, for each i, we can find $\{c^i\} \in M$ satisfying (4.4.8) such that a subsequence of $\{c^i\}$, which we again call $\{c^i\}$, converges.

Now with the c^i chosen, we consider the corresponding x_i defined by (4.4.6), that is,

$$(4.4.9)\quad x_i(t) = \begin{cases} c^i, & \text{for } t \leq t_0 \\ U(t)c^i + U(t)\int_{t_0}^{t} U^{-1}(s)b\big(s,x_i(s-(h/i)\big)\,ds, & \text{for } t \in \Delta, \end{cases}$$

and

$$(4.4.10)\qquad Tx_i = r.$$

Furthermore, since

$$U(t)c^i = T_A^{-1}r - T_A^{-1}TU(t)\int_{t_0}^{t} U^{-1}(s)b\ s,x_i(s-(h/i))\ \ ds,$$

we have

$$(4.4.11)\qquad \|x_i\| \le \|T_A^{-1}r\| + \left(1 + \|T_A^{-1}T\|\right)\cdot \left\| \int_{t_0}^{t} U(t)U^{-1}(s)b\left(s, x_i(s - (h/i))\right)\, ds \right\|$$

$$\le \|T_A^{-1}r\| + \left(1 + \|T_A^{-1}T\| \sup_{t,s} \|U(t)U^{-1}(s)\|\right)\int_{\Delta} v(s)\, ds.$$

Hence, $\{x_i(t)\}$ are uniformly bounded. Moreover, it is easy to see that $\{x_i(t)\}$ are equicontinuous and by Ascoli's theorem a subsequence of $\{x_i(t)\}$ which we again call $\{x_i(t)\}$ converges to a function $y(t)$. From (4.4.9) and (4.4.10), it is clear that $y(t)$ is a solution of the BVP (4.4.1), (4.4.2).

We now prove the result for the original case. First we obtain a bound on $\sup_{t,s} \|U(t)U^{-1}(s)\|$. Immediately we have

$$U(t)U^{-1}(s) = I + \int_s^t A(\theta)U(\theta)U^{-1}(s)\, d\theta, \qquad t,\ s \in \Delta.$$

Since $\|I\| = n$, it follows that

$$\|U(t)U^{-1}(s)\| \le n + \int_s^t \|A(\theta)\|\ \|U(\theta)U^{-1}(s)\|\, d\theta.$$

From the Gronwall inequality, we observe that

$$\|U(t)U^{-1}(s)\| \le n \exp \int_s^t \|A(\theta)\|\, d\theta$$

and from (b) we see

$$\sup_{t,s} \|U(t)U^{-1}(s)\| \le n \exp \int_{\Delta} \mu(\theta)\, d\theta.$$

Using (4.4.11), it follows by letting $i \to \infty$

$$(4.4.12) \qquad \|y\|_0 \le \|T_A^{-1}r\| + (1 + \|T_A^{-1}T\|)n \exp\left(\int_\Delta \mu(\theta)\, d\theta\right) \int_\Delta v(\theta)\, d\theta.$$

Now assume the existence of a k and v_k satisfying (4.4.4) and (4.4.5). Consider the vector defined by

$$\bar{b}(t,x) = \begin{cases} b(t,x), & \|x\| \le k, \\ b(t, kx/\|x\|), & \|x\| > k. \end{cases}$$

Obviously $\bar{b}(t,x)$ satisfies Carathéodary conditions as well as

$$\|\bar{b}(t,x)\| \le v_k(t)$$

for all $x \in R^n$ and $t \in \Delta$. However, we have shown the problem

$$\bar{x}' = A(t)\bar{x} + \bar{b}(t,\bar{x}), \qquad T\bar{x} = r$$

has a solution satisfying (4.4.12) with v replaced by v_k. Because of (4.4.11) and (4.4.5), we have $\|\bar{x}\| \le k$, and thus $\bar{b}(t,\bar{x}) = b(t,\bar{x})$. Consequently, $\bar{x}(t)$ is a solution of the BVP (4.4.1), (4.4.2). This completes the proof of Theorem 4.4.1.

COROLLARY 4.4.1. Assume conditions (a) and (b). Suppose that

(c') $b(t,x)$ satisfies Carathéodary conditions and for each $\rho > 0$, there exists an integrable function $v_\rho(t)$ such that

$$(4.4.13) \qquad \|b(t,x)\| \le v_\rho(t)$$

and

$$(4.4.14) \qquad \lim_{\|x\|\to\infty} b(t,x)/\|x\| = 0.$$

Then the BVP (4.4.1), (4.4.2) has a solution.

Proof: From (4.4.14), it follows that for $\varepsilon > 0$ such that

$$(1 + \|T_A^{-1}T\|)n \exp \int_\Delta \mu(\theta)\, d\theta < h/\varepsilon \tag{4.4.15}$$

there exists $\bar{k} \equiv \bar{k}(\varepsilon) > 0$ for which

$$\|b(t,x)\| \leq \varepsilon\|x\|, \qquad t \in \Delta, \qquad \|x\| \geq \bar{k}.$$

Hence, if we pick $k > \bar{k}$, then

$$\|b(t,x)\| \leq k \quad \text{for} \quad t \in \Delta \quad \text{and} \quad \bar{k} \leq \|x\| \leq k.$$

Furthermore, from (4.4.13), we have

$$\|b(t,x)\| \leq v_{\bar{k}}(t), \qquad t \in \Delta, \quad \|x\| \leq \bar{k}$$

and

$$\|b(t,x)\| \leq v_{\bar{k}}(t) + \varepsilon k, \qquad t \in \Delta, \quad \|x\| \leq k. \tag{4.4.16}$$

Now let k be chosen so that it also satisfies

$$k\big(1 - (1 + \|T_A^{-1}\|)\big)n \exp \int_\Delta \mu(s)\, ds$$
$$> \|T_A^{-1}\| + (1 + \|T_A^{-1}T\|)n \exp \int_\Delta \mu(s)\, ds \int_\Delta v_{\bar{k}}(s)\, ds.$$

This is possible in view of (4.4.15). Hence,

$$k \geq \|T_A^{-1}r\| + (1 + \|T_A^{-1}\|)n \exp\left(\int_\Delta \mu(s)\, ds\right) \times \left(\int_\Delta \mu\, v_{\bar{k}}(s) + \varepsilon k\right) ds. \tag{4.4.17}$$

Inequalities (4.4.16) and (4.4.17) imply that inequalities (4.4.4) and (4.4.5) hold with $v_k(t)$ replaced by $v_{\bar{k}}(t) + \varepsilon k$. Thus we may apply Theorem 4.4.1 to assert the stated result.

EXERCISE 4.4.1. For the BVP (4.4.1) and (4.4.2), let

(i) $\|b(t,x)\| \leq \beta(t) + \gamma(t)\|x\|$ and

(ii) $1 - \|U(t)U^{-1}(s)\| \left(1 + \|T_A^{-1}\| \, \|T\| \, \|U(t)U^{-1}(s)\|\right) \int_{\Delta}\gamma(s)\, ds > 0.$

Suppose also that the conditions (a) and (b) hold. Then the BVP has a solution.

Hint: Reduce the conditions so that the second half of the proof of Theorem 4.4.1 is applicable. Do this by constructing a new function $\bar{b}(t,x)$ which is bounded by an integrable function by using conditions (i) and (ii). Show the solution is bounded by an appropriate number where $\bar{b}$ and b agree.

Recall that our operator T of Section 4.3 could be described as

$$\begin{aligned} Tx &= \left(x_1(t_1), x_2(t_2), \ldots, x_n(t_n)\right) \\ &= (r_1, r_2, \ldots, r_n) \end{aligned}$$

whenever $x \in R^n$. Thus, if $U(t)$ is a fundamental matrix of (4.4.3), then the matrix $T^n U$ is the diagonal matrix

$$T_A \overset{\text{def}}{\equiv} T^n U = \begin{pmatrix} u_{11}(t_1) & 0 & \cdot & \cdot & \cdot \\ 0 & u_{22}(t_2) & \cdot & \cdot & \cdot \\ 0 & 0 & \cdot & \cdot & \cdot \\ 0 & 0 & \cdot & \cdot & \cdot \\ 0 & 0 & \cdot & \cdot & \cdot \\ 0 & 0 & \cdot & \cdot & u_{nn}(t_n) \end{pmatrix}$$

We now look at a more general representation for the operator T. In particular, let us consider the Stieltzes integral

$$\text{(4.4.18)} \qquad Tx = \int_{\Delta} dF\, x = r, \qquad x \in C[\Delta, R^n],$$

where $F = F(t)$ is an $n \times n$ matrix of bounded variation for $t \in \Delta$. Then for any solution u of (4.4.3)

$$T_u = T_A u = T_A Uc = \int_{\Delta} dF\, Uc$$

and thus hypothesis (a) is equivalent to

$$\text{(4.4.19)} \qquad \det \int_{\Delta} dF\, U \neq 0;$$

that is, the matrix $D_A = \int_{\Delta} dF\, U$ has an inverse D_A^{-1}.

Hence we immediately have the following extension of Theorem 4.4.1.

THEOREM 4.4.2. Assume (4.4.19) holds as well as conditions (b) and (c) of Theorem 4.4.1. Then the BVP (4.4.1), (4.4.18) has a solution.

REMARK 4.4.1. In the preceding results, namely Theorems 4.4.1 and 4.4.2, observe that we have assumed that the range of the operator T in (4.4.2) lies in R^n. This is not necessary as the range of T may be a subset of R^m for any m. The proofs are exactly the same where, instead of the condition $\det T^n U \neq 0$, we assume the operator $T_A \equiv T^n U$: $R^n \to R^m$ is one-to-one. In fact, for the case $R^m = R^n$, the solution of the BVP (4.4.1), (4.4.18) is equivalent to finding a solution of the integral equation

$$x(t) = U(t) D_A^{-1} r + \int_{\Delta} G(t,s) b\big(s, x(s)\big)\, ds,$$

where $G(t,s)$ is the Green's matrix defined as

$$G(t,s) = \begin{cases} -U(t)D_A^{-1} \int_s^{t_0+h} dF\, U(t)U^{-1}(s) + U(t)U^{-1}(s), & t < s, \\ -U(t)D_A^{-1} \int_s^{t_0+h} dF\, U(t)U^{-1}(s), & t > s. \end{cases}$$

Of course, a fixed point theorem could be applied directly to this integral equation. However, the proof of Theorem 4.4.1 is quite flexible in that it was not necessary to assume $m = n$.

We again consider the questions posed in Sections 4.1 - 4.3, that is, finding solutions of the BVP

$$x' = A(t,x)x + b(t,x) \tag{4.4.20}$$

satisfying (4.4.18).

Let us now assume that instead of (a), $A(t,x)$ satisfies

(a') $\|A(t,x)\| \leq \mu(t)$, $t \in \Delta$, $x \in R^n$, where $\mu(t)$ is integrable on Δ.

Let $\{t_i\}_{i=1}^{\infty}$, $t_i < t_{i+1}$, be any infinite sequence of points in Δ. Consider the series

$$R(t,s,x) = I + \int_s^t A(t_1,x)\, dt_1 + \int_s^t \int_s^{t_1} A(t_1,x)A(t_2,s)\, dt_1\, dt_2 + \cdots . \tag{4.4.21}$$

[If $A(t,x)$ is independent of x, then $R(t,s,x) = U(t)U^{-1}(s)$]. It is easy to verify that this series converges absolutely and uniformly with respect to (t,s,x), for $t,s \in \Delta$ and $x \in R^n$ and thus $R(t,s,x)$ is a continuous function of (t,s,x).

We have the following result.

THEOREM 4.4.3. Consider the BVP (4.4.20), (4.4.18) where $A(t,x)$ satisfies (a') and $b(t,x)$ satisfies (c″) $\|b(t,x)\| \leq v(t)$ for all $(t,x) \in \Delta \times R^n$, where $v(t)$ is integrable on Δ.

Moreover, assume that

$$(4.4.22) \qquad \inf_{x \in C[\Delta,R^n]} \left| \det \int_\Delta dF\, R(t,t_0,x) \right| = d > 0.$$

Then the BVP (4.4.20), (4.4.18) has a solution.

Proof: Since the proof is similar to that of Theorem 4.4.1 we shall only sketch it.

For a fixed integer i, and for each c define $x \in C[\Delta,R^n]$ as

$$x(t) = c, \qquad t \leq t_0,$$

$$x(t) = R(t,t_0,\bar{x})c + \int_{t_0}^{t} R(t,s,\bar{x})b\; s,\bar{x}(s) \quad ds, \quad \text{where}$$

$$\bar{x}(t) = x\big(t - (n/i)\big), \qquad t \in \Delta.$$

Using (4.4.18), we have

$$\int_\Delta dF\; r(t,t_0,\bar{x})c = r - \int_\Delta dF \int_{t_0}^{t} R(t,s,\bar{x})b\big(s,\bar{x}(s)\big)\; ds$$

which can be written in the form

$$B(c)c = b(c).$$

The matrix $B(c)$ and vector $b(c)$ are continuous in c since the functions x, $\bar{x}$ are.

Now condition (4.4.22) implies that $B(c)$ is bounded and has an inverse. By an application of the Brouwer fixed point theorem we can obtain for each integer i, a set of solutions c^i which lie in a bounded set for all i.

Using the same technique as in Theorem 4.4.1 we have the existence of a $y \in C[\Delta,R^n]$ and $c \in R^n$ such that

$$y = R(t,t_0,y)c + \int_{t_0}^{t} R(t,s,y)b(s,y)\,ds, \qquad \int_{\Delta} dF\, y = r.$$

It is not difficult to show that y is a solution of the BVP (4.4.20), (4.4.18).

We have mentioned that one application of Theorem 4.4.1 is the Nicoletti BVP. We now show an application of Theorem 4.4.3.

EXAMPLE 4.4.1. Consider the equation

$$x^{(k)} = \lambda\varphi(t,x,x^1,\ldots,x^{k-1}),$$

where λ is a parameter. Letting

$$y = \begin{pmatrix} x \\ x^1 \\ \cdot \\ \cdot \\ \cdot \\ x^{(k-1)} \\ \lambda \end{pmatrix}, \qquad A = \begin{pmatrix} 0 & 1 & 0 & 0 & 0 & 0 \\ 0 & 0 & 1 & 0 & 0 & 0 \\ \cdot & \cdot & \cdot & \cdot & \cdot & \cdot \\ \cdot & \cdot & \cdot & \cdot & \cdot & \cdot \\ \cdot & \cdot & \cdot & \cdot & \cdot & \cdot \\ 0 & 0 & 0 & 0 & 0 & 1 \\ 0 & 0 & 0 & 0 & 0 & 0 \end{pmatrix}, \qquad b = \begin{pmatrix} 0 \\ 0 \\ 0 \\ 0 \\ 0 \\ 0 \end{pmatrix},$$

we derive (4.4.20) where $n = k + 1$. In this case we obtain

$$R(t,t_0,x) =$$

$$\begin{pmatrix} 1 & (t-t_0) & \frac{(t-t_0)^2}{2!} & \cdots & \frac{1}{(k-1)!}\int_{t_0}^{t} (t-t_1)^{k-1}\varphi(t_1,x)dt_1 \\ 0 & 1 & (t-t_0) & & \frac{1}{(k-2)!}\int_{t_0}^{t} (t-t_1)^{k-2}\varphi(t_1,x)dt_1 \\ 0 & 0 & 1 & (t-t_0) & \\ \cdot & & & 1 & \cdot \\ \cdot & & 0 & 0 & \cdot \\ \cdot & & & & \cdot \\ 0 & 0 & 0 & 0 & 1 \end{pmatrix}$$

We want the matrix F then to obey the condition

$$\int_{\Delta} dF\, R(t,t_0,x) \neq 0,$$

that is, the only polynomial $P(t)$ of degree less than or equal to k satisfying the condition

$$\int_{\Delta} dF \begin{pmatrix} P(t) \\ P^1(t) \\ \cdot \\ \cdot \\ \cdot \\ P^{(k)}(t) \end{pmatrix} = 0$$

is $P(t) \equiv 0$.

4.5 GENERALIZED SPACES

If $x,y \in R^n$, we say $x \leq y$ if and only if $x_i \leq y_i$, $i = 1,\ldots,n$.

DEFINITION 4.5.1. Let E be a real vector space. A generalized norm for E is a mapping $\|\cdot\|_G\colon E \to R^n$ denoted by

$$\|x\|_G = (\alpha_1(x),\ldots,\alpha_n(x))$$

such that

(a) $\|x\|_G \geq 0$, that is, $\alpha_i(x) \geq 0$ for all i;

(b) $\|x\|_G = 0$ if and only if $x = 0$, that is, $\alpha_i(x) = 0$ for all i if and only if $x = 0$;

(c) $\|\lambda x\|_G = |\lambda|\ \|x\|_G$, that is, $\alpha_i(\lambda x) = |\lambda|\alpha_i(x)$;

(d) $\|x+y|_G \leq \|x\|_G + \|y\|_G$, that is, $\alpha_i(x+y) \leq \alpha_i(x) + \alpha_i(y)$.

For each $x \in E$, and $\varepsilon \in R^n$, $\varepsilon > 0$, let

$B_\varepsilon(x) = [y \in E: \|y - x\|_G < \varepsilon]$. Then $[B_\varepsilon(x): x \in E, \varepsilon \in R^n, \varepsilon > 0]$ is a basis for a topology on E.

REMARK 4.5.1. It is not difficult to see that every generalized normed space $(E, \|\cdot\|_G)$ has an equivalent (ordinary) norm. For example in R^2, $\|x\|_G = (|x_1|, |x_2|)$ and $\|x\| = \max(|x_1|, |x_2|)$ are equivalent. For purely algebraic and topological considerations, it is immaterial whether we view E as a generalized norm space or an ordinary norm space. Such concepts as convexity, closure, completeness and compactness remain the same. We do, however, have more flexibility working with generalized spaces.

We shall need the following terminology.

DEFINITION 4.5.2. An <u>A-matrix</u> is a nonnegative matrix S such that $I - S$ is positive definite.

A <u>positive definite matrix</u> S will be any $n \times n$ matrix such that $x \cdot Sx > 0$ for all $x \in R^n$. We will use the following properties of a positive definite matrix S:

(i) $\det S > 0$,

(ii) all the principal minors of S are positive definite,

(iii) if all the off-diagonal elements of S are nonpositive then S^{-1} is nonnegative,

(iv) if $S \geq 0$, then $\sum_{n=0}^{\infty} S^n$ converges if and only if for some m, $I - S^m$ is positive definite in which case $(I - S)^{-1} = \sum_{n=0}^{\infty} S^n$.

EXERCISE 4.5.1. Prove (i) - (iv).

We now state the Schauder fixed point theorem and contraction mapping theorem in a generalized normed space.

THEOREM 4.5.1. Let E be a generalized Banach space and

let $F \subset E$ be closed and convex. If $T: F \to F$ is completely continuous, then T has a fixed point.

Proof: In view of Remark 4.5.1, we may view E as an ordinary Banach space with an equivalent ordinary norm. Then Theorem 4.5.1. becomes the classical Schauder-Tychonoff theorem.

DEFINITION 4.5.3. Let E be a real vector space. A generalized metric for E is a mapping $d: E \times E \to R^n$ such that

(a) $d(x,y) = d(y,x)$;

(b) $d(x,y) \geq 0$ and $d(x,y) = 0$ if and only if $x = y$;

(c) $d(x,z) \leq d(x,y) + d(y,z)$, where x, y, z are any elements of E.

THEOREM 4.5.2. Let E be a complete generalized metric space and let $T: E \to E$ such that

$$d(Tx,Ty) \leq Sd(x,y),$$

where S is a nonnegative matrix such that for some m, S^m is an A-matrix. Then T has a unique fixed point x^*. Further for any $x \in E$

$$x^* = \lim_{n\to\infty} T^n x$$

and

$$d(x^*,T^n x) \leq S^n(1-S)^{-1}d(Tx,x).$$

We leave the proof as an exercise.

EXERCISE 4.5.2. Prove Theorem 4.5.2.

Hint: Use (iv) to show $T^n x$ is a Cauchy sequence.

Then use the same arguments as in the classical case.

COROLLARY 4.5.1. Let E be a complete generalized metric space and let $T: E \to E$ such that

$$d(Tx, Ty) \leq Sd(x,y),$$

where S is a nonnegative matrix. If there is an $x_0 \in E$ such that $\sum_{n=0}^{\infty} S^n d(Tx_0, x_0)$ converges, then T has a fixed point x^* such that

$$x^* = \lim_{n\to\infty} T^n x_0.$$

4.6 INTEGRAL EQUATIONS

Let $J = [a,b]$ be a fixed interval and let $p = (p_1,\ldots,p_n)$ be a vector such that $1 \leq p_i \leq \infty$. We will consider the space

$$L_p[J] \equiv L_p = L_{p_1}[J] \times \cdots \times L_{p_n}[J],$$

where $L_{p_i}[J]$ is the space of Lebesgue p_i-integrable functions. L_p is a generalized Banach space with the generalized norm

$$\|x\|_p = (\|x_1\|_{p_1},\ldots,\|x_n\|_{p_n}), \qquad \text{where } x = (x_1,x_2,\ldots,x_n).$$

To avoid confusion, if $f(t) = \int_a^b g(t,s)\,ds$ and $f \in L_p$, then we write $\|\int_a^b g(t,s)\,ds\|_p \equiv \|f\|_p$. We wish to consider a vector integral equation

$$(4.6.1) \qquad F[x](t) = \int_a^b K\big(t,s,x(s)\big)\,ds + b(t),$$

where $K: J \times J \times R^n \to R^n$ and $b: J \to R^n$. Let K be the operator defined by

$$K[x](t) = \int_a^b K(t,s,x(s))\,ds.$$

We will denote the ith coordinate of $K(t,s,x)$ by $K_i(t,s,x)$; hence $K(t,s,x) = (K_1(t,s,x),\ldots,K_n(t,s,x))$. A similar notation will be employed for other vector-valued functions.

We shall need the following assumptions:

(H_1) $K_i(t,s,x)$ satisfies the Carathéodory conditions; that is $K_i(t,s,x)$ is continuous in x for each fixed (t,s) and measurable in (t,s) for each fixed x.

(H_2) $\|K(t,s,x)\|_G \leq M(t,s)\|x\|_G + r(t,s)$, where $M\colon J \times J \to R^{n^2}$ is a nonnegative measurable matrix-valued function and $r\colon J \times J \to R^n$ is a nonnegative measurable vector-valued function.

(H_3) for some $p = (p_1,\ldots,p_n)$ the operator M defined by $M[x](t) = \int_a^b [M(t,s)x(s) + r(t,s)]\,ds$ maps L_p into L_p.

(H_4) $\|\int_a^b M(t,s)x(s)\,ds\|_p \leq S\|x\|_p$, $x \in L_p$, where S is a nonnegative matrix such that for some m, S^m is an A-matrix.

(H_5) (i) $\|K(t,s,x) - K(t,s,y)\|_G \leq M(t,s)\|x-y\|_G$,

(ii) $\|K(t,s,0)\|_G \leq r(t,s)$.

LEMMA 4.6.1. If $K(t,s,x)$ satisfies (H_1) - (H_3), then the operator K maps L_p into L_p.

EXERCISE 4.6.1. Prove Lemma 4.6.1.

Hint: First assume $x(s)$ is a simple function and show that the assertion holds. Then assume $x(s)$ is an arbitrary measurable function and show that $K(t,s,x(s))$ is measurable. Let $x \in L_p$ and using (H_2), (H_3) conclude that $K_i(t,s,x(s)) \in L_{p_i}(J^2)$. An immediate application of Fubini's theorem yields the result.

THEOREM 4.6.1. Assume that (H_1) - (H_4) are satisfied and

(i) $b(t) \in L_p$;

(ii) the operator $K: L_p \to L_p$ is completely continuous.

Then F has a fixed point in L_p. Further any fixed point x of F in L_p satisfies

$$\|x\|_p \le (I - S)^{-1}\left[\left\|\int_a^b r(t,s)\, ds\right\|_p + \|b\|_p\right].$$

Proof: We will apply Theorem 4.5.1 to obtain our result. Since $F[x](t) = K[x](t) + b(t)$, $F: L_p \to L_p$ is completely continuous from (ii); it is sufficient to show that there exists a closed convex set which is invariant under F.

Now we have from (H_2) - (H_4)

$$\begin{aligned}\|F[x]\|_p &\le \|K[x]\|_p + \|b\|_p \\ &\le \left\|\int_a^b M(t,s)\|x(s)\|_G\, ds\right\|_p + \left\|\int_a^b r(t,s)\, ds\right\|_p + \|b\|_p \\ &\le S\|x\|_p + \eta,\end{aligned}$$

where $\eta = \|\int_a^b r(t,s)\, ds\|_p + \|b\|_p$. By property (iv) in Section 4.5, $I - S$ is invertible and $(I - S)^{-1} \ge 0$. Let $\sigma = (I - S)^{-1}\eta$. Then $S\sigma + \eta = \sigma$. Define

$$B = [x \in L_p: \|x\|_p \le \sigma].$$

Obviously B is closed and convex. For $x \in B$, we have

$$\|F[x]\|_p \le S\|x\|_p + \eta \le S\sigma + \eta = \sigma.$$

Thus $F: B \to B$ and, by Theorem 4.5.1, F has at least one fixed point in B.

Moreover, if x is any fixed point of F, we obtain that

$$\|x\|_p = \|F[x]\|_p \le S\|x\|_p + \eta,$$

which implies

$$(I - S)\|x\|_p \le \eta,$$

and since $(I - S)^{-1} \ge 0$, we conclude that

$$\begin{aligned}\|x\|_p &\le (I - S)^{-1}\eta, \\ &\le (I - S)^{-1}\left[\left\|\int_a^b r(t,s)\, ds\right\|_p + \|b\|_p\right]\end{aligned}$$

THEOREM 4.6.2. Assume that $(H_1), (H_3) - (H_5)$ are satisfied and that $b(t) \in L_p$. Then F has a unique fixed point $x^* \in L_p$. Further for any $x \in L_p$, we have

$$x^*(t) = \lim_{n\to\infty} F^n[x](t)$$

and

$$\|x^* - F^n[x]\|_p \le S^n(I - S)^{-1}\|x - F[x]\|_p. \tag{4.6.2}$$

Proof: First observe that from (H_5)

$$\begin{aligned}\|K(t,s,x)\|_G &\le \|K(t,s,x) - K(t,s,0)\|_G + \|K(t,s,0)\|_G \\ &\le M(t,s)\|x\|_G + r(t,s).\end{aligned}$$

Thus (H_2) is satisfied. By Lemma 4.6.1, $K[x]$ and hence F maps L_p into L_p. Further for any $x, y \in L_p$, we see, using (H_5) and (H_4),

$$\|F[x] - F[y]\|_p = \left\|\int_a^b \left[K(t,s,x(s)) - K(t,s,y(s))\right] ds\right\|_p \le$$

$$\leq \left\| \int_a^b M(t,s) \|x(s) - y(s)\|_G \, ds \right\|_p$$

$$\leq S \|x - y\|_p,$$

where $S \geq 0$ and for some m, S^m is an A-matrix. Hence by Theorem 4.5.2, F has a unique fixed point which may be obtained by successive approximations. The estimate in Theorem 4.5.2 implies (4.6.2) holds thus concluding the proof of Theorem 4.6.2.

4.7 APPLICATION TO EXISTENCE AND UNIQUENESS

In this section we wish to consider the BVP

$$x' = f(t,x) \tag{4.7.1}$$

$$x_i(t_i) = x_i^0, \qquad t_i \in J \equiv [a,b], \quad i = 1,\dots,n, \tag{4.7.2}$$

where $f\colon J \times R^n \to R^n$. By a solution of (4.7.1) and (4.7.2), we mean an absolutely continuous function which satisfies (4.7.1) almost everywhere and passes through the points (t_i, x_i^0).

If we let $x_0 = (x_1^0, x_2^0, \dots, x_n^0)$, then the substitution $y = x - x_0$ transforms (4.7.1) and (4.7.2) into

$$y' = f(t, y + x_0) \equiv g(t,y) \tag{4.7.3}$$

$$y_i(t_i) = 0, \qquad t_i \in [a,b], \quad i = 1,\dots,n. \tag{4.7.4}$$

Solving (4.7.3) and (4.7.4) is equivalent to finding a fixed point of the integral operator $G[y](t)$ given by

$$G_i[y](t) = \int_{t_i}^t g_i\big(s, y(s)\big) \, ds, \qquad i = 1,\dots,n. \tag{4.7.5}$$

LEMMA 4.7.1. The following identity holds.

$$G[y](t) = \int_a^b K\big(t,s,y(s)\big)\,ds, \tag{4.7.6}$$

where

$$K_i(t,s,y) = \psi_i(t,s)g_i(s,y),$$

in which

$$\psi_i(t,s) = \begin{cases} 1, & t_i \le s \le t, \\ -1, & t \le s < t_i, \\ 0, & \text{otherwise.} \end{cases}$$

Proof: From (4.7.5),

$$\begin{aligned} G_i[y](t) &= \int_{t_i}^{t} g_i\big(s,y(s)\big)\,ds \\ &= \operatorname{sgn}(t - t_i)\int_{\min(t_i,t)}^{\max(t_i,t)} g_i\big(s,y(s)\big)\,ds \\ &= \int_{\min(t_i,t)}^{\max(t_i,t)} \psi_i(t,s)g_i\big(s,y(s)\big)\,ds \\ &= \int_a^b \psi_i(t,s)g_i\big(s,y(s)\big)\,ds \\ &= \int_a^b K_i\big(t,s,y(s)\big)\,ds. \end{aligned}$$

Hence (4.7.6) holds.

We now state our main existence result, a special case of which is the Nicoletti BVP considered in Section 4.3.

THEOREM 4.7.1. Assume that $f(t,x)$ satisfies the

Carathéodory conditions and that

(i) $|f_i(t,x)| \leq \sum_{j=1}^n a_{ij}(t)|x_j| + c_i(t)$, where $a_{ij}(t) \geq 0$ and contained in $L_{q_j}[a,b]$, $c_i(t) \in L_{p_i}[a,b]$, $c_i(t) \geq 0$, $1 \leq p_i \leq \infty$ and $(1/p_i) + (1/q_i) = 1$;

(ii) there is a matrix $S = (\sigma_{ij})$ such that for some m, S^m is an A-matrix and

$$\left\| \left| \int_{t_i}^{t} a_{ij}(s)^{q_j} \, ds \right|^{1/q_j} \right\|_{p_i} \leq \sigma_{ij}. \tag{4.7.7}$$

Then the BVP (4.7.1), (4.7.2) has a solution $x^* \in L_p$. Moreover any solution x^* satisfies the estimate

$$\|x^* - x_0\|_p \leq (I - S)^{-1}\eta, \tag{4.7.8}$$

where $\eta \in R_+^n$ and

$$\eta_i = \left\| \left(\sum_{j=1}^{n} |x_j^0| \right) \left| \int_{t_i}^{t} a_{ij}(s) \, ds \right| + (b - a)c_i(t) \right\|_{p_i}.$$

<u>Proof</u>: We first observe that $\int_{t_i}^{t} a_{ij}(s)^{q_j} \, ds$ is a continuous function of t so that the left-hand side of (4.7.7) is defined.

We will apply Theorem 4.6.1 to show that the operator G given by

$$G[y](t) = \int_a^b K(t,s,y(s)) \, ds,$$

where $K(t,s,y)$ is defined in Lemma 4.7.1, has a fixed point in L_p. By our previous remarks this will imply that the BVP (4.7.1), (4.7.2) has a solution.

We now show hypotheses (H_1) - (H_4) hold as well as (i) and (ii) of Theorem 4.6.1.

Verification of (H_1). We have

$$K_i(t,s,y(s)) = \psi_i(t,s)g_i(s,y(s)) = \psi_i(t,s)f_i(s,y(s) + x_0).$$

Since f_i satisfies the Carathéodory conditions so does $g_i(t,y) = f_i(t, y+x_0)$. Moreover $\psi_i(t,s)$ is easily seen to be measurable. Hence $K_i(t,s,y)$ satisfies the Carathéodory conditions, verifying (H_1).

Verification of (H_2). We observe that

$$\begin{aligned} |K_i(t,s,y)| &= |\psi_i(t,s)f_i(s,y+x_0)| \\ &\leq |\psi_i(t,s)| \left| \left[\sum_{j=1}^{n} a_{ij}(s)|y_j + x_j^0| + c_i(s) \right] \right| \\ &\leq \sum_{j=1}^{n} |\psi_i(t,s)|a_{ij}(s)|y_j| \\ &\quad + \sum_{j=1}^{n} |\psi_i(t,s)|a_{ij}(s)|x_j^0| + |\psi_i(t,s)|c_i(s). \end{aligned}$$

Thus

$$\|K(t,s,y)\|_G \leq M(t,s)\|y\|_G + r(t,s),$$

where

$$M_{ij}(t,s) = (|\psi_i(t,s)|a_{ij}(s))$$

and

$$r_i(t,s) = \sum_{j=1}^{n} M_{ij}(t,s)|x_j^0| + |\psi_i(t,s)|c_i(s).$$

Hence (H_2) is verified.

Verification of (H_3). Let $y \in L_p$, $p = (p_1,\ldots,p_n)$;

then

$$(4.7.9) \qquad \sum_{j=1}^{n} M_{ij}(t,s)y_j(s) + r_i(t,s) = |\psi_i(t,s)|\left(\sum_{j=1}^{n} a_{ij}(s)[y_j(s) + |x_j^0|] + c_i(s)\right).$$

Applying Tonelli's theorem to the positive and negative parts of (4.7.9), we see that $\int_a^b \sum_{j=1}^n M_{ij}(t,s)y_j(s) + r_i(t,s)\ ds$ is a measurable function of t. Furthermore,

$$\left|\int_a^b \left(\sum_{j=1}^{n} M_{ij}(t,s)y_j(s) + r_i(t,s)\right) ds\right|$$

$$\le \int_a^b |\psi_i(t,s)|\left[\sum_{j=1}^{n} a_{ij}(s)\left(|y_j(s)| + |x_j^0|\right) + c_i(s)\right] ds$$

$$\le \int_a^b \left[\sum_{j=1}^{n} a_{ij}(s)\left[|y_j(s)| + |x_j^0|\right] + c_i(s)\right] ds$$

$$\le \sum_{j=1}^{n} \|a_{ij}\|_{q_j} \|y_j\|_{p_j} + |x_j^0| \int_a^b \sum_{j=1}^{n} a_{ij}(s)\ ds + \int_a^b c_i(s)\ ds$$

from Holder's inequality. Since $a_{ij}(s) \in L_{q_j}[a,b]$, $q_j \ge 1$, we have $a_{ij}(s) \in L_1[a,b]$. Therefore our estimates yield

$$\left|\int_a^b \left(\sum_{j=1}^{n} M_{ij}(t,s)y_j(s) + r_i(t,s)\right) ds\right| \le N_i + \int_a^b c_i(s)\ ds,$$

where $N_i = \sum_{j=1}^n \|a_{ij}\|_{q_j} \|y_j\|_{p_j} + |x_j^0| \int_a^b a_{ij}(s)\ ds$ is a constant depending only on $\|y\|_p$. Since $c_i \in L_{p_i}$, we conclude

$$\int_a^b \left(\sum_{j=1}^{n} M_{ij}(t,s)y_j(s) + r_i(t,s)\right) ds \in L_{p_i}.$$

Thus (H_3) holds.

Verification of (H_4). Let $S = (\sigma_{ij})$ be the matrix given in (ii). We wish to show

$$\left\| \int_a^b M(t,s)y(s)\, ds \right\|_p \le S\|y\|_p,$$

that is,

$$\left\| \int_a^b \sum_{j=1}^n M_{ij}(t,s)y_j(s)\, ds \right\|_{p_i} \le \sum_{j=1}^n \sigma_{ij}\|y_j\|_{p_j}.$$

For a fixed i, let

$$\|a_{ij}\|_{q_j}(t) = \begin{cases} \left| \int_{t_i}^t a_{ij}(s)^{q_j}\, ds \right|^{1/q_j}, & q_j < \infty, \\ \operatorname*{ess\,sup}_{[t_i,t]} |a_{ij}(s)|, & q_j = \infty; \end{cases}$$

$$\|y_j\|_{p_j}(t) = \begin{cases} \left(\int_{t_i}^t |y_j(s)|^{p_j}\, ds \right)^{1/p_j}, & p_j < \infty, \\ \operatorname*{ess\,sup}_{[t_i,t]} |y_j(s)|, & p_j = \infty. \end{cases}$$

Then

$$\left\| \int_a^b \sum_{j=1}^n M_{ij}(t,s)y_j(s)\, ds \right\|_{p_i}$$

$$= \left\| \int_a^b \sum_{j=1}^n |\psi_i(t,s)| a_{ij}(s)y_j(s)\, ds \right\|_{p_i}$$

$$\le \sum_{j=1}^n \left\| \int_a^b |\psi_i(t,s)| a_{ij}(s)y_j(s)\, ds \right\|_{p_i} \le$$

$$\leq \sum_{j=1}^{n} \left\| \int_{t_1}^{t} a_{ij}(s) y_j(s)\, ds \right\|_{p_i}$$

$$\leq \sum_{j=1}^{n} \left\| \|a_{ij}\|_{q_j}(t) \|y_j\|_{p_j}(t) \right\|_{p_i}$$

(by Holder's inequality)

$$\leq \sum_{j=1}^{n} \left\| \|a_{ij}\|_{q_j}(t) \|y_j\|_{p_j} \right\|_{p_i}$$

$$\leq \sum_{j=1}^{n} \left\| \|a_{ij}\|_{q_j}(t) \right\|_{p_i} \|y_j\|_{p_j}$$

$$\leq \sum_{j=1}^{n} \left\| \left| \int_{t_i}^{t} a_{ij}(s)^{q_j} ds \right|^{1/q_j} \right\|_{p_i} \|y_j\|_{p_j}$$

$$\leq \sum_{j=1}^{n} \sigma_{ij} \|y_j\|_{p_j}.$$

Hence (H_4) is satisfied.

All that remains is to verify conditions (i) and (ii) of Theorem 4.6.1. We observe immediately that $b(t) \in L_p$ since in the case under consideration $b(t) \equiv 0$. Hence (i) holds.

Verification of (ii). We show that the operator $G: L_p \to L_p$ defined by

$$G[y](t) = \int_a^b K\big(t, s, y(s)\big)\, ds$$

is completely continuous. To prove this we will use Helley's first theorem which we state here.

HELLEY'S FIRST THEOREM. Let $\mathfrak{F}$ be an infinite uniformly bounded family of functions on $[a,b]$ whose members are uniformly bounded in variation; that is, there is an $M > 0$ such that for each $f \in \mathfrak{F}$, $\bigvee_a^b [f] \leq M$, where $\bigvee_a^b [f]$ is the total variation of f on $[a,b]$. Then there is an infinite sequence in $\mathfrak{F}$ which converges pointwise on $[a,b]$.

Now let

$$B = [y \in L_p : \|y\|_p \leq \alpha],$$

where α is an arbitrary element in R_+^n, that is, $\alpha = (\alpha_1, \alpha_2, \ldots, \alpha_n)$, $\alpha_i \geq 0$, $i = 1, \ldots, n$. Then for each $y \in B$

$$\begin{aligned} K_i[y](t) &= \int_a^b K_i\big(t,s,y(s)\big)\, ds \\ &= \int_a^b \psi_i(t,s) f_i\big(s, y(s) + x_0\big)\, ds \\ &= \int_{t_i}^t f_i\big(s, y(s) + x_0\big)\, ds; \end{aligned}$$

and hence

$$\begin{aligned} \bigvee_a^b K_i[y] &\leq \int_a^b \left| f_i\big(s, y(s) + x_0\big)\right| ds \\ &\leq \sum_{j=1}^n \int_a^b a_{ij}(s) |y_j(s) + x_0^j|\, ds + \int_a^b c_i(s)\, ds \\ &\leq \sum_{j=1}^n \|a_{ij}\|_{q_j} \left[\|y_j\|_{p_j} + |x_0^j| (b-a)^{1/p_j}\right] + \int_a^b c_i(s)\, ds \\ &\leq \sum_{j=1}^n \|a_{ij}\|_{q_j} \left[\alpha_j + |x_0^j| (b-a)^{1/p_j}\right] + \int_a^b c_i(s)\, ds \\ &\equiv M_i, \end{aligned}$$

where M_i only depends on α. Thus for each $y \in B$, $K_i[y]$

are uniformly bounded in variation on $[a,b]$. Since $K_i[y](t_i) = 0$ for each i, we conclude that $|K_i[y](t)| \leq M_i$; that is, $K_i[y]$ are uniformly bounded for $y \in B$.

Let $y_n(t)$ be any sequence in B and apply Helly's first theorem to select a subsequence, again denoted by $y_n(t)$, such that $K[y_n](t)$ converges pointwise on $[a,b]$ to some function $z(t)$. Since $|K_i[y_n](t)| \leq M_i$, we have by the Lebesgue dominated convergence theorem that

$$\|K[y_n] - z\|_p \to 0 \qquad \text{as} \quad n \to \infty.$$

Thus $K: L_p \to L_p$ is completely continuous and hence (ii) is satisfied.

Hence we apply Theorem 4.6.1 to obtain a solution $y(t)$ of (4.7.3), (4.7.4). Thus $x(t) = x_0 + y(t)$ is a solution of (4.7.1), (4.7.2).

We also have from Theorem 4.6.1 that

$$\|x^* - x_0\|_p = \|y^*\|_p \leq (I - S)^{-1}\eta,$$

where

$$\eta_i = \left\| \sum_{j=1}^{n} |x_j^0| \left| \int_{t_i}^{t} a_{ij}(s)\, ds \right| + \int_{t_i}^{t} c_i(s)\, ds \right\|_{p_i}.$$

This completes the proof of Theorem 4.7.1.

REMARK 4.7.1. Our proof of verifying (ii) essentially handles the case when $p_i < \infty$ for all i. If all $p_i > 1$, then all $q_i < \infty$ and we may apply Ascoli's theorem to verify (ii). Then the general case in which $1 \leq p_i \leq \infty$ follows by combining the cases $1 \leq p_i < \infty$ and $1 < p_i \leq \infty$.

We now apply Theorem 4.6.2 to obtain a unique solution of (4.7.1), (4.7.2).

THEOREM 4.7.2. Assume that $f(t,x)$ satisfies the Carathéodory conditions and that

(i) $|f_i(t,x) - f_i(t,y)| \le \sum_{j=1}^n a_{ij}(t)|x_j - y_j|$, where $a_{ij} \in L_{q_j}$, $f_i(t,x_0) \in L_{p_i}$, such that

$$(1/q_i) + (1/p_i) = 1;$$

(ii) there exists a matrix $S = (\sigma_{ij}) \ge 0$ such that

$$\left\| \left| \int_{t_i}^t a_{ij}(s)^{q_j} \, ds \right|^{1/q_j} \right\|_{p_i} \le \sigma_{ij},$$

where for some m, S^m is an A-matrix. Then the BVP (4.7.1), (4.7.2) has a unique solution $x^* \in L_p$. Moreover, x^* can be realized by successive approximations and satisfies

$$\|x^* - x_0\|_p \le (I - S)^{-1}\|\eta\|_p,$$

where

$$\eta_i(t) = \int_{t_i}^t f_i(s,x_0) \, ds.$$

Proof: We wish to apply Theorem 4.6.2 to the operator G given by

$$G[y](t) = \int_a^b K\big(t,s,y(s)\big) \, ds.$$

Since much of the argument is the same as in Theorem 4.7.1 we will only sketch the proof.

As in Theorem 4.7.1, $K(t,s,y)$ satisfies (H_1). Moreover (H_3) and (H_4) follow also as in Theorem 4.7.1.

Verification of (H_5). We first show (i) of (H_5) holds.

Observe that

$$\begin{aligned}&|K_i(t,s,y) - K_i(t,s,x)|\\&\quad = |\psi_i(t,s)f_i(s,x + x_0) - \psi_i(t,s)f_i(s,y + x_0)|\\&\quad \leq |\psi_i(t,s)| \sum_{j=1}^{n} a_{ij}(s)|x_j - y_j|;\end{aligned}$$

hence (i) of (H_5) is satisfied with

$$M(t,s) = [|\psi_i(t,s)|a_{ij}(s)].$$

We now show (ii) of (H_5) holds. Since

$$|K_i(t,s,0)| = |\psi_i(t,s)f_i(s,x_0)|$$

we immediately have that (ii) of (H_5) is satisfied with

$$r_i(t,s) = |\psi_i(t,s)f_i(s,x_0)|.$$

Observe that (H_2) follows from (H_5), (H_1), and (i). Thus Theorem 4.6.2 may be applied to the operator G to guarantee a unique solution y^* of (4.7.3) and (4.7.4). Hence $x^*(t) = x_0(t) + y^*(t)$ is the unique solution of (4.7.1), (4.7.2) and $x^* \in L_p$. Moreover x^* can be obtained by successive approximations. Finally, by letting $y_0 = 0$ we have from Theorem 4.6.2 that

$$\|y^* - 0\|_p \leq (I - S)^{-1}\|G[0] - 0\|_p$$

which implies

$$\|x^* - x_0\|_{p_i} \leq (I - S)^{-1}\left\|\int_{t_i}^{t} f_i(s,x_0)ds\right\|_{p_i}.$$

This completes the proof of Theorem 4.7.2.

In both Theorems 4.7.1 and 4.7.2 we observe that if

$f(t,x)$ is continuous in both t and x, then we obtain a $C^{(1)}$ solution of the BVP.

We now study the BVP

$$x'' + f(t,x,x') = 0, \tag{4.7.10}$$

$$x(a) = x(b) = 0, \tag{4.7.11}$$

where $f\colon [a,b] \times R^2 \to R$.

Using the Green's function we obtain some interesting estimates on the solutions of (4.7.10), (4.7.11) by applying the previous techniques of this section.

Recall that the Green's function $G(t,s)$ associated with $x'' = 0$, $x(a) = x(b) = 0$ is

$$G(t,x) = \begin{cases} (s-a)(b-t)/h, & s \le t, \\ (t-a)(b-s)/h, & s \ge t, \end{cases}$$

where $h = b - a$. Then a solution $x(t)$ of

$$x'' + f(t) = 0 \tag{4.7.12}$$

satisfying (4.7.11) has the form

$$x(t) = \int_a^b G(t,s)f(s)\,ds. \tag{4.7.13}$$

We now obtain an existence result for the BVP (4.7.10), (4.7.11) as well as estimates on the L_1 norm and L_2 norm of the solutions (in Chapter 1 we discussed the L_∞ norm).

THEOREM 4.7.3. Let $f(t,x,y) \in C\big[[a,b] \times R^2, R\big]$ and let α and β be nonnegative numbers such that

$$(\alpha h^2/4) + \beta h < 1.$$

(a) If

$$|f(t,x,y)| \leq \alpha|x| + \beta|y| + r(t),$$

where $r(t) \in L_1[a,b]$, then (4.7.10), (4.7.11) has a solution. Further every solution satisfies

$$\|x\|_1 \leq \frac{h^2}{4 - \alpha h^2 - 4\beta h} \|r\|_1, \tag{4.7.14}$$

and

$$\|x'\|_1 \leq \frac{4h}{4 - \alpha h^2 - 4\beta h} \|r\|_1. \tag{4.7.15}$$

(b) If

$$|f(t,x_1,y_1) - f(t,x_2,y_2)| \leq \alpha|x_1 - x_2| + \beta|y_1 - y_2|,$$

then (4.7.10), (4.7.11) has a unique solution satisfying (4.7.14) and (4.7.15) with $r(t) \equiv f(t,0,0)$.

EXERCISE 4.7.1. Prove Theorem 4.7.3 by using Theorem 4.6.1 in (a) and Theorem 4.6.2 in (b). Here our generalized space E is the space of L_1 functions $x(t) = \big(x_1(t),x_2(t)\big)$ with norm $\|x\|_1 = \big(\|x_1\|_1, \|x_2\|_1\big)$.

A similar theorem is now presented for the $\|\cdot\|_2$ norm.

THEOREM 4.7.4. Let $f(t,x,y)$ be continuous and let α and β be nonnegative constants such that

$$(\alpha h^2/\pi^2) + (\beta h/\pi) < 1.$$

(a) If

$$|f(t,x,y)| \leq \alpha|x| + \beta|y| + r(t),$$

where $r(t) \in L_2[a,b]$, then (4.7.10), (4.7.11) has a solution. Every solution satisfies, moreover,

$$\|x\|_2 \leq \frac{h^2}{\pi^2 - \alpha h^2 - \beta\pi h} \|r\|_2 \tag{4.7.16}$$

and

$$\|x'\|_2 \leq \frac{\pi h}{\pi^2 - \alpha h^2 - \beta\pi h} \|r\|_2. \tag{4.7.17}$$

(b) If

$$|f(t,x_1,y_1) - f(t,x_2,y_2)| \leq \alpha|x_1 - x_2| + \beta|y_1 - y_2|,$$

then (4.7.10), (4.7.11) has a unique solution which satisfies (4.7.16) and (4.7.17) with $r(t) \equiv f(t,0,0)$.

EXERCISE 4.7.2. Prove Theorem 4.7.4.

We now wish to consider the BVP

$$x'' + A^2x = f(t,x,x'), \tag{4.7.18}$$

$$x(a) = x(b) = 0, \tag{4.7.19}$$

where $f: [a,b] \times R^{2n} \to R^n$ and A is a constant matrix of order n. Using the techniques of Chapter 1, we consider the associated nonhomogeneous linear problem

$$x'' + A^2x = f(t) \tag{4.7.20}$$

and express the solution of (4.7.18), (4.7.19) in terms of the Green's function for the problem

$$x'' + A^2x = 0 \tag{4.7.21}$$

and (4.7.19).

In order to guarantee the uniqueness of solutions of (4.7.18) and (4.7.19), and hence the existence of a Green's function, we assume the eigenvalues of A, λ_j, $j = 1,\ldots,n$ have the property that

$$\lambda_j \neq k\pi/(b - a), \qquad k = 0, \pm 1, \pm 2, \ldots . \tag{P}$$

This allows us to deduce that $\sin A(b - a)$ is invertible and we may obtain, after some computation, that the Green's function is

$$G(t,s) = \begin{cases} A^{-1}(\sin Ah)^{-1} \sin A(b-t) \sin A(s-a), & s \le t, \\ A^{-1}(\sin Ah)^{-1} \sin A(b-s) \sin A(t-a), & s > t. \end{cases}$$

Hence, the unique solution of (4.7.20), (4.7.19) is

$$x(t) = \int_a^b G(t,s)f(s)\, ds;$$

further,

$$x'(t) = \int_a^b G_t(t,s)f(s)\, ds.$$

To obtain an existence theorem for the BVP (4.7.18), (4.7.19) we need the following preliminary computations. If $B = (b_{ij})$ is a matrix of order n, then let $|B| = (|b_{ij}|)$. If $|B| \le |C|$, then

$$\begin{aligned} |\sin B| &= \left| \sum_{k=0}^{\infty} \frac{B^{2k+1}(-1)^{2k+1}}{(2k+1)!} \right|. \\ &\le \sum_{k=0}^{\infty} \frac{|C|^{2k+1}}{(2k+1)!} \\ &= \frac{1}{2}\left(\sum_0^{\infty} \frac{|C|^k}{k!} - \sum_0^{\infty} \frac{|C|^k(-1)^k}{k!} \right) \\ &= \frac{e^{|C|} - e^{-|C|}}{2}. \end{aligned}$$

Similarly,

$$|\cos B| \le (e^{|C|} + e^{-|C|})/2.$$

We now present an existence theorem.

THEOREM 4.7.5. Let $f(t,x,y)$ be continuous and assume that

(a) $\|f(t,x,y)\|_\infty \leq P\|x\|_\infty + Q\|y\|_\infty + r(t)$, where P and Q are nonnegative constant matrices and $r \in C\left[[a,b],R_+^n\right]$.

Assume that the eigenvalues of A have property (P). Let

$$S = \begin{pmatrix} s_{11} & s_{12} \\ s_{21} & s_{22} \end{pmatrix}$$

be a nonnegative $2n \times 2n$ matrix in which

$$s_{11} = |A^{-1}[\sin Ah]^{-1}| \frac{(e^{|A|h} - e^{-|A|h})^2}{4} hP,$$

$$s_{12} = |A^{-1}(\sin Ah)^{-1}| \frac{e^{|Ah|} - e^{-|A|h}}{4} hQ,$$

$$s_{21} = |(\sin Ah)^{-1}| \frac{e^{2|A|h} - e^{-2|A|h}}{4} hP,$$

$$s_{22} = |(\sin Ah)^{-1}| \frac{e^{2|A|h} - e^{-2|A|h}}{4} hQ.$$

Moreover, assume for some m, S^m is an A-matrix. Then the BVP (4.7.18), (4.7.19) has a solution. Furthermore any solution $x(t)$ of (4.7.18) and (4.7.19) satisfies

$$\begin{pmatrix} \|x\|_\infty \\ \|x'\|_\infty \end{pmatrix} \leq (I - S)^{-1} \begin{pmatrix} \|r\|_\infty \\ \|r'\|_\infty \end{pmatrix}.$$

Recall here $r(t) \equiv f(t,0,0)$.

EXERCISE 4.7.3. Prove Theorem 4.7.5.

<u>Hint</u>: A solution $x(t)$ satisfies

$$x(t) = \int_a^b G(t,s)f\big(s,x(s),x'(s)\big)\, ds,$$

$$x'(t) = \int_a^b G_t(t,s)f\big(s,x(s),x'(s)\big)\, ds.$$

Apply Theorem 4.6.1 where now the operator $F = (F_1, F_2)$ is defined as

$$F_1[z](t) = \int_a^b G(t,s)f\big(s,z(s)\big)\, ds,$$

and

$$F_2[z](t) = \int_a^b G_t(t,s)f\big(s,z(s)\big)\, ds.$$

Show the hypotheses of Theorem 4.6.1 are satisfied in a manner similar to that used to prove Theorem 4.6.1. The only new point is the verification of (H_4). In this case show

$$M(t,s) = \begin{pmatrix} |G(t,s)|P & |G(t,s)|Q \\ |G_t(t,s)|P & |G_t(t,s)|Q \end{pmatrix}$$

and

$$r(t,s) = \begin{pmatrix} |G(t,s)| & 0 \\ 0 & |G_t(t,s)| \end{pmatrix} \begin{pmatrix} |r(s)| \\ |r(s)| \end{pmatrix}.$$

This then implies $|M(t,s)| \le (1/h)\, S$ and thus

$$\left\| \int_a^b M(t,s)z(s)\, ds \right\|_\infty \le S\|z\|_\infty.$$

The conclusion of the theorem follows readily.

4.8 METHOD OF A PRIORI ESTIMATES

Here we shall present theorems on the existence of solutions of boundary value problems, with both linear and nonlinear boundary conditions, using the method of "a priori estimates." This technique can be described as follows.

Suppose that it is required to prove the existence of a solution, where the solution is chosen from some topological space H; in other words, it is required to show that a certain subset $G \subseteq H$ is nonempty (the subset of solutions of the problem). To do this, we construct a sequence of problems of a similar type, that is, a sequence of sets $G_p \subseteq H$ $(p = 1,2,\ldots,)$ approximating the original problem in the sense that $g_{p_k} \in G_{p_k}$ $(p_k \to \infty)$, $g_{p_k} \to g$ implies that $g \in G$. We further construct a scalar function $w(h) \geq 0$, $h \in H$, which will give an estimate such that the condition $\sup_{g_p \in G_p} w(g_p) < \infty$ will imply the existence of a convergent subsequence $g_{p_k} \to g$. Finally, it is shown that each of the approximating problems has at least one solution g_p satisfying $\sup_p w(g_p) < \infty$. This implies that G is nonempty.

Suppose in some linear topological space H we have a set $G \subseteq H$ and a mapping L of the space H into R^n. We are interested in showing the existence of a solution of the problem

$$g \in G, \quad Lg = a, \tag{4.8.1}$$

where $a \in R^n$ is given. In applying the method to boundary value problems, G is the set of solutions of the given differential equation. Assume that there exists a scalar lower semicontinuous function w in H satisfying the inequalities

$$\begin{aligned} &0 \leq w(h) \leq \infty, \qquad w(h_1 + h_2) \leq w(h_1) + w(h_2), \\ &w(ah) \leq w(h) \qquad \text{whenever} \quad 0 < a < 1. \end{aligned} \tag{4.8.2}$$

Suppose that the following conditions are satisfied:

(i) The operator L can be written as the sum of two continuous operators $L^0 + L^1$;

(ii) there exists a set $G^0 \subseteq H$ for which the problem

$$v \in G^0, \qquad L^0 v = a \tag{4.8.3}$$

has a unique solution v_0 for any $a \in R^n$ and the solution depends continuously on a;

(iii) there exists a "comparison problem"

$$y \in G^0, \qquad L^* y = b,$$

where L^* is a continuous operator mapping H into R^n whose solution y_b exists and is unique for any $b \in R^n$ and for any sphere $X \subset R^n$, we have $\sup_{y_b \in G^0} w(y_b) < \infty$.

The preceding conditions guarantee the existence of an operator A defined as

$$Aa \equiv L^* v_a$$

which is a continuous, one-to-one mapping of R^n into itself and is hence a homeomorphism. Thus

$$v_a = y_{Aa}, \qquad y_b = v_{A^{-1}b}, \qquad \text{and} \qquad L^0 y_b = Bb,$$

where $B \equiv A^{-1}$.

(iv) The set G can be approximated by sets G_p such that $g_{p_k} \in G_{p_k}$ and $g_{p_k} \to g$ as $p_k \to \infty$ implies that $g \in G$; and for any $b \in R^n$ each of the problems

$$s \in G_p, \qquad L^* s = b$$

has a unique solution s_b^p depending continuously on b and

$$\Omega = \sup_{\substack{p \in R \\ b \in R^n}} w(s_b^p - y_p) < \infty.$$

(v) The condition $\sup_p w(g_p) < \infty$ implies the existence of a convergent subsequence $g_{p_k} \to g$.

(vi) For any $M > 0$ and any $a \in R^n$, there exists a sphere $X \subset R^n$ such that for any $b \in X$, $s \in G$, $w(s - y_b) < M$, we have

$$A(L^0 y_b - Ls + a) \in X.$$

THEOREM 4.8.1. If conditions (i) - (vi) are satisfied, then the problem (4.8.1) has at least one solution.

Proof: Let $M = 1 + \Omega$ and for the given a, choose the sphere X according to (vi). Define mappings T_p $(p = 1,2)$ of the space R^n into itself by the formula

$$T_p b = A(L^0 y_b - Ls_p^b + a)$$

and set $X^p = T_p X$. Let X_ε, $\varepsilon > 0$, be a sphere having the same center as X with a radius ε times greater. We claim, for any $\varepsilon > 1$, that there exists a p_ε such that $X^p \subseteq X_\varepsilon$ for $p > p_\varepsilon$. Suppose the claim is false. Then there exists an $\varepsilon > 1$ and a sequence $b_k \in X$ such that $b_k \to b_0$ as $k \to \infty$ and $d_k \equiv T_{p_k} b_k \notin X_\varepsilon$. Set $s_k = s_{b_k}$, then

$$\begin{aligned}\sup_k w(s_k) &\leq \sup_k \left(w(s_k - y_{b_k}) + w(y_{b_k})\right)\\ &\leq \sup_{p,b} w(s_b^p - y_b) + \sup_{b \in X} w(y_b) < \infty\end{aligned}$$

by condition (iv). Then (iv) and (v) imply the existence of a subsequence s_{k_i} of s_k, $s_{k_i} \to s \in G$. Hence

$$d_{k_i} = A\left(L^0 y_{n_{k_i}} - Ls_{b_{k_i}}^{p_{k_i}} + a\right) \to A\left(L^0 y_{b_0} - Ls + a\right) \in X,$$

since the lower semicontinuity of w implies that

$$w(s - y_{b_0}) \leq \inf_i w(s_{b_{k_i}}^{p_{k_i}} - y_{b_{k_i}}) < M.$$

However, then, for I sufficiently large, $d_{k_i} \in X_\varepsilon$ for all $i \geq I$. This is a contradiction and the claim is proved.

Hence $X^p \subseteq X_\varepsilon$ for p sufficiently large. Since $T_p \colon X \to X^p \in X_\varepsilon$, we have by the Brouwer fixed point theorem that for each p there exist $b_p \in X$ and ε_p such that

$$T_p b_p = b_p + \varepsilon_p.$$

Moreover, $\varepsilon_p \to 0$ as $p \to \infty$. Hence, the equality

$$A(L^0 y_{b_p} - Ls_{b_p}^p + a) = b_p + \varepsilon_p$$

implies that

$$\begin{aligned} Ls_{b_p}^p &= a + L^0 y_{b_p} - B(b_p + \varepsilon_p) \\ &= \varepsilon + Bb_p = B(b_p + \varepsilon_p) \to a \quad \text{as} \quad p \to \infty. \end{aligned}$$

Since $\sup_p w(s_{b_p}^p) < \infty$, we obtain from (v) the existence of a subsequence $\{s_{b_{p_i}}^{p_i}\}$ of $s_{b_p}^p$ such that $s_{b_{p_i}}^{p_i} \to g$, $Ls_{b_{p_i}}^{p_i} \to a$ and thus g is a solution of (4.8.1).

REMARK 4.8.1. If the operator L^0 is linear, the set G^0 is linear and n-dimensional, and $v^1, v^2, \ldots, v^n$ is some basis for G^0, then the statement that the problem (4.8.3) is uniquely solvable and that the solution depends continuously on a is equivalent to the statement that the determinant

$$\Delta = \det(L^0 v^1, L^0 v^2, \ldots, L^0 v^n)$$

is nonzero.

We apply now the results to the boundary value problem

$$(4.8.4)\quad x_i^{(n_i)} + f_i\left(t, x_1, \ldots, x_m^{(n_m-1)}\right) = 0 \qquad (i = 1,\ldots,m, \quad t_1 \le t \le t_2)$$

$$L_k(x_1,\ldots,x_m) = a_k \qquad (k = 1,\ldots,n, \quad \text{where} \quad n = n_1 + \cdots + n_m),$$

where f_i are defined for $t \in [t_1,t_2]$, $-\infty < x_1,\ldots,x_m^{(n_m-1)} < \infty$, all $n_i > 0$, and the functionals L_k are defined on the set of functions $x(t) = \{x_i(t)\}$, $t_1 \le t \le t_2$, for which all the derivatives $x_i^{(n_i-1)}$ are absolutely continuous. We assume that f_i satisfy the Carathéodory conditions; that is, f is measurable in t for fixed $x_1,\ldots,x_m^{(n_m-1)}$, continuous with respect to $(x_1,\ldots,x_m^{(n_m-1)})$ for each fixed t, and for any $M > 0$ there exists an integrable function f_M on (t_1,t_2) such that $|f_i| \le f_M(t)$ for all i and $t \in (t_1,t_2)$, and $\|x\| < M$,

We may write problem (4.8.4) in the form

$$(4.8.5)\qquad Dx = 0, \qquad Lx = a.$$

Here H is the space of functions $x(t) = \{x_i(t)\}$, $t_1 \le t \le t_2$, for which all the derivatives $x_i^{(n_i-1)}$ are absolutely continuous and the norm on H is

$$\|x\| = \max_{[t_1,t_2]} \sum_{i=1}^{m} \left(\sum_{j=0}^{n_i-1} |x_i^{(j)}(t)| \right).$$

Let G be the set of solutions of $Dx = 0$, where

$$w(x) = \|x\|_1 \equiv \sum_{i=1}^{m} \left(\sum_{j=0}^{n_i-1} |x_i^{(j)}(t_1)| + \int_{t_1}^{t_2} |x_i^{(n_i)}(\tau)| \, d\tau \right).$$

The operator D^0 is obtained from D by setting all $f_i = 0$, then G^0 is the set of all solutions of the equation $D^0 x = 0$.

Let the comparison problem be the Cauchy problem

$$D^0 y = 0$$

$$y_1(t_1) = b_1,\ y_2(t_1) = b_2, \ldots, y_m^{(n_m-1)}(t_1) = b_n \qquad (L^* y = b).$$

We thus have the following result.

THEOREM 4.8.2. Suppose that the following conditions are satisfied:

(i) Let $|f_i| \le f_0(t)$ for $t \in [t_1,t_2]$ and all $x_i, \ldots, x_m^{(n_m-1)}$, where f_0 is integrable on $[t_1,t_2]$;

(ii) The operator L can be written in the form $L = L^0 + L^1$, where L^0 and L^1 are continuous operators in H;

(iii) The boundary value problem $D^0 v = 0$, $L^0 v = a$ has a unique solution for any $a \in R^n$, depending continuously on a;

(iv) For any $M > 0$ and for any $a \in R^n$, there exists a sphere $X \subset R^n$ such that for any $b \in X$, $s \in G$, and

$w(s - y_b) < M$, we have

$$A(L^0 y_b - Ls + a) \in X$$

(see the notation in Theorem 4.8.1).

Then the BVP (4.8.5) has at least one solution.

Proof: We use Theorem 4.8.1 to prove the result; it thus suffices to verify conditions (iv) and (v) of Theorem 4.8.1, since conditions (i) - (iii), (vi) follow immediately.

Observe that each f_i can be approximated by f_{i_p} which is Lipschitz and satisfies an estimate of the form $|f_{i_p}| \leq f_0(t)$, where $f_0(t)$ is integrable on $[t_1, t_2]$. Then conditions (iv) and (v) follow from the Lipschitzian properties of f_{i_p}, the definition of $w(x)$, and the continuous dependence of differential equations on initial data. This concludes the proof of Theorem 4.8.2. Similarly the following results may be obtained.

THEOREM 4.8.3. Suppose conditions (i) - (iii) of Theorem 4.8.2 hold. If L^0 is positively homogeneous $(L^0 \alpha x = \alpha^{\mu} Lx$, $\alpha > 0$, $\mu = \text{constant} > 0)$ and for any $M > 0$, we have $\lim_{a \to \infty} \sup_{w(x) < M} \|a^{-\mu} L^1 ax\| = 0$, then the BVP (4.8.5) has at least one solution.

EXERCISE 4.8.1. Prove Theorem 4.8.3.

THEOREM 4.8.4. Suppose that conditions (i) and (ii) of Theorem 4.8.2 hold. If the operator L^0 is linear, $\Delta \neq 0$, and for any $M > 0$ we have $\lim_{a \to \infty} \sup_{w(x) \leq M} \|a^{-1} L^1 ax\| = 0$, then the BVP (4.8.5) has at least one solution.

EXERCISE 4.8.2. Prove Theorem 4.8.4.

REMARK 4.8.2. Observe that if the boundary operator L is linear, then we may assume $L^0 = L$ and $L^1 = 0$. Then Theorem 4.8.4 essentially says that the BVP (4.8.5) has a solution if the problem $D^0x \equiv 0$, $L^0x = 0$ has only the trivial solution. This result has been proved using other techniques in Section 4.3.

4.9 BOUNDS FOR SOLUTIONS IN ADMISSIBLE SUBSPACES

In this section our treatment of the previously considered BVP

$$x' = A(t)x + f(t,x), \tag{4.9.1}$$

$$Lx = r, \tag{4.9.2}$$

is quite different from that of earlier sections. In particular, the treatment rests on a characterization of the class of linear differential equations

$$x' = A(t)x + b(t) \tag{4.9.3}$$

for which there exists a solution satisfying (4.9.2).

Let J be any compact interval and we may think of L as a continuous linear mapping from $C[J]$ onto R^n. Denote by V the inverse image of any $r \in R^n$. We see that V is a closed linear variety of codimension m in C.

We first examine under what conditions (4.9.3) has at least one solution belonging to V whenever $b \in C \equiv C[J,R^n]$.

Corresponding to some fixed $t_0 \in J$, define $\varphi\colon C \to C$ to be the mapping which associates with each $b \in C$ the particular solution of (4.9.3) that equals 0 at t_0; that is,

$$\varphi\big(b(t)\big) = \int_{t_0}^{t} U(t)U^{-1}(s)b(s)\,ds,$$

where $U(t)$ is a fundamental matrix of $x' = A(t)x$. Let $\Psi: R^n \to C$ be the mapping whose value at each $s \in R^n$ is the solution of $x' = A(t)x$ which equals s at t_0; that is

$$\psi(s) = U(t)U^{-1}(t_0)s.$$

Thus every solution x of (4.9.3) has the unique representation $x = \psi(s) + \varphi(b)$ where $s = x(t_0)$. Hence, it follows that (4.9.3) has a solution in V if and only if the equation

$$L \circ \psi(s) = r - L \circ \varphi(b) \tag{4.9.4}$$

has a solution in R^n.

Let $Y = L \circ \psi(R^n)$, $\Phi = -L \circ \varphi$ and define

$$B \equiv \Phi^{-1}(-r + Y).$$

Then B is either empty or a closed linear variety in C. In the latter case, every equation (4.9.3) with $b \in B$ has at least one solution in V, that is, satisfying (4.9.2).

Observe that the null space X_0 of $L \circ \psi$ is a closed linear subspace of R^n. Let X_1 be the complement of X_0 and let P be the projection of R^n onto X_0. Moreover, the restriction of $L \circ \psi$ to X_1 is an isomorphism of X_1 onto Y. In addition, there is a constant $\lambda > 0$, in which for each $z \in Y$, there is a unique $x_1 \in X_1$ such that $L \circ \psi(x_1) = z$ and $\|x_1\| \leq \lambda\|z\|$. We essentially have proved the following lemma.

LEMMA 4.9.1. If B is nonempty, there exist positive constants α, β, γ such that given any $x_0 \in X_0$, (4.9.3) has for every $b \in B$ a unique solution $x \in V$ with $Px(t_0) = x_0$ for which

$$\|x\| \leq \alpha\|x_0\| + \beta\|r\| + \gamma\|b\|. \tag{4.9.5}$$

From Lemma 4.9.1, we see that if B is nonempty, (4.9.3) induces a mapping $\sigma: (x_0,b) \to x$ of $x_0 \times B$ into V which is continuous. For, consider (x_0,b_1), $(\hat{x}_0,b_2)$ in $X_0 \times B$, then the mapping $w = \sigma(x_0,b_1) - \sigma(\hat{x}_0,b_2)$ is a solution of (4.9.3) with $b = b_1 - b_2$ such that $pw(t_0) = x_0 - \hat{x}_0$ and $L(w) = 0$. This implies from (4.9.5) that

$$(4.9.6) \qquad \|\sigma(x_0,b_1) - \sigma(\hat{x}_0,b_2)\| \leq \alpha\|x_0 - \hat{x}_0\| + \gamma\|b_1 - b_2\|.$$

As we pointed out before, if $r \in Y$, then B is a closed linear variety. This occurs if and only if $x' = A(t)x$ has at least one solution belonging to V. For $r = 0$ this is always true.

LEMMA 4.9.2. If $r \in Y$ and $L \circ \varphi \circ \psi$ is a continuous mapping of R^n onto R^n, then B is a topological direct summand of C; that is $C = A \oplus B$, where A and B are closed linear varieties.

<u>Proof</u>: Since $L \circ \varphi \circ \psi$ is a continuous linear mapping of R^n onto R^n whose null space is a topological direct summand of R^n, there exists a continuous linear injection $M: R^n \to R^n$ which is the right inverse of $L \circ \varphi \circ \psi$. Define the injection $\Psi = -\psi \circ M$ of R^n into C and let Q be the projection of R^n onto the supplement of Y. We claim that $\tau = \Psi \circ Q \circ \Phi$ is a continuous projection of C for which $\tau(B) = 0$, that is $\tau^{-1}(0) = B$. Clearly τ is continuous and is a projection because $\Phi \circ \Psi$ is the identity mapping. Moreover, $\tau(c) = 0$ for some $c \in C$, and this is equivalent to $Q \circ \Phi(c) = 0$, which is equivalent to $c \in \Phi^{-1}(Y)$. Since $r \in Y$ implies $B = \Phi^{-1}(Y)$, it follows that $\tau^{-1}(0) = B$. This completes the proof.

We now consider the differential equation (4.9.1). For convenience introduce, for every $b \in V$, the injection g_b: $t \to (t,b(t))$ of J into $J \times R^n$, so we may write $f \circ g_b$ for the continuous mapping $t \to f(t,b(t))$ of J into R^n. The constants α, β, γ will always be those referred to in Lemma 4.9.1.

A necessary condition for (4.9.1) to have a solution belonging to V is that B be nonempty and we assume this throughout.

We now attempt to obtain sufficient conditions for the existence of solutions of (4.9.1) belonging to V, that is, satisfying (4.9.2).

Suppose there is a closed ball $\mathcal{X}$ in V such that $f \circ g_b \in B$ for every $b \in \mathcal{X}$. Then (4.9.1) gives rise to the mapping

$$\Sigma: (x_0, b) \to \sigma(x_0, f \circ g_b)$$

of $X_0 \times B$ into V. Of course $\Sigma(x_0, b)$ is the unique solution x of the linear differential equation

$$x' = A(t)x + f(t,b(t)) \tag{4.9.7}$$

which belongs to V and satisfies $Px(t_0) = x_0$. The continuity of Σ follows from the continuity of the mapping $b \to f \circ g_b$ of $\mathcal{X}$ into B and the continuity of $\sigma\colon X_0 \times B \to V$. Clearly every $u \in \mathcal{X}$ for which $\Sigma(x_0, u) = u$ for some $x_0 \in X_0$ is a solution of (4.9.1) in V with $Pu(t_0) = x_0$. Thus we need conditions for which the mapping $b \to \Sigma(x_0, b)$ of $\mathcal{X}$ into V has a fixed point for some $x_0 \in X_0$.

THEOREM 4.9.1. Let $a \geq 0$, $s > 0$, $\rho > 0$ be constants such that $\alpha a + \|r\| + \gamma\rho \leq s$ and denote by $\mathcal{X}_0$ the closed ball in X_0 with center at 0 and radius a. If $b \in V$ and $\|b\| \leq r$ implies $f \circ g_b \in B$ and $\|f \circ g_b\| \leq \rho$, then the BVP (4.9.1), (4.9.2) has a solution x for each $x_0 \in \mathcal{X}_0$ with $\|x\| \leq s$ and $Px(t_0) = x_0$.

Proof: The ball $X = [b \in V: \|b\| \leq s]$ is a convex closed subset of C because V is closed. By Lemma 4.9.1 and our assumptions, we have $\Sigma(x_0, X) \subset X$ for any $x_0 \in X$. Hence the set $[v(t): v \in \Sigma(x_0, X)]$ is relatively compact in R^n. Moreover, any $v \in \Sigma(x_0, X)$ is a solution of (4.9.7) for some $b \in X$ and thus satisfies

$$(4.9.8) \qquad \|v(t) - v(t_1)\| \leq s \left| \int_{t_1}^{t} \|A(\tau)\| \, d\tau \right| + \rho |t - t_1|.$$

Hence $\Sigma(X_0, X)$ is equicontinuous and by Ascoli's theorem, relatively compact in C. By Schauder's fixed point theorem, for each $x_0 \in X_0$, there exist a function $x \in X$ such that $\Sigma(x_0, x) = x$. Hence x is a solution of the BVP (4.9.1), (4.9.2). This concludes the proof.

THEOREM 4.9.2. Let $a \geq 0$ and positive constant k, s, ρ be so chosen that $\alpha a + \beta \|r\| + \gamma \rho \leq s$ $(k\gamma < 1)$, $2ks < \rho$, and denote by H the closed ball in R^n with center at 0 and radius s. Suppose

(i) f is Lipschitzian in $I \times H$ for the constant k;

(ii) $f \circ g_b \in B$ for every $b \in V$ with $\|L\| \leq s$;

(iii) there is a $b_0 \in V$ with $\|b_0\| \leq s$ such that $\|f \circ g_{b_0}\| \leq \rho - 2kr$.

Then for each $x_0 \in X_0$ there exist a unique solution u_{x_0} of the BVP (4.9.1), (4.9.2) with $\|u_{x_0}\| \leq s$ and $Pu_{x_0}(t_0) = x_0$. Moreover the mapping $x_0 \to u_{x_0}$ is continuous in X_0.

Proof: The ball $X = [b \in V: \|b\| \leq s]$ is a complete subspace of C. Since by our assumption $\|f \circ g_b\| \leq \rho$ for every $b \in X$, by Lemma 4.9.1 we can define a sequence $\{v_n\}$

of points in X such that $v_0 = b_0$ and $v_n = \Sigma(x_0, v_{n-1})$ for $n \geq 1$. By induction, we may write $v_n \equiv u_n(x_0)$, where each $u_n: X_0 \to X$ is continuous. From (4.9.6) we observe that

$$(4.9.9) \qquad \|u_n(x_0) - u_{n-1}(x_0)\| \leq k\gamma\|u_{n-1}(x_0) - u_{n-2}(x_0)\| \leq 2r(k\gamma)^{n-1}$$

for any $x_0 \in X_0$, and this implies that u_n converges uniformly in X_0. Thus $u_{x_0} = \lim u_n(x_0)$ exists for every $x_0 \in X_0$ and $x_0 \to u_{x_0}$ is continuous in X_0. It follows immediately $u_{x_0} \in X$ and $\Sigma(x_0, u_{x_0}) = u_{x_0}$ for every $x_0 \in X_0$. Clearly u_{x_0} is unique. For if w is another solution of (4.9.1), (4.9.2) satisfying $\|w\| \leq s$ and $Pw(t_0) = x_0$, then $v = w - u_{x_0}$ is a solution of the linear equation (4.9.3) with $b = f \circ g_w - f \circ g_{u_{x_0}}$ such that $Pv(t_0) = 0$ and $T(v) = 0$. Since by Lemma 4.9.1 there is only one solution we conclude $v = \sigma(0,b)$ and this implies from (4.9.6)

$$\|w - u_{x_0}\| \leq \gamma\|f \circ g_w - f \circ g_{u_{x_0}}\| \leq k\gamma\|w - u_{x_0}\|.$$

However, $w = u_{x_0}$ because $k\gamma < 1$, thus completing the proof.

When B is a topological direct summand of C we can say more. For in this case, there exists a continuous projection τ of C with $\tau^{-1}(0) = B$ and, for every $b \in C$, a unique $h_b \in C$ such that $f \circ g_b - h_b \in B$. Here $h_b = \tau(f \circ g_b)$ so that $\|h_b\| \leq \|\tau\| \, \|f \circ g_b\|$ and $\|f \circ g_b - h_b\| \leq 2\|\tau\| \, \|f \circ g_b\|$. Thus, even if $f \circ g_b$ does not belong to B for any $b \in V$, the linear equation

$$(4.9.10) \qquad x' = A(t)x + f\big(t, b(t)\big) - h_b(t)$$

has a unique solution $x(t)$ satisfying (4.9.2) for each $x_0 \in X_0$ with $Px(t_0) = x_0$. This yields immediately the following result.

COROLLARY 4.9.1. Let B be a topological direct summand of C. Let τ be a continuous projection of C with $\tau^{-1}(0) = B$ and let $a \geq 0$, $s > 0$, $\rho > 0$ be constants such that $\alpha a + \beta\|r\| + 2\gamma\rho\|\beta\| \leq s$. If $\|f \circ g_b\| \leq \rho$ for every $b \in V$ with $\|b\| \leq s$, then there exists for each $x_0 \in X_0$ a function u satisfying (4.9.2) with $\|u\| \leq s$, $Pu(t_0) = x_0$ and

$$(4.9.11) \qquad u'(t) = A(t)u(t) + f\big(t,u(t)\big) - hu(t).$$

[Here $h(u) = \tau(f \circ g_u)$].

EXERCISE 4.9.1. State and prove a corollary for Theorem 4.9.2 in a formulation similar to Corollary 4.9.1.

We now apply our previous results to the case in which $J = [0,T]$ and (4.9.2) becomes

$$(4.9.12) \qquad Lx = x(0) - x(T) = 0.$$

Thus we are looking for a solution of (4.9.1) satisfying the periodic boundary conditions $x(0) = x(T)$. Here V is the space of periodic mappings $u \in C$ for which $u(0) = u(T)$. Choose $t_0 = 0$ and let U be the fundamental matrix of $x' = A(t)x$ with $U(0) = I$. Then the mappings $\psi\colon R^n \to C$ and $\varphi\colon C \to C$ can be written as

$$(4.9.13) \qquad \psi(x)\colon t \to U(t)x, \qquad \varphi(b)\colon t \to U(t) \int_0^t U^{-1}(s)b(s)\, ds.$$

Thus X_0 is the null space, and Y the range of the mapping $I - U(T)$. The mapping $L \circ \varphi \circ \psi$ is the bijection $x \to -TU(T)x$ so the assumptions of Lemma 4.9.2 are satisfied. Hence the set B of mappings $b \in C$ for which the linear differential equation (4.9.3) has at least one periodic solution is a topological summand of C.

A topological supplement to B may be constructed by choosing Q (in the proof of Lemma 4.9.2), for example, as the projection of R^n onto the null space Y_0 of the adjoint of $I - U(T)$, which is an orthogonal supplement to the range of $I - U(T)$. Then B is the null space of the continuous projection $\tau = \Psi \circ Q \circ \Phi$ where $\Phi: C \to R^n$ is the mapping

$$\varphi(b) \to U(T) \int_0^T U^{-1}(s)b(s)\, ds$$

and $\Psi: R^n \to C$ such that

$$\Psi(x): t \to (1/T)\, U(t)U^{-1}(T)x.$$

It follows that B is precisely the set of those $b \in C$ for which $Qy_b = 0$ where

$$y_b = (1/T)\, U(T) \int_0^T U^{-1}(s)b(s)\, ds.$$

4.10 LERAY-SCHAUDER'S ALTERNATIVE

Let us now investigate the general existence problems based on different geometric ideas. We shall employ, as the title of the section indicates, the Leray-Schauder's theory of topological degree.

DEFINITION 4.10.1. Let T_0, T_1 be completely continuous operators defined for $u \in S_\rho$, where $S_\rho = [u \in E: \|u\| \leq \rho]$, with values in a Banach space E. Then we shall say that T_0, T_1 are homotopic if there exists an operator $T(u,\lambda)$, that is completely continuous on $E\times[0,1]$ such that $T(u,0) \equiv T_0u$, $T(u,1) = T_1u$ for $u \in S_\rho$ and $T(u,\lambda) \neq u$ for

$\|u\| = \rho$.

We need the following result of Leray-Schauder.

LEMMA 4.10.1. Let T be a completely continuous operator defined on S_ρ with values in E. Suppose that T is homotopic to the operator identically equal to zero. Then there exists at least one solution u of the equation $Tu = u$ such that $\|u\| < \rho$.

Let $\lambda \in [0,1]$ and let $S(\lambda)$ denote the set of functions $x \in C^{(2)}\big[[0,1],R^n\big]$ satisfying

$$x'' = \lambda f(t,x,x'), \tag{4.10.1}$$

where $f \in C\big[[0,1] \times R^n \times R^n, R^n\big]$, and the general boundary conditions

$$x(0) - A_0 x'(0) = 0, \tag{4.10.2}$$

$$x(1) + A_1 x'(1) = 0, \tag{4.10.3}$$

A_0, A_1 being $n \times n$ matrices. Then we can prove the following result which we give as an exercise with generous hints.

EXERCISE 4.10.1. Let A_0, A_1 be positive definite or be identically zero. Suppose further there exists a constant $B > 0$ such that, if $\lambda \in [0,1]$ and $x \in S(\lambda)$, we have

$$\|x(t)\| \le B \quad \text{and} \quad \|x'(t)\| \le B, \qquad t \in [0,1]. \tag{4.10.4}$$

Then show that the set $S(1)$ is nonempty.

Hints: Recall that the trivial solution is the only solution of the homogeneous equation $x'' = 0$ which satisfies (4.10.2), (4.10.3) and that there exists a Green's matrix $G(t,s)$ such that (4.10.1) with (4.10.2) (4.10.3) is equivalent to $x(t) = \lambda \int_0^1 G(t,s) f\big(s,x(s),x'(s)\big)\, ds$. Now use the fact that the map $x(t) \to \int_0^1 G(t,s)f(s,x(s),x'(s))\,ds$ is completely continuous in the Banach space $C^{(1)}[[0,1],R^n]$ with the supremum norm and apply Lemma 4.10.1.

4.11 APPLICATION OF LERAY-SCHAUDER'S ALTERNATIVE

As an application of Leray-Schauder's alternative, we shall, in this section, prove existence of solutions of the differential system

$$x'' = f(t,x,x') \tag{4.11.1}$$

where $f \in C[[0,1]\times R^n \times R^n, R^n]$, subjected to the boundary conditions (4.10.2) and (4.10.3). Let us begin with the following result.

THEOREM 4.11.1. Assume that

(i) $f \in C[[0,1]\times R^n\times R^n,R^n]$ and A_0, A_1 are positive definite;

(ii) $V \in C^{(2)}[[0,1]\times R^n,R^+]$, $V(t,x)$ is positive definite, $g \in C[[0,1]\times R^+ \times R, R^-]$, $g(t,u,v)$ is nonincreasing in u for each (t,v) and for $(t,x) \in [0,1]\times R^n$,

$$V''_f(t,x) \geq g((t,V(t,x),\ V'(t,x)), \tag{4.11.2}$$

where

$$V'(t,x) = V_t(t,x) + V_x(t,x)x',$$

$$V''_f(t,x) = V_{tt}(t,x) + 2V_{tx}(t,x)\cdot x' + V_{xx}(t,x)x'\cdot x' + V_x\cdot f(t,x,x')$$

and

$$V_{tt}(t,x) + 2V_{tx}(t,x)\cdot x' + V_{xx}(t,x)x'\cdot x' \geq 0$$

for $(t,x,x') \in [0,1]\times R^n\times R^n$;

(iii) the conditions (4.10.2), (4.10.3) imply that

$V'(0,x(0)) \geq 0$, $V'(1,x(1)) \leq 0$ and $V(0,x(0)) \leq \alpha V'(0,x(0))$ for some $\alpha > 0$.

(iv) $G \in C[[0,1] \times R^+, R]$ and there exists an $L > 0$ such that for $u \geq L$, $t \in [0,1]$,

$$(1/u)\, g(t,u,v) - (v/u)^2 \geq G(t,v/u)$$

and for any $\tau \in (0,1]$, the left maximal solution $r(t,\tau,0)$ of

$$z' = G(t,z), \qquad z(\tau) = 0$$

satisfies the inequality $r(t,\tau,0) < \alpha_0$, $t \in [0,\tau]$, where $\alpha_0 = \min(\frac{1}{2}, 1/\alpha)$;

(v) the left maximal solution $r(t,1,0)$ and the right minimal solution $\rho(t,0,0)$ of

$$v' = g(t,2L,v)$$

exists on $[0,1]$.

Assume also that for each solution $x_0(t)$ of (4.11.1) either $x_0(t)$ is defined for all $t \in [0,1]$ or $\|x_0(t)\|$ is unbounded. Then there exists a solution of (4.11.1) satisfying (4.10.2) and (4.10.3).

Proof: By Lemma 1.14.1 and the positive definiteness of V, we have for each $(\lambda,x) \in S$,

$$\|x(t)\| \leq B^*, \qquad 0 \leq t \leq 1. \tag{4.11.3}$$

We shall show that S is compact in $[0,1] \times C^{(1)}[[0,1],R^n]$ with respect to norm $\|\varphi\|_1 = \max[\|\varphi\|_0, \|\varphi'\|_0]$, where $\|\varphi\|_0 = \sup_{0 \leq t \leq 1} \|\varphi(t)\|$ and $\varphi \in C^{(1)}[[0,1],R^n]$. If not, there exists a sequence $\{\lambda_i, x_i(\cdot)\}$ with no limit in S. Since $\|x_i(0)\| \leq B^*$, we may assume that λ_i and $x_i(0)$ are convergent to some λ_0 and x_0 as $i \to \infty$. Let $y_i = x_i'(0)$ for all i. Since A_0 is assumed to be positive definite,

$y_i = A_0^{-1}x_i(0)$, by (4.10.2), which converges to some $y_0 = A_0^{-1}x_0$. From the standard convergence theorem, there is a solution $x_0(t)$ of (4.10.1) with $\lambda = \lambda_0$ defined on an interval $I = [0,T) \subset [0,1]$ such that

(a) $x_0(0) = x_0$, $x'(0) = y_0$ and either $I = [0,T]$ or $x_0(t)$ cannot be defined continuously on any larger interval as well as at T;

(b) $x_i(t) \to x_0(t)$ and $x_i'(t) \to x_0'(t)$ uniformly, as $i \to \infty$, on I.

By (4.11.3), we must have $\|x_0(t)\| \leq B^*$, $t \in I$. Since $x_0(t)$ is bounded, the hypothesis of the theorem implies $I = [0,1]$. At $t = 1$, applying (b) above, it follows that $x_0(t)$ satisfies (4.10.3). Hence $(\lambda_0, x_0) \in S$, contradicting our assumption that (λ_i, x_i) has no limit point in S. The set S is therefore compact as claimed.

We now consider the derivative evaluation function $F: [0,1] \times S \to R^n$ given by $F(t,x(t)) \to x'(t)$. This function is continuous and its domain is compact. Hence $\|F\|$ is bounded by some $B_1 > 0$, which, in turn, implies that

$$\|x'(t)\| \leq B_1, \qquad 0 \leq t \leq 1 \text{ and } (\lambda,x) \in S.$$

Let $B = \max[B^*, B_1]$. An application of Exercise 4.10.1 proves the stated result.

Notice that the assumption A_0 is positive definite is crucial in the proof of Theorem 4.11.1. Whether the result is true, if A_0, A_1 are identically zero, remains an open question.

COROLLARY 4.11.1. Assume that $f \in C[[0,1] \times R^n \times R^n, R^n]$ and for some $k \geq 0$,

$$\|y\|^2 + x \cdot f(t,x,y) \geq -k[1 + \|x\| + |x \cdot y|]$$

for $(t,x,y) \in [0,1] \times R^n \times R^n$. Let the matrices A_0, A_1 be positive definite. Suppose that for each solution $x_0(t)$ is defined for all $t \in [0,1]$ or $\|x_0(t)\|$ is unbounded. Then the BVP (4.11.1), (4.10.2), and (4.10.3) has a solution.

Next we shall give another proof of Theorem 1.14.1 which depends on Leray-Schauder's alternative.

Alternate proof of Theorem 1.14.1. Let

$$S = \big[(\lambda,x)\colon \lambda \in [0,1] \quad \text{and} \quad x \in S(\lambda)\big],$$

where $S(\lambda)$, as in Section 4.10, denotes the set of functions x satisfying (4.10.1), (4.10.2), and (4.10.3). The set S is nonempty, since if $x \equiv 0$, then $(0,x) \in S$. Let $(\lambda,x) \in S$ and define $m(t) = V\big(t,x(t)\big)$, so that because of assumption (c), we have (1.14.16) as before. Also since $(\lambda,x) \in S$, $0 \le \lambda \le 1$, $g(t,u,v) \le 0$ and $U(t,x,x') \ge 0$, it is easily verified, as in the previous case, that

$$V''_{\lambda f}(t,x) \ge g\big(t,V(t,x),V'(t,x)\big) + \sigma\|\lambda f(t,x,x')\|$$

which again leads to the differential inequality (1.14.18). The rest of the proof is exactly the same as that of the proof of Theorem 1.14.1 as given in Section 14 of Chapter 1, until we arrive at the inequality (1.14.19). The stated conclusion follows, in the present case, by Exercise 4.10.1 and this completes the proof.

COROLLARY 4.11.2. Let hypothesis (a) of Theorem 1.14.1 hold. Assume that for some $k > 0$ and $\sigma > 0$,

$$\|y\|^2 + x\cdot f(t,x,y) \ge -k\big[1 + \|x\| + |x\cdot y|\big] + \sigma\|f(t,x,y)\|$$

for $(t,x,y) \in [0,1] \times R^d \times R^d$. Then there exists a solution for the boundary value problem (4.11.1), (4.10.2), and (4.10.3).

COROLLARY 4.11.3. Assume that hypothesis (a) of Theorem 1.14.1 is satisfied. For some $k > 0$, $\sigma > 0$, let

$$\|y\|^2 + x \cdot f(t,x,y) \geq -k[1 + \|x\| + |x \cdot y|] + \sigma \|y\|$$

for $(t,x,y) \in [0,1] \times R^d \times R^d$. Then, if f satisfies Nagumo's condition, there exists a solution for the problem (4.11.1), (4.10.2), and (4.10.3).

We can also give another proof of Theorem 1.14.2.

Alternate proof of Theorem 1.14.2. Let $S, (\lambda, x)$ and $m(t)$ be as in the alternate proof of Theorem 1.14.1. Distinguishing the two cases $\sigma \geq 1$ and $0 < \sigma < 1$, and arguing as in the proof of Theorem 1.14.2, we obtain

$$V''_{\lambda f}(t,x) \geq g\big(t, V(t,x), V'(t,x)\big) - 1 + \|x'\|$$

or

$$V''_{\lambda f}(t,x) \geq g\big(t, V(t,x), V'(t,x)\big) - 1 + \sigma \|x'\|.$$

These inequalities imply the further inequalities

$$m''(t) \geq -(N + 1) + \|x'(t)\|, \qquad m''(t) \geq -(N + 1) + \sigma \|x'(t)\|$$

and consequently, as before, we arrive at $\|x'(t)\| \leq \gamma(M)$, $0 \leq t \leq 1$, using Lemma 1.12.1. This then implies the estimates (1.14.19) and Exercise 4.10.1 then concludes the proof.

4.12 PERIODIC BOUNDARY CONDITIONS

This section provides sufficient conditions for the solutions of the second-order differential system

$$x'' = f(t,x,x') \tag{4.12.1}$$

satisfying the periodic boundary conditions

$$x(0) = x(T), \qquad x'(0) = x'(T). \tag{4.12.2}$$

A solution of the BVP (4.12.1), (4.12.2) will be called a <u>periodic solution</u>.

The approach is to establish existence results for the boundary condition

$$(4.12.3) \qquad x(0) = y = x(T)$$

and then study the vector field

$$(4.12.4) \qquad U(y) = x'(0,y) - x'(T,y),$$

where $x(t,y)$ is the unique solution of (4.12.1), (4.12.3). To solve the BVP (4.12.1), (4.12.2), it is sufficient to prove the existence of a y such that $U(y) = 0$. We shall assume that $f \in C[[0,T] \times R^n \times R^n, R^n]$.

LEMMA 4.12.1. Let D_n be the closed n-ball with radius one, that is, $D_n = [x: \|x\| \leq 1]$. Assume for each $y \in D_n$, (4.12.1), (4.12.3) has a unique solution $x(t;y)$. Further, let there exist a constant N [depending on D_n^0 (interior of D_n) and f] such that

$$|x(t,y)| \leq N, \qquad |x'(t,y)| \leq N \qquad \text{for} \quad 0 \leq t \leq T, \quad y \in D_n.$$

Then $x(t,y)$ and $x'(t,y)$ are continuous in y for each $t \in I$.

EXERCISE 4.12.1. Prove Lemma 4.12.1 by applying Ascoli's theorem.

Before introducing our main result, we need the following information on degree theory. Because the ideas of degree theory are strongly connected with algebraic topology we shall only mention, without proof, the essential theory needed.

The idea is to obtain for each $f: S^n \to S^n$ an integer (positive, negative, or zero) called its <u>degree</u>. Here S^n is the boundary of D_{n+1}. We shall always assume f is continuous.

For $n = 1$ the degree of $f: S^1 \to S^1$ is simply the number of times the image point $f(z)$ rotates around S^1 when z performs one oriented rotation of S^1. In particular, for each $k = 0, \pm 1, \pm 2, \ldots,$ the map $z \to z^k$ has degree k.

The definition of the degree of $f: S^n \to S^n$ for $n > 1$ is a generalization of the case $n = 1$. We shall omit the details. Denote the degree of f by $D(f)$.

Let X, Y be two spaces and $I = [0,1]$. Recall that two continuous maps $f,g: X \to Y$ are called <u>homotopic</u> if there exists a continuous $\Phi: X \times I \to Y$ such that $\Phi(x,0) = f(x)$, $\Phi(x,1) = g(x)$ for each $x \in X$ and $\varphi(x,\cdot)$ never vanishes.

LEMMA 4.12.2. Let $n \geq 0$. If $f,g: S^n \to S^n$ are homotopic, then $D(f) = D(g)$. Let $n \geq 0$ and $f: S^n \to S^n$ have the property that $f(-x) = -f(x)$. Then $D(f)$ is odd; in particular, there exists an $x \in D_{n+1}$ such that $f(x) = 0$.

We now state our main result.

THEOREM 4.12.1. Let the hypotheses of Lemma 4.12.1 hold. Assume that Ω is a convex homeomorphic image of D_n^0 and let $\bar{\Omega}$ be symmetric about a point $z \in \Omega$. For each $y \in \bar{\Omega}$ define $U(y) = x'(0,y) - x'(T,y)$. Let A be the continuous mapping of $\partial\Omega$ (the boundary of Ω) onto itself which maps each $y \in \partial\Omega$ onto the point which is symmetric to y with respect to z. Further assume that for all $y \in \partial\Omega$ for which $U(y) \neq 0$, $U(y)$ and $U(Ay)$ do not have the same direction [that is, there exists no $c > 0$ such that $U(y) = cU(Ay)$]. Then there exists a solution of the BVP (4.12.1), (4.12.2).

<u>Proof</u>: If there exists a $y \in \partial\Omega$ such that $U(y) = 0$, then the proof is complete. Assume therefore that the vector field U does not vanish on $\partial\Omega$. Since $\bar{\Omega}$ is convex and symmetric about z, there exists a homeomorphism g of D_n

onto $\overline{\Omega}$ such that

$$g(0) = z, \qquad g^{-1}Ag(r) = -r,$$

for all $r \in \partial D_n$, that is, $g^{-1}Ag$ is the antipodal map of S^{n-1} onto itself.

Define the vector field φ on D_n by

$$\varphi(r) = U(g(r)).$$

By Lemma 4.12.1, U is continuous, thus implying φ is continuous. Since U does not vanish on $\partial\Omega$, φ will not vanish on ∂D_n. Observe that

$$\varphi(r) = U(g(r))$$

and

$$\varphi(-r) = U(g(-r)) = U(Ag(r)).$$

From the hypothesis on U, we see that $\varphi(r)$ and $\varphi(-r)$ will have different directions on ∂D_n; that is

$$\varphi(-r)/|\varphi(-r)| \neq \varphi(r)/|\varphi(r)|.$$

Hence the vector field

$$\psi(r,\lambda) = \varphi(r)/|\varphi(r)| - \lambda\, \varphi(-r)/|\varphi(-r)|$$

never vanishes on ∂D_n for $0 \leq \lambda \leq 1$. Since $\psi(r,0)$ and $\psi(r,1)$ are homotopic, their topological degrees are identical by Lemma 4.12.2. Since $\psi(-r,1) = -\psi(r,1)$ we have by Lemma 4.12.3 that the degree of $\psi(r,1)$ is an odd integer. Hence the degree of $\psi(r,0)$ is an odd integer which implies there exists $\overline{r} \in D_n^0$ such that $\psi(\overline{r},0) = 0$. This implies $\varphi(\overline{r}) = 0$ and hence $U(g(\overline{r})) = 0$ where $g(\overline{r}) \in \Omega$. This proves the existence of a solution of the BVP (4.12.1), (4.12.2).

In our first application, we shall assume for each $M > 0$, there exists an $N > 0$ such that whenever $x(t)$ is a solution of (4.12.1) defined on $I = [0,T]$ with $\|x(t)\| \leq M$, then $\|x'(t)\| \leq N$. Further there exists a constant $p \geq N$ such that if $x(t)$ is a solution defined on I of the perturbed equation

$$(4.12.5) \qquad x'' = f(t,x,x') + \varepsilon x, \qquad 0 < \varepsilon \leq \varepsilon_0,$$

with $\|x(t)\| \leq M$, then $\|x'(t)\| \leq p$.

We further adopt the following convention. If x, $\bar{x}$, y, $\bar{y}$ are vectors in R^n, then let $\Delta x = x - \bar{x}$, $\Delta y = y - \bar{y}$, $\Delta f = f(t,x,y) - f(t,\bar{x},\bar{y})$.

THEOREM 4.12.2. Let there exist a positive constant R such that

$$(4.12.6) \qquad x \cdot f(t,x,y) + \|y\|^2 \geq 0 \qquad \text{if} \quad x \cdot y = 0,$$

$\|x\| = R$, and

$$(4.12.7) \qquad \Delta x \cdot \Delta f + \|\Delta y\|^2 \geq 0$$

for any x, $\bar{x} \in R^n$, $\|x\|$, $\|\bar{x}\| \leq R$, $x \neq \bar{x}$ and y, $\bar{y} \in R^n$ with $\Delta x \cdot \Delta y = 0$. Then there exists a solution $x(t)$ of (4.12.1), (4.12.2) with $\|x\| \leq R$.

Proof: Let $\varepsilon > 0$ be given with $\varepsilon \leq \varepsilon_0$. Consider the perturbed equation (4.12.5) and let $F(t,x,x',\varepsilon) \equiv f(t,x,x') + \varepsilon x$. Then, from (4.12.5), it follows that

$$(4.12.8) \qquad x \cdot F + \|y\|^2 = x \cdot f + \varepsilon\|x\|^2 + \|y\|^2 \geq \varepsilon\|x\|^2 > 0$$

if $x \cdot y = 0$ and $\|x\| = R$.

Furthermore, from (4.12.7), we observe that

$$(4.12.9) \qquad \Delta x \cdot \Delta F + \|\Delta y\|^2 = \Delta x \cdot \Delta f + \varepsilon\|\Delta x\|^2 + \|\Delta y\|^2 \geq \varepsilon\|\Delta x\|^2 > 0,$$

if $\Delta x \cdot \Delta y = 0$ and $\Delta x \neq 0$.

We have seen in Section 1.14 that conditions (4.12.8) and (4.12.9) imply that the BVP

$$x'' = F(t,x,x',\varepsilon) \tag{4.12.10}$$

$$x(0) = z = x(T) \tag{4.12.11}$$

has a unique solution $x(t,z,\varepsilon)$ with

$$\|x(t,z,\varepsilon)\| \leq R, \tag{4.12.12}$$

for any z, $\|z\| \leq R$.

Let $\Omega = [x\colon \|x\| < R]$. Then, by assumption there exists a $p > 0$ such that $\|x'(t,z,\varepsilon)\| \leq p$ for any $z \in \bar{\Omega}$. Thus by Lemma 4.12.1, the vector field

$$U(z,\varepsilon) = x'(0,z,\varepsilon) - x'(T,z,\varepsilon)$$

is continuous on $\bar{\Omega}$. Again assume that $U(z,\varepsilon)$ does not vanish for $z \in \partial\Omega$. Letting $r(t) = \frac{1}{2}\|x(t,z,\varepsilon)\|^2$, we find

$$r'(t) = x(t,z,\varepsilon)\cdot x'(t,z,\varepsilon), \tag{4.12.13}$$

and

$$r''(t) = x\cdot F + \|x\|^2. \tag{4.12.14}$$

From (4.12.8) and (4.12.12) - (4.12.14), it follows that

$$r'(0) < 0 < r'(T)$$

for any $z \in \partial\Omega$ which implies $z\cdot U(z,\varepsilon) < 0$ for $z \in \partial\Omega$. Therefore,

$$U(z,\varepsilon)/\|U(z,\varepsilon)\| \neq U(-z,\varepsilon)/\|U(-z,\varepsilon)\|, \qquad z \in \partial\Omega,$$

that is, $U(z,\varepsilon)$ and $U(-z,\varepsilon)$ cannot have the same direction. By Theorem 4.12.1, we conclude $U(z,\varepsilon)$ will have a zero in Ω for every ε, $0 < \varepsilon < \varepsilon_0$, that is, there exists a solution $x(t,\varepsilon)$ of the BVP (4.12.10), (4.12.11). By a standard application of Ascoli's theorem, there exists a sequence $\varepsilon_i \to 0$

such that $x(t,\varepsilon_i) \to \bar{x}(t)$ as $i \to \infty$ and $\bar{x}(t)$ is a solution of the BVP (4.12.1), (4.12.2). This completes the proof of Theorem 4.12.1.

We now use the theory of differential inequalities discussed in Chapter 1 together with Theorem 4.12.1 to derive some results. We shall assume that in (4.12.1) f is independent of x'. In R^n, consider the usual partial ordering $x \leq y$ if and only if $x_i \leq y_i$, $i = 1,\ldots,n$ and $x < y$ if and only if $x_i < y_i$, $i = 1,\ldots,n$. Recall that a function $\alpha \in C^{(2)}[I,R^n]$ is called a <u>lower solution</u> of (4.12.1) if

$$\alpha''(t) \geq f(t,\alpha(t)), \qquad t \in I. \tag{4.12.15}$$

Similarly, $\beta \in C^{(2)}[I,R^n]$ is called an <u>upper solution</u> if

$$\beta''(t) \leq f(t,\beta(t)), \qquad t \in I. \tag{4.12.16}$$

Further, we will assume that $f(t,x)$ is quasimonotone increasing in x.

THEOREM 4.12.3. Let there exist lower and upper solutions α and β of (4.12.1) with $\alpha(t) \leq \beta(t)$, $t \in I$ and let f be quasimonotone increasing in x on the set

$$w = \{(t,x): \alpha(t) \leq x \leq \beta(t),\ t \in I\}.$$

Moreover, let α and β be such that

$$\begin{aligned} \alpha(0) &= \alpha(T), & \beta(0) &= \beta(T), \\ \alpha'(0) &\geq \alpha'(T), & \beta'(0) &\leq \beta'(T). \end{aligned} \tag{4.12.17}$$

Further, assume that for every y, with $\alpha(0) \leq y \leq \beta(0)$ the BVP

$$x'' = f(t,x), \tag{4.12.18}$$

and (4.12.3), has at most one solution $x(t)$ such that $(t,x(t)) \in w$. Then there exists a solution of the BVP (4.12.18), (4.12.2).

<u>Proof</u>: Let $[\alpha,\beta] = [y: \alpha(0) \leq y \leq \beta(0)]$. Then by Theorem 1.11.1 and the hypotheses of Theorem 4.12.3 there exists a unique solution $x(t,y)$ of (4.12.18) such that

$$x(0,y) = x(T,y)$$

and $\alpha(t) \leq x(t,y) \leq \beta(t)$ for every $y \in [\alpha,\beta]$. Further there exists an $N > 0$ depending on $[\alpha,\beta]$ such that $\|x(t,y)\|$, $\|x'(t,y)\| \leq N$ for all $y \in [\alpha,\beta]$. The vector field

$$U(y) = x'(0,y) - x'(T,y)$$

is continuous on $[\alpha,\beta]$. If $\alpha(0) = \beta(0)$, it follows from (4.12.17), that $U(\alpha(0)) = 0$ and the proof is complete. In fact, if there exists an i, $1 \leq i \leq n$ such that $\alpha_i(0) = \beta_i(0)$, then it follows again from (4.12.17) that $(U(y))_i$ is zero for all $y \in [\alpha,\beta]$. Therefore we need only to consider the components $(U(y))_j$ for those j in which $\alpha_j(0) < \beta_j(0)$. We assume now $\alpha_j(0) < \beta_j(0)$ for $j = 1,\ldots,n$, the contrary situation will follow by using a similar argument in a lower dimensional setting.

Thus let $\Omega = [y: \alpha(0) < y < \beta(0)]$. If there exists $y \in \partial\Omega$ such that $U(y) = 0$, the proof is done. Otherwise assume U does not vanish on $\partial\Omega$. Let A be the mapping on $\partial\Omega$ mapping y into the point symmetric to y about $[\alpha(0) + \beta(0)]/2$.

Let h_i be a continuous function on $[\alpha_i(0),\beta_i(0)]$ such that $h_i(\beta_i(0)) < 0 < h_i(\alpha_i(0))$ and let $\varepsilon > 0$ be given. Consider the vector field defined for $y \in \overline{\Omega}$ given by

$$\Phi(y,\varepsilon) = U(y) + \varepsilon(h_i(y_i),\ldots,h_n(y_n)).$$

Then if y^* is on the face $y_i = \alpha_i(0)$, we have

$$x_i'(0,y^*) \geq \alpha_i'(0) \geq \alpha_i'(T) \geq x_i'(T,y^*)$$

and thus $(U(y^*))_i \geq 0$; therefore

$$\Phi(y^*,\varepsilon)_i = U(y^*)_i + \varepsilon h_i(\alpha(0)) > 0,$$

and if, $\bar{y} = Ay^*$, then $\bar{y}$ lies on the face $y_i = \beta_i(0)$ and we obtain

$$\Phi(\bar{y},\varepsilon)_i = U(\bar{y})_i + \varepsilon h_i(\beta_i(0)) \leq \varepsilon h_i(\beta_i(0)) < 0.$$

Thus $\Phi(y,\varepsilon)$ and $\Phi(Ay,\varepsilon)$ do not have the same direction, which implies $\Phi(y,\varepsilon)$ must have a zero in Ω. Pick a sequence $\varepsilon_1 > \varepsilon_2 > \varepsilon_3 \cdots$ converging monotonically to zero and let y^n be a sequence of $\Phi(y,\varepsilon_n)$ in Ω. This sequence has a convergent subsequence converging to a zero of $U(y)$ in $\bar{\Omega}$. This concludes the proof of Theorem 4.12.3.

COROLLARY 4.12.1. Assume that there exists constant vectors α and β, $\alpha < \beta$ such that

$$f(t,\alpha) \leq 0 \leq f(t,\beta)$$

and f is quasimonotone increasing on $\{x: \alpha \leq x \leq \beta\}$ for each $t \in I$. Let $f(t,x)$ satisfy

$$\|f(t,x) - f(t,y)\| \leq L\|x - y\|$$

for some $L > 0$ where $\alpha \leq x$, $y \leq \beta$. Then the BVP (4.12.1), (4.12.2) has a solution provided $L < 8/T^2$.

EXERCISE 4.12.2. Prove Corollary 4.12.1.

<u>Hint</u>: Apply Theorem 4.12.3 by showing there exist at most one solution of (4.12.18) and (4.12.3). This may be done by setting up an integral equation and showing the operator is a contraction mapping with constant $T^2L/8$.

EXAMPLE 4.12.1. Consider the two-dimensional second-order system

$$(4.12.19) \qquad x'' = x^3 - y + p(t), \qquad y'' = -x + y^3 + q(t),$$

where $|p(t)| \leq a$, $|q(t)| \leq a$. Choose $b > 0$ so that $b^3 - b - a \geq 0$. Then letting $\beta = (b,b)$, $\alpha = (-b,-b)$, we find that β is an upper solution and α is a lower solution of (4.12.19). The right-hand side of (4.12.19) is quasimonotone increasing in x and y for $-b \leq x, y \leq b$ and is Lipschitz-continuous in the region. Thus, we may conclude for T sufficiently small the hypotheses of Corollary 4.12.1 are satisfied. Thus there exists a solution of (4.12.19), (4.12.2) for T sufficiently small.

4.13 SET-VALUED MAPPINGS AND FUNCTIONAL EQUATIONS

We wish to present here an existence and uniqueness result for nonlinear functional equations, which will be stated in terms of the theory of set-valued mappings. The approach is topological in nature and it permits us to establish the existence of solutions provided a criterion of uniqueness is fulfilled.

Let E be a Banach space and let $n(E)$ denote the family of all nonempty subsets of E. For a set A in $n(E)$, a mapping $H: A \to n(E)$ is called <u>upper semicontinuous</u> if its graph $[(x,y): y \in H(x)]$ is closed in $A \times E$. The map H is said to be compact if, for any bounded subset B of A, the closure of the set $\bigcup_{x \in B} H(x)$ is compact in E. The map H is called <u>completely continuous</u> if it is upper semicontinuous and compact.

For a single-valued mapping $h: A \to E$, the upper semi-continuity of the mapping $H: x \to h(x)$ is equivalent to the continuity of h, the compactness of H is equivalent to that of h and the complete continuity of H means the complete continuity of h.

We now state our fixed point theorem.

THEOREM 4.13.1. Let U be a neighborhood of 0 in the space E and let $H: U \to n(E)$ be a completely continuous mapping such that

(4.13.1) $$x \in H(x),\ x \in U \ \text{implies}\ x = 0.$$

Then, for any continuous mapping $h: E \to E$, the condition

(4.13.2) $$h(x) - h(y) \in H(x - y) \quad \text{for} \quad x - y \in U$$

implies that the equation

(4.13.3) $$x = h(x)$$

has exactly one solution.

Proof: Denote by $\mathcal{U}(x)$ the neighborhood of x defined by $\mathcal{U}(x) = \mathcal{U} + x = [y + x: y \in \mathcal{U}]$ and by $K(x)$ the ball with center at x and of radius ε chosen in such a way that $\mathcal{U}(x)$ contains the ball with center at x of radius 2ε.

Assumption (4.13.1) implies that the mapping $T = I - h$, where I denotes the identity mapping, is one-to-one when restricted to the neighborhood $\mathcal{U}(x)$, for every $x \in E$, which, in its turn, shows that

(4.13.4) $$T(x_1) = T(x_2),\quad x_1 \neq x_2 \Rightarrow K(x_1) \cap K(x_2) = \emptyset.$$

Let $S(x)$ denote the boundary of the ball $K(x)$. We claim that there exists a $\delta > 0$, independent of x, such that

(4.13.5) $$y \in S(x) \Rightarrow \|T(y) - T(x)\| > \delta.$$

Suppose that it is false. Then there exist sequences $\{x_n\}$, $\{y_n\}$ such that $\lim_{n\to\infty} \big(T(y_n) - T(x_n)\big) = 0$ and $\|y_n - x_n\| = \varepsilon$ for $n = 1,2,\ldots$. Setting $r_n = T(y_n) - T(x_n)$, we have

$$y_n - x_n - r_n = h(y_n) - h(x_n), \qquad n = 1,2,\ldots,$$

and consequently, by (4.13.2),

(4.13.6) $$y_n - x_n - r_n \in H(y_n - x_n), \qquad n = 1,2,\ldots .$$

Since by assumption the closure of the set $\bigcup_{\|x\| \leq \varepsilon} H(x)$ is compact, we may suppose that the sequence $\{y_n - x_n - r_n\}$ or, which means the same, the sequence $\{y_n - x_n\}$ is convergent to an element z, such that $\|z\| = \varepsilon$. Relation (4.13.6) and the upper semicontinuity of H imply that $z \in H(z)$, which is a contradiction to (4.13.1).

From the condition (4.13.2), it follows that the continuous mapping h is completely continuous. Thus, by (4.13.5), for every $x \in E$, we have

(4.13.7) $$[y: \|y - T(x)\| \leq \delta \subset T(k(x))].$$

This shows, in turn, that the set $T(E)$ is open in E.

It is easily seen that $T(E) = E$. Indeed, suppose that $T(E) \neq E$, and let y be an arbitrary element on the boundary of $T(E)$. Since $T(E)$ is open, y does not belong to $T(E)$. On the other hand, for a point $T(x)$ lying in the neighborhood of y of radius δ, we see, by (4.13.7), that $y \in T(E)$.

Relations (4.13.4), (4.13.7) and the fact $T(E) = E$ imply that the pair (E,T) is a covering space for E. Since E is simply connected, T is a homeomorphism, and this completes the proof of the theorem.

REMARK 4.13.1. It is easy to see that if a mapping h satisfies condition (4.13.2), then for every $a \in E$, so does the mapping $h_a: x \to h(x) - a$. Therefore, Theorem 4.13.1 may be stated in an equivalent form as follows: the mapping $I - h$ is a homeomorphism of E onto itself.

If h is a mapping of the form $h(x) = Ax + b$, where $b \in E$ and $A: E \to E$ is a completely continuous linear operator, Theorem 4.13.1 yields the essential part of the first theorem of

Fredholm. It suffices to set $H(x) = \{Ax\}$ and to observe that condition (4.13.1) means the uniqueness of solution of the linear homogeneous equation $x = Ax$. We also note that in the Fredholm theorem the map A may be noncontractive in general.

The assumption that the map h is continuous, in Theorem 4.13.1 may be dropped by strengthening the conditions imposed on the mapping H. This is the content of the next theorem.

THEOREM 4.13.2. Let U be a neighborhood of 0 in the space E and let $H: U \to n(E)$ be a completely continuous mapping such that the implication (4.13.1) holds true and

$$H(0) = 0. \tag{4.13.8}$$

Then for any mapping $h: E \to E$ satisfying (4.13.2), Eq. (4.13.3) has exactly one solution.

Proof: It is enough to show that h is necessarily continuous. To this end, assume that a sequence $\{x_n\}$ is a convergent to x_0. Then, for n sufficiently large, we have

$$h(x_n) - h(x_0) \in H(x_n - x_0). \tag{4.13.9}$$

Suppose now that $\{h(x_n)\}$ does not converge to $h(x_0)$. By the compactness of H, there exists a subsequence $\{h(x_{k_n})\}$ which converges to $y_0 \neq h(x_0)$. On passing to the limit in (4.13.9), we see that

$$y_0 - h(x_0) \in H(0)$$

and this implies, by (4.13.8), that $y_0 - h(x_0) = 0$. This contradiction proves the theorem.

We wish to point out that without the assumption (4.13.8), Theorem 4.13.2 is not true. For example, in R^2 let

$$H(x) = \begin{cases} [\frac{3}{4}x, 1] & \text{for } x > 0, \\ [-1, 1] & \text{for } x = 0, \\ [-1, \frac{3}{4}x] & \text{for } x < 0, \end{cases}$$

and

$$h(x) = \frac{3}{4}(x - \tfrac{1}{2} - [x - \tfrac{1}{2}]).$$

It is easy to verify that all the assumptions of Theorem 4.13.2 are satisfied except (4.13.8). The function h, however, is not continuous and the equation $x = h(x)$ does not have any solution.

4.14 GENERAL LINEAR PROBLEMS

As an application of the topological method developed in the preceding section, we shall consider the question of existence of solutions of general linear problems. In general, one compares the linear problem under consideration with another homogeneous problem suitably chosen and from the uniqueness of solutions of the second, concludes the existence of solutions of the first. In the present case this comparison problem involves either an equation with set-valued right-hand side or some differential inequality.

Let $cf(R^n)$ denote the family of all nonempty, closed, and convex subsets of R^n. For $A \in cf(R^n)$, we set $\|A\| = \sup[\|p\|: p \in A]$. Let J be a compact interval of R and let $C = C[J, R^n]$ with the norm of uniform convergence, $\|x\| = \max_{t \in J} \|x(t)\|$. Let us assume that

(1) F is a mapping of $J \times R^n$ into $cf(R^n)$. For each $t \in J$, $F(t,x)$ is upper semicontinuous with respect to x, and for each compact set $K \subset R^n$ there exists a function $m(t)$, summable on J, such that

$$\|F(t,x)\| \leq m(t), \qquad t \in J, \quad x \in K; \tag{4.14.1}$$

(2) f is a mapping of $J \times R^n$ into R^n. For each $t \in J$, $f(t,x)$ is continuous relative to x, and for fixed $x \in R^n$, $f(t,x)$ is summable with respect to t;

(3) L is a linear continuous mapping of C into R^n. Given the mappings F, f, and L, we consider the equation

$$x'(t) \in F(t, x(t)) \tag{4.14.2}$$

with the homogeneous linear condition

$$Lx = 0 \tag{4.14.3}$$

and the ordinary differential equation

$$x'(t) = f(t, x(t)) \tag{4.14.4}$$

with the linear condition

$$Lx = r, \qquad r \in R^n. \tag{4.14.5}$$

An absolutely continuous function $x \in C$ will be called a <u>solution</u> of (4.14.2) [respectively, of (4.14.4)] if it satisfies (4.14.2) [respectively, (4.14.4)] almost everywhere on J.

THEOREM 4.14.1. Assume that the functions F, f, and L satisfy conditions (1) - (3) and for $t \in J$, $p, q \in R^n$,

$$f(t,q) - f(t,p) \in F(t, q-p). \tag{4.14.6}$$

Suppose further that the function $x \equiv 0$ is the unique solution of (4.14.2) satisfying (4.14.3). Then for every $r \in R^n$, there exists one and only solution of (4.14.4) satisfying (4.14.5).

<u>Proof</u>: We wish to apply Theorem 4.13.1. Consider the mapping H of $E = C \times R^n$ into $cf(R^n)$ such that for every point (x,p) its image $H(x,p)$ is a set of all pairs (y,q) given by the formula

$$y(t) = \int_a^t u(s)\, ds + p, \qquad q = p - Lx$$

for $u(s) \in F(s,x(s))$ and the mapping h of E into itself such that for every point (x,p) its image $h(x,p)$ is a pair (y,q) given by

$$y(t) = \int_a^t f(s,x(s))\, ds + p, \qquad q = p - Lx + r$$

where a is a fixed point in J. From (4.14.1), (4.14.6), it follows that for every compact set $K \subset R^n$, the function $\sup|\,\|f(t,p)\|\colon p \in K|$ is bounded by a summable function $m(t) + \|f(t,0)\|$.

It is easy to see that the maps H, h satisfy conditions (4.13.1), (4.13.2) and that the existence of solutions of problem (4.14.4), (4.14.5) is equivalent to the existence of solutions of the functional equation (4.13.3). Evidently the map H is compact and the map h is continuous. Thus to complete the proof, it is sufficient to show that H is upper semicontinuous.

Suppose that

$$(y_k, q_k) \in H(x_k, p_k), \quad (x_k, p_k) \to (x,p), \quad (y_k, q_k) \to (y,q) \quad \text{as} \quad k \to \infty.$$

Then we have

$$(4.14.7) \qquad y_k(t) = \int_a^t u_k(s)\, ds + p_k, \qquad u_k(t) \in F(t, x_k(t)),$$
$$q_k = p_k - Lx_k$$

and consequently

$$y_k'(t) \in F(t, x_k(t)).$$

This, together with the upper semicontinuity of F, implies

$$\lim_{k\to\infty} \rho\big(y_k'(t),\ F(t,x(t))\big) = 0,$$

where $\rho(p,A)$ is the distance function. By Lemma 4.14.1 (which follows below) it then follows that

$$y'(t) \in F\big(t,x(t)\big). \tag{4.14.8}$$

On passing to the limit in (4.14.7), we obtain, for $t = a$,

$$y(a) = p, \qquad q = p - Lx \tag{4.14.9}$$

and, as a result,

$$y(t) = \int_a^t y'(s)\,ds + p. \tag{4.14.10}$$

From relations (4.14.8), (4.14.9), and (4.14.10) it follows that $(y,q) \in H(x,p)$ and the proof is complete by Theorem 4.13.1.

We now prove Lemma 4.14.1 which we used in the preceding proof.

LEMMA 4.14.1. Let G be a mapping on J into $cf(R^n)$ and let a sequence $\{v_k\}$ of absolutely continuous functions, $v_k\colon J \to R^n$ $(k = 1,2,\ldots)$ satisfy the conditions

$$\lim_{k\to\infty} v_k(t) = v(t) \qquad \text{for } t \in J \text{ a.e.},$$

$$|v_k'(t)| \le h(t), \qquad \text{a.e.} \int_J h(s)\,ds < \infty, \tag{4.14.11}$$

$$\lim_{k\to\infty} \rho\big(v_k'(t),G(t)\big) = 0 \quad \text{a.e.} \tag{4.14.12}$$

Then the function v is absolutely continuous, and

$$v'(t) \in G(t) \ \text{ a.e. in } J. \tag{4.14.13}$$

<u>Proof</u>: Define

(4.14.14) $$H(t) = \bigcap_{i=1}^{\infty} \operatorname{conv} \bigcap_{k=i}^{\infty} v_k'(t),$$

where $\operatorname{conv} A$ denotes the smallest convex and closed set containing A. In view of (4.14.12),

(4.14.15) $$H(t) \subset G(t) \quad \text{a.e.} \quad t \in J.$$

Clearly $H(t)$ is compact, convex, nonempty a.e. $t \in J$, measurable in t and

(4.14.16) $$\|H(t)\| \leq h(t) \quad \text{a.e.} \quad t \in J.$$

In the proof we shall use the following property

(4.14.17) $$(t-s)^{-1} \int_s^t r\big(H(u),\ H(s)\big)\, du \to 0 \quad \text{as} \quad t \to s,\ t \neq s,$$

for almost every s, where $r(A,B) = \max\big(\rho(A,B), \rho(B,A)\big)$, $\rho(A,B) = \sup_{x \in A} \rho(x,B), A, B \subset R^n$. If $H(t)$ reduces to a (integrable) vector function, this property is a classical result. For the general case, it can be proved by using the standard real variable methods.

The function $v(t)$ is absolutely continuous because of (4.14.11). Hence the derivative $v'(t)$ exists a.e. in J.

We shall now show that

(4.14.18) $$v'(s) \in H(s) \quad \text{if} \quad v'(s) \text{ exists}$$

and if (4.14.17), (4.14.12) for $t = s$ are satisfied. The inclusion (4.14.18) is equivalent to

(4.14.19) $$\rho\left(\frac{v(t) - v(s)}{t-s},\ H(s)\right) \to 0 \quad \text{as} \quad t \to s, \quad t > s,$$

because $v'(s)$ exists. By virtue of (4.14.11), (4.14.16),

$$\rho\big(v_k'(t), H(t)\big) \leq 2h(t)$$

and by (4.14.14)

$$\rho\big(v_k'(t), H(t)\big) \to 0 \quad \text{a.e.} \quad \text{in } J.$$

Hence

$$\int_s^t \rho\big(v_k'(u), H(u)\big)\, du \to 0 \qquad \text{as} \quad k \to \infty. \tag{4.14.20}$$

Since $H(s)$ is compact and convex, there exists a unique point $p_k(u) \in H(s)$ such that

$$\rho\big(v_k'(u), H(s)\big) = \|v_k'(u) - p_k(u)\|, \tag{4.14.21}$$

the vector $p_k(u)$ is integrable and

$$g_k(t) = (t-s)^{-1} \int_s^t p_k(u)\, du \in H(s). \tag{4.14.22}$$

We have

$$\rho\left(\frac{v(t) - v(s)}{t-s}, H(s)\right) = \lim_{k\to\infty} \rho\left(\frac{v_k(t) - v_k(s)}{t-s}, H(s)\right). \tag{4.14.23}$$

It follows from (4.14.22) that

$$\begin{aligned} q_k(t) &= \rho\left(\frac{v_k(t) - v_k(s)}{t-s}, H(s)\right) \\ &\le \left\| \frac{v_k(t) - v_k(s)}{t-s} - g_k(t) \right\|. \end{aligned} \tag{4.14.24}$$

Therefore, we find that

$$\begin{aligned} q_k(t) &\le \left\| (t-s)^{-1} \int_s^t \big(v_k'(t) - p_k(u)\big)\, du \right\| \\ &\le (t-s)^{-1} \int_s^t \|v_k'(t) - p_k(u)\|\, du \\ &= (t-s)^{-1} \int_s^t \rho\big(v_k'(u), H(s)\big)\, du, \end{aligned}$$

because of (4.14.21), and consequently

$$q_k(t) \le (t-s)^{-1} \int_s^t \rho\big(v_k'(u) - H(u)\big)\, du$$
$$+ (t-s)^{-1} \int_s^t r\big(H(u),H(s)\big)\, ds.$$

By virtue of (4.14.20), (4.14.23), (4.14.24), we obtain

$$\rho\left(\frac{v(t)-v(s)}{t-s}, H(s)\right) \le (t-s)^{-1} \int_s^t r\big(H(u),H(s)\big)\, du$$

and by (4.14.17) we obtain (4.14.19) and (4.14.18). The conclusion (4.14.13) easily follows because of (4.14.15). The proof is complete.

Theorem 4.14.1 assumes a particularly simple form, if Eq. (4.14.2) reduces to a differential inequality

$$\|x'\| \le w(t, \|x\|), \tag{4.14.25}$$

where $w\colon J \times R^+ \to R^+$ is a function satisfying the condition

(4) for each $t \in J$, $w(t,u)$ is continuous with respect to u and for each $u \in R^+$, $w(t,u)$ is measurable with respect to t; the functions $\sup[w(t,u)\colon u \le k]$, $k = 1,2,\ldots,$ are summable.

COROLLARY 4.14.1. If the functions f, w and L satisfy the conditions (2) - (4) and, in addition, the inequality

$$\|f(t,p) - f(t,q)\| \le w\big(t, \|p-q\|\big)$$

holds and if the problem (4.14.3), (4.14.4) has only the trivial solution $x \equiv 0$, then there exists one and only solution of the problem (4.14.4), (4.14.5).

For the proof it is enough to set $F(t,p) = \big[q\colon \|q\| \le w\big(t,\|p\|\big)\big]$ and then apply Theorem 4.14.1.

Let us demonstrate Corollary 4.14.1 by an example.

Consider Eq. (4.14.4) with the boundary condition

$$(4.14.26) \qquad x(a) + \lambda x(b) = r, \qquad \lambda > 0, \quad J = [a,b].$$

It is easily verified that every solution of the differential inequality $\|x'\| < \varphi(t)\|x\|$ where $\varphi(t)$ is a positive summable function such that $\int_a^b \varphi(s)\, ds < \pi$, satisfying the homogeneous boundary condition $x(a) + \lambda x(b) = 0$ is necessarily trivial. Indeed, it is well known that for an absolutely continuous function $x(t)$ satisfying the differential inequality $\|x'\| < \varphi(t)\|x\|$, either $x(t) \equiv 0$ or $x(t)$ is never 0 on J. Suppose that $x(t)$ is never 0. Then setting $z(t) = x(t)/\|x(t)\|$, we arrive at the following contadiction

$$\pi \leq \int_a^b \|z'(s)\|\, ds \leq \int_a^b \frac{\|x'(s)\|}{\|x(s)\|}\, ds \leq \int_a^b \varphi(s)\, ds.$$

Thus from Corollary 4.14.1 the next corollary follows immediately.

COROLLARY 4.14.2. If the function f satisfies hypothesis (2) and the inequality

$$\|f(t,p) - f(t,q)\| \leq \varphi(t)\|p - q\|,$$

where $\varphi(t)$ is positive summable function verifying $\int_a^b \varphi(s)\, ds < \pi$, then for each $r \in R^n$, the problem (4.14.4), (4.14.26) has exactly one solution.

4.15 GENERAL RESULTS FOR SET-VALUED MAPPINGS

In what follows, we wish to construct a general theory of linear problems for set-valued differential equations or some times called contingent equations, by utilizing fixed point theorems for set-valued mappings. With this motive, we introduce necessary tools and prove some general results concerning set valued mappings. We shall first state the Kakutani-Ky Fan fixed point theorem.

THEOREM 4.15.1. Let L be a locally convex topological linear space and K a compact convex set in L. Let $cf(K)$ be the family of all closed convex nonempty subsets of K. Then for any upper semicontinuous point to set transformation f from K into $cf(K)$, there exists a point $x_0 \in K$ such that $x_0 \in f(x_0)$.

We shall now consider some preliminary results. We shall continue to use the same notation of Section 4.14. Let L^p $(1 \leq p \leq \infty)$ denote the Banach space of all real-valued functions p-summable on $J = [a,b]$; L_n^p will denote its nth cartesian power. In both L^p, L_n^p, the norm will be denoted by $\|\cdot\|_p$.

For a linear normed space E, let, as before, $cf(E)$ denote the set of all closed, convex, nonempty subsets of E. A mapping $G: J \to cf(R^n)$ is called measurable if for every $p \in R^n$ the distance from p to $G(t)$ is measurable on J. The following lemmas are needed in our main results.

LEMMA 4.15.1. If sequences $\{w_i\} \subset L_n^p$, $\{v_i\} \subset L^p$ $(1 \leq p < \infty)$, and functions $v \in L^p$ and w satisfy the conditions: $|w_i(t)| \leq v_i(t)$ a.e. on J, $w_i(t) \to w(t)$ a.e. on J and $\|v_i - v\|_p \to 0$, then $w \in L_n^p$ and $\|w_i - w\|_p \to 0$.

The proof is left to the reader as an easy exercise.

LEMMA 4.15.2. For a sequence $\{w_i\} \subset L_n^1$ and a function $\varphi \in L^1$ satisfying $|w_i(t)| \leq \varphi(t)$ a.e. on J, there exists a double sequence $\{\lambda_{ik}\}$ $(i = 1,2,\ldots;\ k = i, i+1,\ldots)$ of real nonnegative numbers such that $\Sigma_{k=1}^{\infty} \lambda_{ik} = 1$, $\lambda_{ik} = 0$ for sufficiently large k (depending on i), and the sequence

$$\tilde{w}_i = \sum_{k=i}^{\infty} \lambda_{ik} w_k \qquad (i = 1,2,\ldots)$$

converges a.e. on J to a function $w \in L_n^1$.

Proof: Setting $y_i(t) = w_i(t)/[1 + \varphi(t)]^{\frac{1}{2}}$, we have

$$\|y_i\|_2 = \int_J \frac{\|w_i(t)\|^2}{1 + \varphi(t)} \, dt \leq \int_J \varphi(t) \, dt < \infty$$

and consequently, we can find a sequence $\{y_{\beta i}\}$ and a function $y_0 \in L_n^2$ such that $\lim_{i\to\infty} y_{\beta i} = y_0$ in the sense of weak convergence in L_n^2. By virtue of the Banach-Saks theorem, one can find a conveniently chosen partial sequence such that

$$\lim_{n\to\infty} \left\| \frac{1}{n^2} \sum_{i=1}^{n^2} y_{\beta i} - y_0 \right\|_2 = 0,$$

in place of the sequence $\{y_{\beta i}\}$. As the sequences

$$\left\| \frac{1}{n^2} \sum_{i=1}^{n-1} y_{\beta i} \right\|_2, \qquad \left\| \frac{1}{n^2} \sum_{i=n^2+1}^{n^2+n-1} y_{\beta i} \right\|_2$$

tend to zero because of (4.15.1), we obtain

$$\lim_{n\to\infty} \left\| \sum_{i=n}^{n^2+n-1} y_{\beta i} - y_0 \right\|_2 = 0.$$

By Riesz's theorem, there exists a sequence of $\sigma_n \geq n$ such that

$$\lim_{n\to\infty} \frac{1}{\sigma_n^2} \sum_{i=\sigma_n}^{\alpha_n} y_{\beta i}(t) = y_0(t), \qquad \alpha_n = \sigma_n^2 + \sigma_n - 1$$

for all $t \in J$. It is now easy to see that

$$\tilde{w}_i(t) = \frac{1}{\sigma_n^2} \sum_{i=\sigma_n}^{\alpha_n} y_{\beta i}(t)$$

converges for all values on J to the function $w = y_0/[1 + \varphi]^{\frac{1}{2}}$. Thus in order to show the conclusion it is sufficient to set $\lambda_{ik} = 1/\sigma_i^2$ if there exists an index i for which $k = \beta_i$, $\sigma_n \leq \beta_i \leq \alpha_n$, and set $\lambda_{ik} = 0$ in the contrary case. The proof is complete.

LEMMA 4.15.3. Let a map $G: J \to cf(R^n)$ be measurable and let $\|G(t)\| \leq \varphi(t)$ a.e. on J, where $\varphi \in L^1$. Then there is a function $y \in L_n^1$ such that $y(t) \in G(t)$ a.e. on J.

Proof: It is enough to take for $y(t)$ the unique point of intersection $G(t) \cap K(t)$, where $K(t)$ is the ball with center at the origin 0 of R^n and radius equal to the distance of 0 to $G(t)$, and apply the known theorem that the intersection of two measurable functions is also measurable.

Let $F: J \times R^n \to cf(R^n)$ be such that

(i) for every fixed $x \in R^n$, the function $F(t,x)$ is measurable on J;

(ii) for every fixed $t \in J$, the function $F(t,x)$ is upper semicontinuous on R^n.

(iii) there exist functions $\alpha \in L^{p_2}$, $\beta \in L^s$, where $s = p_1p_2/(p_1 - qp_2)$ and $0 \leq p_2 q < p_1$, $1 \leq p_1, p_2 < \infty$ (in case $p_1 = \infty$, $1 \leq p_2 < \infty$, $0 \leq q < \infty$, $s = p_2$ is not excluded), such that

$$\|F(t,x)\| \leq \alpha(t) + \beta(t)\|x\|^q, \qquad (t,x) \in J \times R^n.$$

THEOREM 4.15.2. Let F satisfy assumptions (i) - (iii). For a given function $x \in L_n^{p_1}$, let $Q(x)$ denote the set of all measurable functions $y: J \to R^n$ such that

$$y(t) \in F(t,x(t)) \quad \text{a.e. on } J.$$

Then the correspondence $x \to Q(x)$ defines a bounded mapping of $L_n^{p_1}$ into $cf(L_n^{p_2})$.

Proof: We have to show that for every $x \in L_n^{p_1}$

(a) $Q(x)$ is nonempty;

(b) $Q(x)$ is convex;

(c) $Q(x) \subset L_n^{p_2}$;

(d) $Q(x)$ is closed;

(e) for every $k > 0$, there exists an $M > 0$ such that $\|x\|_{p_1} \le k$ implies $\|y\|_{p_2} \le M$ for every $y \in Q(x)$.

We would prove only (a), because (b) is trivially true, (e) follows immediately from assumption (iii) and obviously implies (c), and (d) will follow from Theorm 4.15.2 below.

Let $\{x_i\} \subset L_n^{p_1}$ be a sequence of measurable functions admitting only a finite number of different values, convergent to a function $x \in L_n^{p_1}$ a.e. on J. By assumption (i) the functions $F(t,x_i(t))$ $(i = 1,2,\ldots)$ are measurable on J and therefore, by Lemma 4.15.3, there are functions $y_i \in L_n^1$ such that $y_i(t) \in F(t,x_i(t))$ a.e. on J. For

$$w_i(t) = y_i(t)\left[\alpha(t) + \beta(t)\|x_i(t)\|^q\right]^{-1},$$

we have, by assumption (iii), $\|w_i(t)\| \le 1$ a.e. on J. Now a straightforward application of Lemma 4.15.2 gives a sequence

$$\tilde{w}_i(t) = \sum_{k=i}^{\infty} \lambda_{ik} w_k(t), \qquad i = 1,2,\ldots,$$

which converges to a function $w \in L_n^1$ a.e. on J. It is easily seen that the corresponding sequence

$$\tilde{y}_i(t) = \sum_{k=i}^{\infty} \lambda_{ik} y_k(t), \qquad i = 1,2,\ldots$$

converges a.e. on J to a function

$$y(t) = w(t)\left[\alpha(t) + \beta(t)\|x(t)\|^q\right].$$

On the other hand, by assumption (ii), for almost every fixed $t \in J$ and any $\varepsilon > 0$ there is an integer $N(\varepsilon,t)$ such that

$$\begin{aligned} F\big(t,x_i(t)\big) &\subset F(t,x(t) \;\; + k_\varepsilon \\ &= \left[u + v\colon u \in F\big(t,x(t)\big), \|v\| \le \varepsilon\right] \end{aligned}$$

for $i \ge N(\varepsilon,t)$. Thus

$$y_i(t) \in F\big(t,x(t)\big) + k_\varepsilon, \qquad i \ge N(\varepsilon,t)$$

and, by the convexity of $F\big(t,x(t)\big)$,

$$y_i(t) \in F\big(t,x(t)\big) + k_\varepsilon, \qquad i \ge N(\varepsilon,t),$$

so that $y(t) \in F\big(t,x(t)\big)$ a.e. on J. This proves the theorem.

THEOREM 4.15.3. If Γ is a linear transformation of $L_n^{p_2}$ into a Banach space E, then ΓQ maps $L_n^{p_1}$ into $cf(E)$ and is upper semicontinuous.

<u>Proof</u>: It is clear that for every $x \in L_n^{p_1}$ the set $\Gamma Q(x)$ is convex. Thus it remains to prove that for sequences $\{x_i\} \subset L_n^{p_1}$, $\{z_1\} \subset E$, and $x \in L_n^{p_1}$, $z \in E$, the conditions

$$\|x_i - x\|_{p_i} \to 0, \qquad \|z_i - z\| \to 0, \qquad z_i \in \Gamma Q(x_i), \quad i = 1,2,\ldots,$$

imply $z \in \Gamma Q(x)$. For this purpose, define $z_i = \Gamma y_i$, $y_i \in Q(x_i)$.

As in the proof of Theorem 4.15.2 [one can assume, without loss of generality, that $x_i(t) \to x(t)$ a.e. on J] an application of Lemma 4.15.2 yields a convergent sequence

$$\tilde{y}_i(t) = \sum_{k=1}^{\infty} \lambda_{ik} y_k(t) \to y(t)$$

such that $y(t) \in F(t,x(t))$ a.e. on J, that is, $y \in Q(x)$.

By assumption (iii), we have

$$\|\tilde{y}_i(t)\| \le \sum_{k=1}^{\infty} \lambda_{ik}[\alpha(t) + \beta(t)\|x_k(t)\|^q]$$

$$= \alpha(t) + \beta(t) \sum_{k=i}^{\infty} \lambda_{ik}\|x_k(t)\|^q.$$

Denoting by $v_i(t)$ the last sum and setting $v(t) = \alpha(t) + \beta(t)\|x(t)\|^q$, we find easily that $\|v_i - v\|_{p_2} \to 0$ and then, applying Lemma 4.15.2, we obtain $\|y_i - y\|_{p_2} \to 0$. On the other hand,

$$\Gamma\tilde{y}_i = \sum_{k=i}^{\infty} \lambda_{ik}\Gamma y_i = \sum_{k=i}^{\infty} \lambda_{ik} z_k \to z$$

so that $z = \Gamma y$ and therefore $z \in \Gamma Q(x)$. The proof is complete.

4.16 SET-VALUED DIFFERENTIAL EQUATIONS

Having developed necessary results concerning set-valued mappings in the preceding section, we are now ready to employ the Kakutani-Ky Fan theorem to prove an existence theorem for boundary value problems of set-valued differential equations. The results we present may be considered as extensions of classical theorems on boundary value problems.

Consider a system

(4.16.1) $$x'(t) \in A(t)x(t) + F(t,x(t))$$

and a condition

(4.16.2) $$Lx = r.$$

We shall assume that

(a) $A: J \to A^*$, where A^* is the algebra of $n \times n$ matrices, which are measurable and integrable in J;

(b) $F: J \times R^n \to cf(R^n)$ such that

(b_1) for each $x \in R^n$, the function $F(t,x)$ is measurable, that is, for each $p \in R^n$, the distance from p to the set $F(t,x)$ is a measurable function in J;

(b_2) for each $t \in J$, the function $F(t,x)$ is upper semicontinuous on R^n, that is, $x_i \to x_0$, $y_i \to y_0$, $y_i \in F(t,x_i)$ implies $y_0 \in F(t,x_0)$;

(b_3) there exist function α, β which are measurable and integrable in J such that

$$\|F(t,x)\| \leq \alpha(t) + \beta(t)\|x\|, \qquad (t,x) \in J \times R^n.$$

(c) L is a linear continuous operator from $C = C[J,R^n]$ into R^m, $r \in R^m$.

(d) A, L, and r are such that the problem

(4.16.3) $$x'(t) - A(t)x(t) = f(t), \qquad Lx = r,$$

admits solutions for all f belonging to a linear manifold Φ_r in L^1_n;

(e) for every $x \in C$, $Q(x)$ is the set of functions $y: J \to R^n$, measurable, such that $y(t) \in F(t,x(t))$ a.e. on J, and suppose that $Q(v) \in cf(\Phi_r)$, for each $v \in C$ with $Lv = r$.

Observe that condition (a) implies the existence of a unique function $U: J \times J \to A^*$, which is continuous and satisfies the integral equation

$$U(t,s) = I_n + \int_s^t A(\tau)U(\tau,s)\,d\tau$$

where I_n is the identity of A^*.

Also $U(\cdot,s)$ is a linear operator, continuous (and compact), from R^n into C, while the composition product

$$L_U = L\cdot U(\cdot,s)$$

is a linear operator from R^n into R^m and hence can be represented by an $m\times n$ matrix. Furthermore, it is known that every $m\times n$ matrix has a generalized inverse and consequently, there exist matrices L_U^* of type $n\times m$ such that

$$L_U L_U^* L_U = L_U.$$

For a fixed $s\in J$, let us define the linear operator

$$\Gamma: f\to \Gamma f = -U(t,s)L_U^*L\int_s^t U(t,\tau)f(\tau)\,d\tau + \int_s^t U(t,\tau)f(\tau)\,d\tau,$$

from L_n^1 into C. It is easy to prove by Ascoli's theorem that this is a compact operator.

Let us fix a solution $c\in R^n$ of $L_U c = 0$ and write

$$Hr = U(t,s)(c + L_U^* r).$$

We claim that if $f\in \Phi_r$, the function

$$x = \Gamma f + Hr$$

is a solution of problem (4.16.3). Indeed, denoting by D the linear operator $d/dt - A(t)$ from C into L_n^1 we have

$$D(\Gamma f + Hr) = D[-U(t,s)L_U^*L\int_s^t U(t,\tau)f(\tau)\,d\tau] +$$

$$+ D\int_s^t U(t,\tau)f(\tau)\,d\tau + D[U(t,s)(c + L_U^* r]$$

$$= D\int_s^t U(t,\tau)f(\tau)\,d\tau = f(t).$$

Concerning the second equation in (4.16.3), we notice that $f \in \Phi_r$ means that there exists an $\eta \in R^n$ such that

$$L_U\eta = r - L\int_s^t U(t,\tau)f(\tau)\,d\tau$$

and hence

$$\begin{aligned}
L(\Gamma f + Hr) &= -L_U L_U^* L\int_s^t U(t,\tau)f(\tau)\,d\tau \\
&\quad + L\int_s^t U(t,\tau)f(\tau)\,d\tau + L_U c + L_U L_U^* c \\
&= L_U L_U^*[r - L\int_s^t U(t,\tau)f(\tau)\,d\tau] + L\int_s^t U(t,\tau)f(\tau)\,d\tau \\
&= L_U L_U^* L_U\eta + L\int_s^t U(t,\tau)f(\tau)\,d\tau \\
&= r - L\int_s^t U(t,\tau)f(\tau)\,d\tau + L\int_s^t U(t,\tau)f(\tau)\,d\tau \\
&= r,
\end{aligned}$$

which proves our claim.

By Theorem 4.15.2, the correspondence $x \to Q(x)$ defines a bounded transformation from C into $cf(L_n^1)$ and consequently, by Theorem 4.15.3, ΓQ transforms C into $cf(C)$ and is upper semicontinuous. Hence

$$T\colon x \to T(x) = \Gamma Q(x) + \{Hr\}$$

is an upper semicontinuous transformation from C into $cf(C)$.

Let us denote by V_r, the linear manifold in C

$$V_r = [v \in C: Lv = r].$$

To prove the existence of solutions of the boundary value problem (4.16.1), (4.16.2), under hypotheses (a) - (e), it is enough to show the existence of a function $v \in V_r$ such that $v \in T(v)$.

By (e), T transforms V_r into $cf(V_r)$. Indeed, the transformation Q maps V_r into $cf(\Phi_r)$, by (e); hence, for each $f \in \Phi_r$, $\Gamma f + Hr$ is a solution of (4.16.3). Consequently, $\Gamma f + Hr \in V_r$ or Γ transforms Φ_r into $V_r - \{H_r\}$. As Γ is linear, $\Gamma Q(x)$ is convex for each $v \in V_r$ and since ΓQ is upper semicontinuous, it follows that ΓQ is closed for each $v \in V_r$. This shows that T transforms V_r into $cf(V_r)$.

For $x \in C$, $z \in T(x)$, we have by assumption (b_3),

$$\|z\| \leq \|\Gamma\|(\alpha_0 + \beta_0)\|x\| + \|Hr\|, \tag{4.16.4}$$

where $\alpha_0 = \int_J \alpha(s)\,ds$, $\beta_0 = \int_J \beta(s)\,ds$, and $\|\Gamma\|$ is the norm of Γ. Let us suppose that $\beta_0\|\Gamma\| < 1$ and set

$$K_\rho = [x \in C: \|x\| \leq \rho,$$
$$\rho = \left(\alpha_0\|\Gamma\| + \|Hr\|\right)\left(1 - \beta_0\|\Gamma\|^{-1}\right) + a,$$

where $a > 0$ is sufficiently large so that $K_\rho \cap V_r \neq \emptyset$.

It follows easily from (4.16.4) that $T(x) \subset K_\rho$ for each $x \in K_\rho$, that is, $T(K_\rho) \subset K_\rho$, where $T(K_\rho) = \bigcup_{x \in K_\rho} T(x)$. Hence, we have $T(K_\rho \cap V_r) \subset K \cap V_r$.

As $K_\rho \cap V_r$ is bounded, by Theorem 4.15.2, $Q(K_\rho \cap V_r)$ is bounded and as Γ is compact, it follows that $T(K_\rho \cap V_r)$ is relatively compact in C.

Let us consider the convex closure of $T(K_\rho \cap V_r)$, which we denote by $\overline{co}\ T(K_\rho \cap V_r)$; then we readily obtain

$$\overline{co}\ T(K_\rho \cap V_r) \subset \overline{co}(K_\rho \cap V_r) = K_\rho \cap V_r$$

as V_r, being the inverse image of the point r by the continuous operator L, is closed and $K_\rho \cap V_r$ is closed and convex. Hence

$$T\left(\overline{co}\ T(K_\rho \cap V_r)\right) \subset T(K_\rho \cap V_r) \subset \overline{co}\ T(K_\rho \cap V_r).$$

Therefore the upper semicontinuous map T transforms the compact and convex set $\overline{co}\ T(K_\rho \cap V_r)$ into itself and consequently an application of Ky Fan fixed point Theorem 4.15.1 shows the existence of $v \in \overline{co}\ T(K_\rho \cap V_r)$ such that $v \in V_r$ and $v \in T(v)$. We have thus proved the following theorem.

THEOREM 4.16.1. Under hypotheses (a) - (e), problem (4.16.1), (4.16.2) has a solution if $\beta_0 = \int_J \beta(s)\, ds$ is sufficiently small.

If $L_U = m \leq n$, problem (4.16.3) has a solution for each $f \in L_n^1$ and each r and conversely. Let us call this hypothesis (f). If (f) holds, $\Phi_r = L_n^1$ for each $r \in R^m$ and in this case (d) is automatically satisfied. Hence from Theorem 4.16.1, we obtain the next corollary.

COROLLARY 4.16.1. Under hypotheses (a) - (c), (f), problem (4.16.1), (4.16.2) has a solution for any $r \in R^m$ if $\beta_0 = \int_J \beta(s)\, ds$ is sufficiently small.

If $L_U = m = n$, there exists only one matrix L_U^* which is the inverse of L_U, for which the only solution of $L_U c = 0$ is $c = 0$. Thus denoting by (h) the hypothesis that problem (4.16.3) has a unique solution for each $f \in L_n^1$ and each $r \in R^n$, we have the following corollary.

COROLLARY 4.16.2. Under hypotheses (a) - (c), (h), problem

(4.16.1), (4.16.2) has a solution for each $r \in R^n$, if $\beta_0 = \int_J \beta(s)\, ds$ is sufficiently small.

In order to illustrate the results, consider the following problem analogous to the classical problem of Nicoletti,

$$(4.16.5) \quad x' \in F(t,x), \quad x_i(t_i) = r_i, \quad t_i \in J, \quad r_i \in R, \; i = 1,2,\ldots,n.$$

Since the corresponding homogeneous problem

$$x' = 0, \qquad x_i(t_i) = 0, \qquad i = 1,2,\ldots,n,$$

has a unique solution $x(t) \equiv 0$, an application of Corollary 4.16.2 with $A(t) \equiv 0$ and $Lx = \big(x(t_1), x(t_2), \ldots, x(t_n)\big)$ yields the existence of at least one solution of (4.16.5) for arbitrary $(r_1, r_2, \ldots, r_n)$, provided F satisfies hypothesis (b) and $\beta_0 = \int_J \beta(s)\, ds < 1$.

In the case of the Cauchy problem

$$(4.16.6) \quad x' \in F(t,x), \qquad x(t_0) = r, \qquad t_0 \in J, \quad r \in R^n,$$

condition $\beta_0 < 1$ is superfluous. It is easy to verify that replacing the usual norm in C by

$$\|x\| = \max_J \|x(t)\| \exp\left[-2 \int_{t_0}^{t} \beta(s)\, ds\right]$$

we obtain, instead of (4.16.4)

$$\|z\| \le \tfrac{1}{2}\|x\| + \alpha_0 + \|r\|$$

for every $z \in T(x)$. Also $T(K_\rho) \subset K_\rho$ for $\rho = 2(\alpha_0 + \|r\|)$. This implies that if F satisfies hypothesis (b), then the Cauchy problem (4.16.6) has at least one solution defined on J.

Let us conclude with one more application. Consider the problem

$$(4.16.7) \qquad x' - A(t)x \in F(t,x), \qquad x(T) - x(0) = 0,$$

where $A(t)$ is a periodic $n \times n$ matrix of period T and $F\colon R \times R^n \to cf(R^n)$ such that $F(t+T,x) = F(t,x)$. Let hypotheses (a), (b) hold and let

$$(4.16.8) \qquad x' - A(t)x = f(t), \qquad x(T) - x(0) = 0$$

with $f(t + T) = f(t)$, be the problem similar to problem (4.16.3). Recall that if the homogeneous system has periodic solutions, then the adjoint system has also such solutions. Denoting by Z the space of such solutions, we have

$$\Phi_0 = \left[f \in L_n^1\colon \int_0^T z^*(t)f(t)\,dt = 0,\ z \in Z\right].$$

Assume that $F\big(t,v(t)\big) \in cf(\Phi_0)$ for each $v \in V_0$, where V_0 is the space of functions from R into R^n, continuous and periodic with period T. Then by Theorem 4.16.1, Eq. (4.16.7) admits at least one solution $x(t)$ and therefore the set-valued differential equation (4.16.7) has a least one periodic solution of period T.

4.17 NOTES AND COMMENTS

Sections 4.1 - 4.3 are concerned with the results of Opial [2]. For related work, see Lasota and Opial [2] and Whyburn [1]. The contents of Section 4.4 are based on Conti [1,2]. For further references, see Conti's survey paper [2]. The results contained in Sections 4.5 - 4.7 are adapted from Perov and Kibenko [2]. See Lepin and Nyshkis [1] for the results in Section 4.8. The work in Section 4.9 is due to Antosiewicz [1]. Sections 4.10 and 4.11 consist of the results of Lasota and Yorke [13]. See, for more general results, Bernfeld et al. [3]. Section 4.12 contains the work of Schmitt [7]. Related results on periodic solutions may be found in Schmitt [1,6,8] and Mawhin [1-3]. Sections 4.13 and 4.14 deal with the results of Lasota and Opial [7] and Lasota [9]. For Lemma 4.14.1, see Plís [1]. For the results of Section 4.15, see Lasota and Opial [3]. The contents of Section 4.16 are taken from Grandolfi [1].

For further results, see Lasota [5], Lasota and Olech [6], Ky Fan [1] and Chow and Lasota [1].

Chapter 5
EXTENSIONS TO FUNCTIONAL DIFFERENTIAL EQUATIONS

5.0 INTRODUCTION

We extend, in this chapter, some of the results of the preceding chapters, to boundary value problems associated with functional differential equations. We begin by proving existence theorems in the small. We then extend the theory of differential inequalities and the modified function approach, to establish existence results in the large. A shooting type method is presented to derive existence and uniqueness results. We also consider the question of existence of solutions of quasilinear functional differential equations with nonhomogeneous linear boundary conditions. Finally, a more general problem is treated where the boundary conditions are prescribed in terms of bounded linear operators.

5.1 EXISTENCE IN THE SMALL

Let $J = [a,b]$, $J_a = (-\infty,a]$, $J_b = [b,\infty)$ and let $h \in C[J \times R, R]$. Suppose that $\varphi \in C^{(1)}[J_a,R]$, $\psi \in C^{(1)}[J_b,R]$ are given functions that are bounded together with their derivatives on their respective domains. We shall consider the BVP

$$(5.1.1) \qquad x''(t) = f\big(t,x(t),x(h(t,x(t))),x'(t)\big),$$

$$(5.1.2) \qquad x(t) = \varphi(t) \text{ on } J_a, \qquad x(t) = \psi(t) \text{ on } J_b,$$

where $f \in C[J \times R \times R, R]$. We seek functions $x \in C^{(2)}[J,R]$ satisfying the BVP (5.1.1), (5.1.2).

Consider the Banach space B given by

$$B = C[R,R] \cap BC^{(1)}[J_a,R] \cap BC^{(1)}[J_b,R] \cap C^{(1)}[J,R]$$

with the norm

$$\|x\|_B = \max\left[\sup_{t\in R} |x(t)|, \sup_{t\in J_a} |x'(t)|, \sup_{t\in J_b} |x'(t)|, \sup_{t\in J} |x'(t)|\right].$$

Here $BC^{(1)}[I,R]$ means, as usual the class of functions $C^{(1)}[I,R]$ such that $y \in BC^{(1)}[I,R]$ implies that y, y' are bounded on I.

Recall from Section 1.1.1, that if $G(t,s)$ is the Green's function associated with the BVP

$$x'' = 0, \qquad x(a) = x(b) = 0,$$

then the solution $x(t)$ of the BVP

$$x'' = p(t), \qquad x(a) = x(b) = 0,$$

is of the form $x(t) = \int_a^b G(t,s)p(s)\,ds$. Define $\bar{G}(t,x)$ as

$$\bar{G}(t,x) = \begin{cases} G(t,x), & t \in J, \\ 0, & t \notin J. \end{cases}$$

Also, let $w(t)$ be defined on R such that

$$w(t) = \begin{cases} \varphi(t), & t \in J_a, \\ \psi(t), & t \in J_b, \\ \dfrac{[\psi(b) - \varphi(a)]}{b-a}(t-a) + \varphi(a), & t \in J. \end{cases}$$

Then clearly $w \in B$.

We next define the operator T on B such that for each $x \in B$,

$$Tx(t) = \int_a^b \bar{G}(t,s) f\big(s, x(s), x(h(s,x(s))), x'(s)\big)\,ds + w(t).$$

The following properties of T may be easily established as in Section 1.1.1:

(1) $Tx(t) = \varphi(t)$, $t \in J_a$, $Tx(t) = \psi(t)$, $t \in J_b$;

(2) $Tx(t)$ is twice continuously differentiable on J;

(3) $(Tx)''(t) = f\big(t,x(t),x(h(t,x(t))),x'(t)\big)$ on J;

(4) $T\colon B \to B$ and the fixed points of T are solutions of the BVP (5.1.1) and (5.1.2);

(5) T is a continuous operator.

Let $M,N > 0$ be given such that $|w(t)| \leq M$, $|w'(t)| \leq N$ on R. We consider the closed, convex, bounded subset B_0 of B defined by

$$B_0 = \big[x \in B\colon |x(t)| \leq 2M,\ |x'(t)| \leq 2N\big]$$

and we let

$$q = \big[\sup|f(t,x,y,z)|\colon t \in J,\ |x|,\ |y| \leq 2M,\ |z| \leq 2N\big].$$

If $x \in B_0$, we can compute, as in Theorem 1.1.2, the estimates

$$|Tx(t)| \leq \big((b-a)^2/8\big)\, q + M, \qquad |(Tx)'(t)| \leq \big((b-a)/2\big)\, q + N,$$

to conclude $TB_0 \subset B_0$; provided that $(b-a)^2/8 \leq M$, $\big((b-a)/2\big)\, q \leq N$. We can then use Ascoli's theorem and Schauder's fixed point theorem to complete the proof of the following result concerning the existence in the small.

THEOREM 5.1.1. Let $M,N > 0$ be given numbers and let for $t \in J$, $|x|,|y| \leq 2M$, $|z| \leq 2N$, $|f(t,x,y,z)| \leq q$, where $f \in C[J \times R^3, R]$. Suppose further that $\delta = \min[(8M/q)^{\frac{1}{2}}, 2N/q]$. Then any BVP (5.1.1), (5.1.2) such that $b - a \leq \delta$, $|\varphi(t)|$, $|\psi(t)| \leq M$, $|\varphi'(t)|$, $|\psi'(t)| \leq N$, $|\psi(b) - \varphi(a)/(b-a)| \leq N$, has a solution.

COROLLARY 5.1.1. Assume that $f \in C[J \times R^3, R]$ and is bounded on $J \times R^3$. Then every BVP (5.1.1), (5.1.2) has a solution.

The foregoing results remain valid if we assume that f

and x are vector-valued functions and $|\cdot|$ denotes an appropriate vector norm. In fact, one can suitably state and prove a more general existence result when the derivative x' also contains a deviated argument like $x'(g(t))$. This we give as an exercise.

EXERCISE 5.1.1. State and prove an existence theorem for the BVP $x''(t) = f(t,x(t),x(h(t,x(t))),x'(t),x'(g(t)))$, $x(t) = \varphi(t), x'(t) = \varphi'(t)$ on J_a, $x(t) = \psi(t), x'(t) = \psi'(t)$ on J_b.

Regarding the question of uniqueness of solutions, the following example illustrates the difference between the BVP (5.1.1), (5.1.2) and its associated BVP without derivatives in the arguments.

Consider the BVP

$$x''(t) = -\frac{x(t)}{1 + |x(x(t))|}, \qquad 0 \leq t < \pi$$

$$x \equiv 0 \quad \text{for} \quad t \leq 0 \quad \text{and} \quad t \geq \pi.$$

It is easy to check that $x(t) = c \sin t$ is a solution of this problem for any $c \leq 0$ whereas, the corresponding problem, namely,

$$x''(t) = -x/(1 + |x|), \qquad 0 \leq t \leq \pi,$$

$$x(0) = x(\pi) = 0,$$

has only the trivial solution. Nonetheless, the uniqueness is guaranteed, if f satisfies an appropriate Lipschitz condition, as is common in these type of problems. A verification of this statement is left to the reader.

EXERCISE 5.1.2. Let $k_1, k_2 > 0$ be such that $|\varphi(t)|, |\varphi'(t)| \leq k_1$ on J_a, $|\psi(t)|, |\psi'(t)| \leq k_2$ on J_b. Let $R_0 = [(x,y,z)\colon |x|, |y| \leq k_2, |z| \leq k_2]$ and for $t \in J$, (x,y,z),

$(x_1,y_1,z_1) \in R_0$,

$$|f(t,x,y,z) - f(t,x_1,y_1,z_1)| \leq L_1|x - x_1| + L_2|y - y_1| + L_3|z - z_1|,$$

and for $t \in J$, $|x|, |x_1| \leq k_1$, $|h(t,x) - h(t,x_1)| \leq M|x - x_1|$. Then the BVP (5.1.1), (5.1.2) has at most one solution $x(t)$ with $|x(t)| \leq k_1$, $|x'(t)| \leq k_2$ provided that $\eta \max[(b-a)^2/2, (b-a)/2] < 1$ where $\eta = L_1 + L_2(1 + k_2M) + L_3$.

Let us now single out for later use, a special case of the BVP (5.1.1), (5.1.2) in which f is of retarded type. More specifically, let $h(t,x) \equiv t - \tau$, $\tau > 0$, so that $J_a = [a - \tau, a]$ and J_b, a single point. The corresponding BVP may be stated as

$$(5.1.3) \qquad x''(t) = f\big(t, x(t), x(t-\tau), x'(t)\big),$$

$$(5.1.4) \qquad x(t) = \varphi(t), \quad t \in [a-\tau, a], \qquad x(b) = B.$$

5.2 EXISTENCE IN THE LARGE

Let us consider the second-order functional differential equation (5.1.1) together with the boundary conditions (5.1.2), with respect to which, we shall define upper and lower solutions.

DEFINITION 5.2.1. A function $\alpha \in B$ having a continuous second derivative on J will be called a <u>lower solution</u> of (5.1.1), (5.1.2), if

(i) $\alpha''(t) \geq f\big(t, \alpha(t), \alpha(h(t,\alpha(t))), \alpha'(t)\big)$, $t \in J$,

(ii) $\alpha(t) \leq \varphi(t)$ on J_a, $\alpha(t) \leq \psi(t)$ on J_b.

An upper solution $\beta(t)$ of (5.1.1), (5.1.2) is defined similarly by reversing inequalities (i) and (ii).

We shall establish the following result.

THEOREM 5.2.1. Consider the BVP (5.1.1), (5.1.2), where f satisfies the conditions:

(i) $f(t,x,y,z)$ is nonincreasing in y for each (t,x,z);

(ii) $|f(t,x,y,z)| \leq q$ on $J \times R^3$,

(iii) there exist constant functions α, β which are lower and upper solutions, respectively, of (5.1.1), (5.1.2) such that $\alpha \leq \beta$. Then there exists a solution $x(t)$ of (5.1.1), (5.1.2) with $\alpha \leq x(t) \leq \beta$, $t \in R$.

<u>Proof</u>: Let us define a modified function F by

$$F(t,x,y,x') = \begin{cases} f(t,\beta,\bar{y},x') + \dfrac{x-\beta}{1+x^2} & \text{if } x > \beta, \\ f(t,x,\bar{y},x') & \text{if } \alpha \leq x \leq \beta, \\ f(t,\alpha,\bar{y},x') + \dfrac{x-\alpha}{1+x^2} & \text{if } x < \alpha, \end{cases}$$

where

$$\bar{y} = \begin{cases} \beta & \text{if } y > \beta, \\ y & \text{if } \alpha \leq y \leq \beta, \\ \alpha & \text{if } y < \alpha. \end{cases}$$

Since F is bounded, because of (ii), by Corollary 5.1.1, the modified BVP $x''(t) = F\big(t,x(t),x(h(t,x(t))),x'(t)\big)$, (5.1.2) has a solution $x(t)$. Also, by the definition of F and assumption (iii), we have

$$F(t,\alpha,\alpha,0) \leq 0 \leq F(t,\beta,\beta,0). \tag{5.2.1}$$

We now claim that $\alpha \leq x(t) \leq \beta$, $t \in R$. Suppose that this is false. Then, since $\alpha \leq x(t) \leq \beta$, $t \notin J$, there exist $t_1, t_2 \in (a,b)$ such that either $x(t) > \beta$ for $t_1 < t < t_2$ and $x(t_1) = \beta = x(t_2)$ or $x(t) < \alpha$ for $t_1 < t < t_2$ and $x(t_1) = \alpha = x(t_2)$. We shall deal with the first case, the arguments of the second case being similar. There exists a $t_0, t_1 < t_0 < t_2$, such that $x(t) - \beta$ assumes a positive maximum at t_0, with $x'(t_0) = 0$. Then because of (5.2.1), we obtain

$$x''(t_0) - 0 \geq F\big(t_0,x(t_0),x(h(t_0,x(t_0))),x'(t_0)\big) - F(t_0,\beta,\beta,0).$$

By condition (i) and the definition of F, we find that

$$F\big(t_0,x(t_0),x(h(t_0,x(t_0))),x'(t_0)\big) \geq f(t_0,\beta,\beta,0) + \frac{x(t_0)-\beta}{1+x^2(t_0)} .$$

Hence $x''(t_0) \geq \big(x(t_0)-\beta\big)/\big(1+x^2(t_0)\big) > 0$. This contradicts the fact that $x(t)-\beta$ assumes a positive maximum at t_0. Therefore $x(t) \leq \beta$, $t \in R$. Similarly, we conclude $\alpha \leq x(t)$ on R. This, however, implies that

$$F\big(t,x(t),x(h(t,x(t))),x'(t)\big) \equiv f\big(t,x(t),x(h(t,x(t))),x'(t)\big),$$

which assures that $x(t)$ is actually a solution of (5.1.1), (5.1.2). The proof is thus complete.

The next result is concerned with the existence of solutions of the BVP (5.1.1), (5.1.2) where assumption (ii) of Theorem 5.2.1 is omitted.

THEOREM 5.2.2. Consider the BVP (5.1.1), (5.1.2) and let (i) and (iii) of Theorem 5.2.1 hold. Assume further that f satisfies Nagumo's condition on $[a,b]$ relative to α, β whenever $\alpha \leq x(t) \leq \beta, \alpha \leq x\big(h(t,x)\big) \leq \beta$. Then the BVP (5.1.1), (5.1.2) has a solution such that $\alpha \leq x(t) \leq \beta$ on R.

<u>Proof</u>: We first define the function F_0 by

$$F_0(t,x,y,x') = f(t,\bar{x},\bar{y},x')$$

where, as before,

$$\bar{x} = \begin{cases} \beta & \text{if } x > \beta, \\ x & \text{if } \alpha \le x \le \beta, \\ \alpha & \text{if } x < \alpha, \end{cases} \qquad \bar{y} = \begin{cases} \beta & \text{if } y > \beta, \\ y & \text{if } \alpha \le y \le \beta, \\ \alpha & \text{if } y < \alpha. \end{cases}$$

Then F_0 satisfies Nagumo's condition. Hence, by Theorem 1.4.1, there is an $N > 0$ such that if $x(t)$ is a solution of $x''(t) = F_0\big(t,x(t),x(h(t,x(t))),x'(t)\big)$ and $\alpha \le x(t) \le \beta$, $\alpha \le x\big(h(t,x(t))\big) \le \beta$, then $|x'(t)| \le N$ on J. Now define the function F by

$$F(t,x,y,z) = \begin{cases} F_0(t,x,y,N) & \text{if} \quad x' > N, \\ F_0(t,x,y,x') & \text{if} \quad |x'| \le N, \\ F_0(t,x,y,-N) & \text{if} \quad x' < -N. \end{cases}$$

Then F also satisfies the Nagumo's condition and furthermore, F is bounded on $J \times R^3$. Also $F(t,\alpha,\alpha,0) \le 0 \le F(t,\beta,\beta,0)$. We thus conclude by Theorem 5.2.1 that the functional differential equation

$$x''(t) = F\big(t,x(t),x(h(t,x(t))),x'(t)\big)$$

together with the boundary conditions (5.1.2) has a solution $x(t)$ satisfying $\alpha \le x(t) \le \beta$ on R. Hence also $|x'(t)| \le N$. By the definition of F, this implies that $x(t)$ is actually a solution of (5.1.1), (5.1.2), completing the proof.

If the deviating argument is independent of the solution itself, that is, $h(t,x) \equiv h(t)$ only, upper and lower solutions may be assumed to be functions of t instead of constants. We shall merely state as exercises results corresponding to that situation. In fact, we shall restrict ourselves to the BVP (5.1.3), (5.1.4).

EXERCISE 5.2.1. With respect to the BVP (5.1.3), (5.1.4), let hypotheses (i) and (ii) of Theorem 5.2.1 hold. Furthermore suppose that there exist lower and upper solutions $\alpha(t),\beta(t)$ relative to (5.1.3), (5.1.4) such that $\alpha(t) \le \beta(t)$ on R. Then there is a solution $x(t)$ of (5.1.3), (5.1.4) satisfying $\alpha(t) \le x(t) \le \beta(t)$ on R.

EXERCISE 5.2.2. Relative to the BVP (5.1.3), (5.1.4), let all the hypotheses of Theorem 5.2.2 hold except that α,β are not assumed to be constant lower and upper solutions. Then there exists a solution $x(t)$ of (5.1.3), (5.1.4) such that $\alpha(t) \le x(t) \le \beta(t)$ on R.

5.3 SHOOTING METHOD

Let us consider the second-order delay differential equation

$$x''(t) = f\big(t,x(t),x(t-h(t))\big), \qquad 0 \le t \le T, \tag{5.3.1}$$

where $h \in C\big[[0,T],R^+\big]$, subject to the boundary conditions

$$x(t) = \varphi(t), \quad t \in [-c,0], \quad x(T) = A, \tag{5.3.2}$$

where $\varphi \in C\big[[-c,0],R\big]$ and $-c = \min[t-h(t)\colon 0 \le t \le T]$. We assume that

(1) $f \in C\big[[0,T]\times R^2,R\big]$, $f(t,x,y)$ is nonincreasing in y for each (t,x);

(2) $P,P_0,q \in C\big[[0,T],R\big]$, $q(t) \ge 0$ such that

$$-P(t)(x-\bar{x}) - q(t)(y-\bar{y}) \le f(t,x,y) - f(t,\bar{x},\bar{y}) \le -P_0(t)(x-\bar{x}),$$

if $x \le \bar{x}$, $y \le \bar{y}$ and $t \in [0,T]$;

(3) the solution $u(t)$ of the initial value problem (IVP)

$$u''(t) + P(t)u(t) + q(t)u\big(t-h(t)\big) = 0, \tag{5.3.3}$$

$$u(t) \equiv 0, \quad -c \le t \le 0, \qquad u'(0+) = 1, \tag{5.3.4}$$

is positive on $(0,T]$.

LEMMA 5.3.1. Under assumptions (1) - (3), the IVP (5.3.1) with the initial conditions $x(t) = \varphi(t)$, $t \in [-c,0]$, $x'(0+) = s$ has a unique solution.

The conclusion of Lemma 5.3.1 is immediate because hypothesis (2) implies that f satisfies a uniform Lipschitz condition with respect to x, y, which in turn yields the existence and uniqueness of solutions of IVP's.

LEMMA 5.3.2. Under assumptions (1) - (3), the BVP (5.3.1), (5.3.2) has at most one solution.

<u>Proof</u>: Let $x(t)$, $y(t)$ be two different solutions of (5.3.1), (5.3.2) and set $m(t) = x(t) - y(t)$. By Lemma 5.3.1, $x'(0+) \neq y'(0+)$. Without loss of generality, we may therefore suppose that there exists a $t_1, 0 < t_1 \leq T$ such that $m(t) > 0$, $0 \leq t \leq t_1$ and $m(0) = m(t_1) = 0$. Also $m(t) \equiv 0$ on $[-c,0]$. Because of (2), we then arrive at the differential inequality

$$m''(t) + p(t)m(t) + q(t)m\big(t - h(t)\big) \geq 0, \qquad 0 \leq t \leq t_1.$$

In view of assumption (3), it is possible to choose an $r > 0$ sufficiently large so that $ru(t) > m(t)$, $0 < t \leq t_1$, and since (5.3.3) is linear and homogeneous, $ru(t)$ is again a solution of (5.3.3).

Consider now the BVP

$$u''(t) + p(t)u(t) + q(t)u\big(t - h(t)\big) = 0, \tag{5.3.5}$$

$$u(t) \equiv 0 \quad \text{on} \quad [-c,0], \qquad u(t_1) = 0. \tag{5.3.6}$$

Clearly $m(t)$ is a lower solution and $ru(t)$ is an upper solution of (5.3.5), (5.3.6). Recalling the fact that $q(t) \geq 0$, we conclude, on the basis of Exercise 5.2.2, that there is a solution $y(t)$ of the BVP (5.3.5), (5.3.6) such that

$$m(t) \le y(t) \le ru(t), \qquad 0 \le t \le t_1.$$

However, $y(t)$ is a solution of (5.3.3) and consequently $y(t) = \tilde{r}u(t)$ for some $\tilde{r} < 0$. This contradicts the fact that $u(t_1) > 0$. Hence the proof is complete.

LEMMA 5.3.3. Let $u(t)$, $x(t,s)$, and $v(t)$ be the solutions of the IVP's (5.3.3), (5.3.4), (5.3.1); $x(t) = \varphi(t)$ on $[-c,0]$, $x'(0+) = s$; and $v'' + P_0(t)v = 0$, $v(0) = 0$, $v'(0+) = 1$, respectively. Then for $\tilde{s} \ge s$, we have

$$(5.3.7) \qquad 0 \le (\tilde{s} - s)u(t) \le x(t,\tilde{s}) - x(t,s) \le (\tilde{s} - s)v(t), \quad 0 \le t \le T.$$

The proof of this lemma is similar to the proof of Lemma 5.3.2, and hence we leave it to the reader.

LEMMA 5.3.4. Let $u(t)$, $x(t,s)$, and $v(t)$ be as in Lemma 5.3.3. Let $M[u(T) + v(T)] = 2$ and $g(s) = s - M[x(T,s) - A]$. Then $g(s)$ has a unique fixed point.

Proof: Let $\tilde{s} > s$. Then

$$g(\tilde{s}) - g(s) = \tilde{s} - s - M[x(T,\tilde{s}) - x(T,s)].$$

Using (5.3.7), we then see that

$$|g(s) - g(\tilde{s})| \le \gamma|s - \tilde{s}|$$

where $\gamma = \big(v(T) - u(T)\big)/\big(u(T) + v(T)\big)$. Since $0 \le \gamma < 1$, g is a contraction mapping and therefore has a unique fixed point s^*.

In view of the preceding lemmas, if we note that $x(T,s^*) = A$, we have proved the following result.

THEOREM 5.3.1. Suppose that hypotheses (1) - (3) hold. Then the BVP (5.3.1), (5.3.2) has a unique solution which may be computed by the shooting method.

Assumption (2) above, may be weakened. Let us replace it by the following:

(4) $p, q \in C[[0,T],R]$, $q(t) \geq 0$ and

$$-p(t)(x-\bar{x}) - q(t)(y-\bar{y}) \leq f(t,x,y) - f(t,\bar{x},\bar{y}),$$

if $x \leq \bar{x}$, $y \leq \bar{y}$, $t \in [0,T]$;

(5) the IVP (5.3.1), $x(t) = \varphi(t)$ on $[-c,0]$, $x'(0+) = s$ has a unique solution which extends to $[0,T]$.

Then we have the following theorem.

THEOREM 5.3.2. Assumptions (1),(3) - (5) imply the conclusion of Theorem 5.3.1.

Proof: We observe that in proving Lemma 5.3.2 and the first half of inequality (5.3.7), assumption (4) rather than (2) was employed. Thus the BVP (5.3.1), (5.3.2) has at most one solution and for $\tilde{s} > s$,

$$0 < (\tilde{s} - s)u(t) \leq x(t,\tilde{s}) - x(t,s), \qquad 0 < t \leq T,$$

holds. Let s be fixed and let $\tilde{s} \to \infty$. Then, since $u(T) > 0$, we obtain $x(T,\tilde{s}) \to \infty$. Keeping $\tilde{s}$ fixed and letting $s \to -\infty$, we conclude similarly that $x(T,s) \to -\infty$. Assumption (5) implies that $x(T,s) - A$ is a continuous function of s, which must cover the whole real line by the above argument. Hence there exists an s^* such that $x(T,s^*) = A$ and the proof is complete.

5.4 NONHOMOGENEOUS LINEAR BOUNDARY CONDITIONS

In this section, we shall be concerned with the question of existence of solutions of a quasilinear functional differential equation subjected to nonhomogeneous linear boundary conditions. The equations considered include both functional differential equations of retarded type and of advanced type.

Let $h, k \geq 0$ and let $T > 0$. As before, let $C[[0,T],R^n]$ denote the Banach space with the norm $\|x\|_0 = \sup_{0<t<T} \|x(t)\|$. We need the following function spaces simultaneously and consequently, we shall adopt the notation given below for convenience.

$$
\begin{aligned}
C &\equiv C[[0,T],R^n], & \text{norm} &\equiv \|\cdot\|_0, \\
C_1 &\equiv C[[-h,0],R^n], & \text{norm} &\equiv \|\cdot\|_1, \\
C_2 &\equiv C[[T,T+k],R^n], & \text{norm} &\equiv \|\cdot\|_2, \\
C_3 &\equiv C[[-h,k],R^n], & \text{norm} &\equiv \|\cdot\|_3, \\
C_4 &\equiv C[[-h,T+k],R^n], & \text{norm} &\equiv \|\cdot\|_4.
\end{aligned}
$$

In each of the above cases the norm is the sup norm. For any $x \in C_4$ and $t \in [0,T]$, we define the element $x_t \in C_3$ by the relationship

$$x_t(\theta) = x(t+\theta), \qquad -h \leq \theta \leq k.$$

Let $A(t)$ be a continuous $n \times n$ matrix with domain $[0,T]$ and let $f \in C[[0,T] \times C_3, R^n]$. We consider the functional differential equation

$$x'(t) = A(t)x(t) + f(t,x_t), \qquad 0 \leq t \leq T. \tag{5.4.1}$$

Let $\Phi(t), \psi(t)$ be continuous $n \times n$ matrices with respective domains $[-h,0]$, $[T,T+k]$ such that

$$\Phi(0) = I = \psi(T),$$

where I is the identity matrix. Let L be a continuous linear operator on C into R^n. We are interested in the existence of at least one solution $x(t)$ of (5.4.1) which, for a given $r \in R^n$, satisfies the following additional conditions

$$Lx(t) = r, \qquad 0 \leq t \leq T, \tag{5.4.2}$$

$$x(t) = \varphi(t)x(0), \qquad -h \leq t \leq 0, \tag{5.4.3}$$

$$(5.4.4) \qquad x(t) = \psi(t)x(T), \qquad T \le t \le T + k.$$

If $B(t)$ is a continuous matrix on $[a,b]$, we let $\|B(t)\| = \sup_{\|r\|=1} \|B(t)r\|$, $r \in R^n$, and $\|B\|_0 = \sup_{t\in[a,b]} \|B(t)\|$. Furthermore, we denote by $\|\cdot\|$ the usual norm for a linear operator L.

Recall the fact that the solution of

$$(5.4.5) \quad x' = A(t)x + b(t), \qquad Lx(t) = r, \qquad r \in R^n,$$

for every $b \in C$ is given by

$$(5.4.6) \quad x(t) = U(t)[L^n U(t)]^{-1}\left[r - L(U(t) \int_0^t U^{-1}(s)b(s)\, ds\right] + U(t) \int_0^t U^{-1}(s)b(s)\, ds,$$

and that BVP (5.4.5) has a unique solution for every $r \in R^n$ and every $f \in C$ if and only if the BVP

$$x' = A(t)x, \qquad Lx(t) = 0$$

has only the trivial solution. Here $U(t)$ is the fundamental matrix solution of $x' = A(t)x$ such that $U(0) = I$.

Let us define a mapping $S\colon C_4 \to C_4$ in the following way. For each $y \in C_4$, let

$$Sy(t) = x(t),$$

where

(i) for each $t \in [0,T]$, $x(t)$ is the unique solution of the differential equation

$$x' = A(t)x + f(t,y_t), \qquad 0 \le t \le T,$$

satisfying the boundary condition $Lx(t) = r$;

(ii) for $t \in [-h,0]$, $x(t) = \Phi(t)x(0)$;

(iii) for $t \in [T,T+k]$, $x(t) = \psi(t)x(T)$.

Noticing that for each $y \in C_4$, $f(t,y_t)$ is a continuous function of t for $t \in [0,T]$, it is clear, from the fact mentioned above, that the mapping S is well defined.

We shall next show that the mapping S is continuous on C_4. Let $y,z \in C_4$. Then

$$\|Sy - Sz\|_4 = \max_{0 \leq i \leq 2} [\|Sy - Sz\|_i].$$

Also, from (5.4.6) we obtain

$$\begin{aligned}(5.4.7)\quad \|Sy - Sz\|_0 \leq \|U\|_0 \Big[&\|(L^n U(t))^{-1}\| \, \Big\| L\Big\{U(t) \\ &\times \int_0^t U^{-1}(s)\big(f(s,y_s) - f(s,z_s)\big)\, ds\Big\} \Big\| \\ &+ T\|U^{-1}\|_0 \|f(t,y_t) - f(t,z_t)\|_0 \Big] \\ \leq T\|U\|_0 \|U^{-1}\|_0 &[\|(L^n U)^{-1}\|_0 \; \|L\| \; \|U\|_0 + 1] \\ &\times \|f(t,y_t) - f(t,z_t)\|_0.\end{aligned}$$

Since $U(t)$ is a fundamental matrix solution and L is a continuous linear operator, there exist positive constants $\alpha, \beta, \gamma, \delta$ such that

$$\|U\|_0 \leq \alpha, \qquad \|U^{-1}\|_0 \leq \beta, \qquad \|(L^n U)^{-1}\|_0 \leq \gamma, \qquad \|L\| \leq \delta.$$

Consequently, (5.4.7) implies that

$$\|Sy - Sz\|_0 \leq \alpha\beta T(\alpha\gamma\delta + 1)\|f(t,y_t) - f(t,z_t)\|_0.$$

Moreover, we have,

$$\begin{aligned}\|Sy - Sz\|_1 &\leq \|\Phi\|_1 \|Sy(0) - Sz(0)\| \\ &\leq \|\Phi\|_1 \|Sy - Sz\|_0,\end{aligned}$$

and

$$\begin{aligned}\|Sy - Sz\|_2 &\leq \|\psi\|_2 |Sy(T) - Sz(T)| \\ &\leq \|\psi\|_2 \|Sy - Sz\|_0.\end{aligned}$$

The foregoing three inequalities, then yield

$$\|Sy - Sz\| \leq \max[\|\Phi\|_1, \|\psi\|_2, 1]\alpha\beta T(\alpha\gamma\delta + 1)\|f(t,y_t) - f(t,z_t)\|_0,$$

which, in view of the assumed continuity of f on $[0,T] \times C_3$ shows that S is a continuous operator on C_4.

We shall impose conditions on f to ensure that for some $\rho > 0$, $S(B_\rho) \subseteq B_\rho$, where $B_\rho = [y \in C_4\colon \|y\|_4 \leq \rho]$. Let $y \in B_\rho$; then it follows from (5.4.6) that

$$\begin{aligned}\|Sy\|_0 &\leq \alpha\gamma(\|r\| + \delta\alpha\beta T\|f(t,y_t)\|_0) + \alpha\beta T\|f(t,y_t)\|_0, \\ \|Sy\|_1 &\leq \alpha\gamma(\|r\| + \delta\alpha\beta T\|f(t,y_t)\|_0)\ \|\Phi\|_1, \\ \|Sy\|_2 &\leq [\alpha\gamma(\|r\| + \delta\alpha\beta T\|f(t,y_t)\|_0) + \alpha\beta T\|f(t,y_t)\|_0]\ \|\psi\|_2.\end{aligned}$$

These inequalities imply that

$$\|Sy\|_4 \leq \eta_1 + \eta_2\|f(t,y_t)\|_0,$$

where η_1, η_2 are constants depending on $\alpha, \beta, \gamma, \delta, T$, $\|\Phi\|_0$ and $\|\psi\|_1$. Thus if ρ is such that

$$\rho \geq \eta_1 + \eta_2\|f(t,y_t)\|_0,$$

whenever $y \in B_\rho$, then $S(B_\rho) \subseteq B_\rho$. In particular, ρ may be chosen this way provided f satisfies the condition that for sufficiently large ρ

$$\|f(t,u)\| \leq (\rho - \eta_1)/\eta_2, \tag{5.4.8}$$

whenever $\|u\|_3 \leq \rho$.

We next verify that $S(B_\rho)$ is sequentially compact. Indeed, since $S(B_\rho) \subseteq B_\rho$, $S(B_\rho)$ is uniformly bounded. Furthermore, if $y \in B_\rho$, then for $t \in [0,T]$

$$(Sy)'(t) = A(t)Sy(t) + f(t,y_t)$$

and hence the assumption (5.4.8) on f imply that $\{Sy\}$ is equicontinuous on $[0,T]$. Moreover, as

$$Sy(t) = \Phi(t)Sy(0), \qquad \text{for } -h \le t \le 0,$$

and

$$Sy(t) = \psi(t)Sy(T), \qquad \text{for } T \le t \le T + k,$$

it is evident that $\{Sy\}$ is also equicontinuous on $[-h,0]$ and $[T,T+k]$. The sequential compactness is therefore a consequence of Ascoli's theorem.

We summarize the foregoing considerations in the following existence theorem.

THEOREM 5.4.1. Let $f(t,u)$ be continuous on $[0,T]\times C_3$ and satisfy (5.4.8). Assume that the BVP (5.4.5) has a unique solution for every $r \in R^n$ and every $b \in C$ if and only if the BVP $x' = A(t)x$, $Lx(t) = 0$ **has only the trivial solution. Then** the BVP (5.4.1) - (5.4.4) has at least one solution.

5.5 LINEAR PROBLEMS

In the following section we shall continue our study of boundary value problems for functional differential equations.

Some preliminary notation will be needed. Let C_h denote the Banach space of continuous functions from $[a-h,a]$ into R^n, where for $\varphi \in C_h$

$$\|\varphi\|_0 = \sup\big[\|\varphi(\theta)\|: \theta \in [a-h,a]\big].$$

Let $L(t,\varphi)$ and $f(t,\varphi)$ be continuous mappings from $[a,b]\times C_h \to R^n$ and for each $t \in [a,b]$, let $L(t,\varphi)$ be a bounded linear operator from C_h into R^n. For any continuous function $f(t)$ defined on $[a,b]$, let f_t denote the element of C_h defined by $f_t(\theta) = f(t+\theta)$, $\theta \in [a-h,a]$, $t \in [a,b]$.

We will consider the two point BVP

$$(5.5.1) \qquad y'(t) = L(t,y_t) + f(t,y_t), \qquad t \in [a,b]$$

$$(5.5.2) \qquad My_a + Ny_b = \psi, \qquad \psi \in C_h, \quad b > a + h,$$

where M and N are bounded linear operators on C_h.

We shall assume:

(H_1) There exists a bounded integrable function $\ell(t)$ such that $\|L(t,\varphi)\| \leq \ell(t)\|\varphi\|_0$ for $t \in [a,b]$ and $\varphi \in C_h$.

(H_2) $f(t,\varphi)$ maps closed bounded subsets of $[a,b] \times C_h$ into bounded sets in R^n and satisfies

$$(5.5.3) \quad \lim\sup_{\|\varphi\|_0 \to \infty} \|f(t,\varphi)\|/\|\varphi\|_0 = 0, \qquad \text{uniformly for } t \in [a,b].$$

(H_3) Solutions of the initial value problem (5.5.1) with $y_a = q \in C_h$ exist and are unique.

We will make use of the properties of the linear equations

$$(5.5.4) \qquad x'(t) = L(t,x_t),$$

$$(5.5.5) \qquad x'(t) = L(t,x_t) + g(t),$$

$g \in C[[a,b],R^n]$. For any initial $q \in C_h$, we write $x(q,g)(t)$ as the solution of (5.5.5) satisfying $x_a(q,g) = q$. For each $g \in C[[a,b],R^n]$ and $q \in C_h$ a solution of the initial value problem (5.5.5) exists and is unique. The solution can be represented as

$$(5.5.6) \quad x(q,g)(t) = x(q,0)(t) + x(0,g)(t), \qquad t \geq a-h.$$

From Gronwall's inequality, it follows, using (5.5.5) and (H_1), that

$$(5.5.7) \qquad \|x_t(q,g)\|_0 \le \left(\|q\|_0 + \int_0^t |g(s)|\,ds\right) \exp \int_a^t \ell(s)\,ds.$$

The following elementary result for the linear equation (5.5.4) gives a necessary and sufficient condition for the BVP (5.5.5), (5.5.2) to have a solution.

LEMMA 5.5.1. The two point BVP (5.5.5), (5.5.2) has a solution if and only if $Nx_b(0,g) \in \psi + R(M + NX)$, where R denotes the range and X is an operator on C_h defined by

$$(5.5.8) \qquad Xq = x_b(q,0),$$

which maps $q \in C_h$ into the segment at b of the solution of the initial value problem (5.5.4) with $x_a = q$.

<u>Proof</u>: Suppose $Nx_b(0,g) \in \psi + R(M + NX)$. To show the BVP (5.5.5), (5.5.2) has a solution, it suffices to establish the existence of a solution of the functional equation

$$(5.5.9) \qquad Mq + Nx_b(q,g) = \psi, \qquad \psi \in C_h.$$

Using (5.5.6), we can write (5.5.9) as

$$(5.5.10) \qquad Mq + NXq + Nx_b(0,g) = \psi.$$

This has a solution since $Nx_b(0,g) \in \psi + R(M + NX)$. Conversely, if the boundary value problem (5.5.5) and (5.5.2) has a solution $x(t)$ on $[a-h,b]$, then $x_a = q$ satisfies (5.5.8) and (5.5.9). Consequently, $Nx_b(0,g) \in \psi + R(M + NX)$ completing the proof.

We also need the following result on the operator X.

LEMMA 5.5.2. The operator X is a completely continuous linear operator on C_h.

Proof: For $q_1, q_2 \in C_h$ and scalars α and β, define

$$z(t) = \alpha x(q_1,0)(t) + \beta x(q_2,0)(t) - x(\alpha q_1 + \beta q_2,0)(t).$$

Immediately, we see, due to the linearity of $L(t,\varphi)$, that $z(t)$ satisfies (5.5.4) with $z_a = 0$. By uniqueness $z(t) \equiv 0$, proving X is linear. Moreover, X is continuous due to the fact that solutions depend continuously on initial conditions. It remains to show X is compact. Let $\{q_n\}$ be a bounded sequence in C_h, $\|q_n\|_0 \leq M$. From (5.5.7)

$$\|Xq_n\|_0 = \|x_b(q_n,0)\|_0 \leq \|q_n\|_0 \exp \int_a^b \ell(s)\, ds.$$

Hence $\{Xq_n\}$ is uniformly bounded. Moreover

$$\begin{aligned}\|x_b'(q_n,0)\|_0 &= \sup_{b-h \leq t \leq b} \|L(t,x_t(q_n))\| \\ &\leq \sup_{b-h \leq t \leq b} \ell(t)\|x_t(q_n)\|_0 \\ &\leq \sup_{b-h \leq t \leq b} \ell(t)\|q_n\|_0 \exp \int_a^b \ell(s)\, ds.\end{aligned}$$

Hence $\{Xq_n\}$ are equicontinuous and an application of the Ascoli's theorem proves that X is completely continuous.

Before stating our main results we need two lemma's which provide conditions on an operator so that the range of the operator is the whole space. We omit the proofs.

LEMMA 5.5.3. Let Ψ be a completely continuous mapping of a Banach space B into itself. If

$$\limsup_{\|\varphi\| \to \infty} \|\Psi\varphi\|/\|\varphi\| < 1, \tag{5.5.11}$$

then $R(I - \Psi) = B$. Recall $R(I - \Psi)$ means the range of $I - \Psi$.

LEMMA 5.5.4. Let T_0 be a contraction operator of a

Banach space B into itself. Let T_1 be a completely continuous operator of a Banach space into itself such that

$$\lim_{\|\varphi\|\to\infty} \|T_1\varphi\|/\|\varphi\| = 0. \tag{5.5.12}$$

Then $R(I + T_0 + T_1) = B$.

LEMMA 5.5.5. Let T be a completely continuous operator from a Banach space B into B. Assume that the only solution of the equation

$$(I + T)x = 0$$

is $x \equiv 0$. Then for each $y \in B$, there exists a unique $z \in B$ such that

$$(I + T)z = y.$$

Lemma 5.5.5 is often referred to as the Fredholm alternative.

5.6 NONLINEAR PROBLEMS

We now consider a result for the BVP (5.5.1), (5.5.2).

THEOREM 5.6.1. Assume hypotheses (H_1) - (H_3) hold for Eq. (5.5.1). Let the operator $(M + NX)$, where M and N are given in (5.5.2) and X in (5.5.8), have a continuous inverse. Then the BVP (5.5.1), (5.5.2) has at least one solution.

Proof: For each $q \in C_h$, denote by $y(t,q)$ the solution of the initial value problem (5.5.1) with $y_a = q$. Define $g(q)(t) = f(t,y_t(q))$. Then $y(t,q)$ is a solution of the nonhomogeneous linear equation

$$x'(t) = L(t,x_t) + g(q)(t). \tag{5.6.1}$$

Thus

$$y(t,q) = x(q,g)(t) = x(q,0)(t) + x\big(0,g(q)(t)\big).$$

Using (5.5.6), define a mapping $T\colon C_h \to C_h$ by

$$\text{(5.6.2)} \qquad Tq = y_b(q) \equiv x_b(q,0) + x_b\big(0,g(q)\big).$$

A solution of the BVP is determined by an initial condition which is a solution of the functional equation

$$\text{(5.6.3)} \qquad (M + NT)q = \psi, \qquad \psi \in C_h.$$

This may be rewritten as $\big(M + N\chi + N(T - \chi)\big)q = \psi$.

Let $\Gamma = (M + N\chi)^{-1}$ which exists by hypothesis. Then (5.6.3) can be rewritten as

$$\text{(5.6.4)} \qquad \big(I + \Gamma N(T - \chi)\big)q = \Gamma\psi, \qquad \psi \in C_h.$$

We will use Lemma 5.5.3 to show the existence of solutions by letting $\Psi = \Gamma N(T - \chi)$ and $B = C[0,h]$. Since ΓN is continuous, it is sufficient to show that $T - \chi$ is completely continuous in order for Ψ to be completely continuous.

We now show $T - \chi$ is completely continuous. Let $q \in C_h$ and consider the solution $y(t,q)$ of (5.5.1) with $y_a = q$. Then, by (5.5.3), (5.5.7), we have

$$\|y_t(q)\|_0 \le L\left(\|q\|_0 + M_0 + \int_a^t \|y_s(q)\|\, ds\right),$$

where $L = \exp\big(\int_a^b \ell(s)\, ds\big)$ and M_0 is any constant such that $\|f(t,\varphi)\| \le \|\varphi\|_0 + M_0$ for all $t \in [a,b]$. An application of Gronwall's inequality then leads to

$$\text{(5.6.5)} \qquad \|y_t(q)\|_0 \le L\big(\|q\|_0 + M_0\big)e^{L(b-a)} \equiv K_1\|q\|_0 + K_2.$$

Let $\{q_n\}$ be any bounded sequence in C_h. Observe that from (5.5.7), (5.6.2), and (5.6.5), we obtain

$$\begin{aligned}\|(T-\chi)q_n\|_0 &= \|x_b(0,g(q_n))\|_0 \\ &\le L\int_a^b \|f(s,y_s(q_n))\|\,ds \\ &\le L\int_a^b \left(M_0 + \|y_s(q_n)\|_0\right) ds \\ &\le L(b-a)\left(M_0 + K_1\|q_n\|_0 + K_2\right),\end{aligned}$$

which shows $\{(T-\chi)q_n\}$ is uniformly bounded. In addition, by (5.5.3)

$$(5.6.6)\qquad \begin{aligned}[T-\chi q_n]' &= \|x_b'(0,g(q_n))\|_0 \\ &\le \sup_{a\le t\le b} \left(\ell(t)\|x_t(0,g(q_n)\|_0 + M_0 + \|y_t(q_n)\|_0\right).\end{aligned}$$

Furthermore, for $t \in [a,b]$, we derive from (5.5.6), (5.5.7), and (5.6.5)

$$\begin{aligned}\|x_t(0,g(q_n))\|_0 &\le \|y_t(q_n)\|_0 + \|x_t(q_n,0)\|_0 \\ &\le K_1\|q_n\|_0 + K_2 + L\|q_n\|_0.\end{aligned}$$

From this estimate, (H_1), and (5.6.5), we see, using (5.6.6), that $\|x_b'(0,g(q_n))\|_0$ is uniformly bounded. An application of Ascoli's theorem yields the complete continuity of $T-\chi$.

In order to apply Lemma 5.5.3, it is sufficient to show (5.5.11) holds. Suppose there exists a sequence of functions $q_n \in C_h$, $\|q_n\| \to \infty$ as $n \to \infty$ such that $\beta_n \equiv \sup_{a\le t\le b} \|y_t(q_n)\| \to \infty$ as $n \to \infty$. Let $0 < \varepsilon < 1/K_1$. By (5.5.3), there exists an $r > 0$ such that if $\|\varphi\|_0 > r$, $\|f(t,\varphi)\| \le \varepsilon\|\varphi\|_0$ for all $t \in [a,b]$. Since for each $s \in [a,b]$, $\|y_s(q_n)\|_0 > r$ or $\|y_s(q_n)\|_0 \le r$, we arrive at

$$(5.6.7)\quad \|x_b(0,g(q_n))\|_0 \le L\int_a^b \|f(s,y_s(q_n))\|\,ds \le L(b-a)\max(R,\varepsilon\beta_n),$$

where $R = \max(\|f(s,\varphi)\|: a \leq s \leq b, \|\varphi\| \leq r)$. From (5.6.5), $\beta_n \leq K_1\|q_n\|_0 + K_2$; hence

$$\|(\chi - T)q_n\| \equiv \|x_b(0,g(q_n))\|_0 \leq L(b-a)\max(R,\varepsilon K_1\|q_n\|_0 + \varepsilon K_2).$$

Since ε is arbitrary, (5.5.11) holds with $\Psi = T - \chi$ for any sequence q_n such that the corresponding sequence $\beta_n \to \infty$ as $n \to \infty$. However, if $\{\beta_n\}$ are bounded, then (5.6.7) implies

$$\lim_{\|q_n\|_0 \to \infty} \frac{\|x_b(0,g(q_n))\|_0}{\|q_n\|_0} = 0,$$

so again (5.5.11) holds. This completes the proof of the theorem.

COROLLARY 5.6.1. Assume that the two point BVP (5.5.4), (5.5.2) has only the trivial solution for $\psi \equiv 0$. If M has a closed inverse, then the BVP (5.5.1), (5.5.2) has a solution.

<u>Proof</u>: By the closed graph theorem, M^{-1} is continuous. The relation (5.5.2) may be rewritten as

$$Iy_a + M^{-1}Ny_b \equiv (I + M^{-1}N\chi)y_a = M^{-1}\psi, \qquad \psi \in C_h.$$

Hence it is sufficient to prove $(I + M^{-1}N\chi)^{-1}$ exists in order to use Theorem 5.6.1. To do this we will invoke Lemma 5.5.5. Since χ is completely continuous, so is $M^{-1}N\chi$. Hence by hypothesis the only solution of

$$(I + M^{-1}N\chi)y_a = 0$$

is $y_a \equiv 0$. Thus $(I + M^{-1}N\chi)$ has an inverse and the result follows.

COROLLARY 5.6.2. Suppose that the mappings L and f in (5.5.1) are periodic with period $P = b - a$; that is $L(t+P,\varphi) = L(t,\varphi)$ and $f(t+P,\varphi) = f(t,\varphi)$ for all $t \geq a$ and all $\varphi \in C_h$. If the only P-periodic solution of (5.5.4) is the identically zero solution, then Eq. (5.5.1) has a P-periodic solution on $[a,\infty)$.

Proof: Let $M = I$, $N = -I$, and $\psi \equiv 0$ in (5.5.2) and apply Corollary 5.6.1.

The following exercise provides the uniqueness of solutions of the BVP by assuming a Lipschitz condition on $f(t,\varphi)$ and utilizing the contraction mapping theorem.

EXERCISE 5.6.1. Consider the BVP (5.5.1), (5.5.2) and assume (H_2) holds. Let

$$\|f(t,\varphi_1) - f(t,\varphi_2)\| \leq K\|\varphi_1 - \varphi_2\|_0,$$

for all $\varphi \in C_h$. If the operator $M + N\chi$, where χ is defined in (5.5.8), has a bounded inverse and if $(b - a)$ is sufficiently small, then the BVP (5.5.1), (5.5.2) has a unique solution.

Hint: Consider the operator F defined by

$$Fq = \Gamma\psi - \Gamma N(T - \chi)q, \qquad \psi \in C_h,$$

where $\Gamma \equiv (M + N\chi)^{-1}$ and T is defined by (5.6.2). Show F defines a contraction on C_h by utilizing the Lipschitz condition on f and choosing $b - a$ sufficiently small. Then show a fixed point of F yields a solution to the BVP (5.5.1), (5.5.2).

EXAMPLE 5.6.1. We now exhibit an example which illustrates

Theorem 5.6.1. Consider the functional differential equation

$$(5.6.8) \qquad y'(t) = y(t-1) + f(t,y_t), \qquad t \in [0,2]$$

subject to the boundary conditions

$$(5.6.9) \qquad y_0 = \tfrac{1}{4} y_2, \qquad y_0, y_2 \in C[-1,0].$$

Here $L(t,y_t) = y(t-1)$, $M = I, N = -\frac{1}{4} I$, $a = 0$, $b = 2$, $h = 1$, $\psi = 0$, and $f \in C\left[[0,2]\times C[-1,0], R^n\right]$ satisfying (H_1) and (H_2). For example, let $f(t,\varphi) = \lambda(t)\|\varphi\|_0^\sigma$, $0 < \sigma < 1$, $\lambda \in C[0,2]$. Observe that the initial value problem for the homogeneous linear equation

$$(5.6.10) \qquad x'(t) = x(t-1)$$

can be solved by the method of steps; that is, for any initial function $q \in C[-1,0]$, we have

$$(5.6.11) \quad x(t) = q(0) + \int_0^t q(s-1)\, ds, \qquad 0 \le t \le 1$$

$$= q(0) + \int_0^1 q(s-1)\, ds + q(0)(t-1)$$

$$+ \int_1^t \int_0^{u-1} q(s-1)\, ds\, du, \qquad 1 \le t \le 2.$$

Thus $\chi q(\theta)$ is given by

$$\chi q(\theta) = \int_1^{2+\theta} \int_0^{u-1} q(s-1)\, ds\, du, \qquad \theta \in [-1,0]$$

and we immediately see that $\sup_{\|q\|_0=1} \|\chi q\| < 4$. Hence $(I - \frac{1}{4}\chi)$ has a continuous inverse and an application of Theorem 5.6.1 yields the existence of a solution of the BVP (5.6.8), (5.6.9).

EXERCISE 5.6.2. Prove the existence of a periodic solution of

$$y'(t) = y(t-1) + \|y_t\|_0^\sigma \sin t,$$

where $0 < \sigma < 1$.

Hint: Apply Corollary 5.6.2 to show the existence of a solution of period 2π.

5.7 DEGENERATE CASES

In this section we consider (5.5.1) when the linear part is not a functional on C_h. More specifically, we consider the functional differential equations of the form

$$y'(t) = A(t)y(t) + f(t, y_t), \tag{5.7.1}$$

where f is a continuous mapping from $[a,b] \times C_h$ into R^n and $A(t)$ is a $n \times n$ matrix function continuous in t. We shall assume that (H_2) and (H_3) hold.

We first consider the case when $A(t) \equiv 0$. In this case the operator X reduces to

$$Xq = \bar{q}(a), \tag{5.7.2}$$

where $\bar{q}(a)$ is the function in C_h with the constant value $q(a)$. The following equivalent norm on C_h will be useful. For each $q \in C_h$, define

$$\|q\|_a = \|q(a)\| + \|q\|_0. \tag{5.7.3}$$

For a linear operator T on C_h define

$$\|T\|_a = \sup_{\|q\|_a = 1} (\|Tq\|_a).$$

THEOREM 5.7.1. If $\Gamma_0 = (M+N)^{-1}$ exists, is continuous, and $\|\Gamma_0 N\|_a < 1$, then the BVP (5.7.1), (5.5.2) with $A(t) \equiv 0$ has a solution.

Proof: In view of Theorem 5.6.1, it is enough to show that $(M+NX)^{-1}$ exists and is continuous where X is given by (5.7.2). Consider the functional equation

$$(M+NX)q = \varphi, \; \varphi \in C_h, \tag{5.7.4}$$

which may be rewritten as

$$\big(M + N + N(X - I)\big)q = \varphi, \qquad \varphi \in C_h,$$

which is equivalent to

$$I + \Gamma_0 N(X - I)\; q = \Gamma_0 \varphi. \tag{5.7.5}$$

Hence, by Lemma 5.5.4, it suffices to show that $\Gamma_0 N(X - I)$ is a contraction, for then, the inverse of $M + NX$ would exist. For any $q_1, q_2 \in C_h$

$$\begin{aligned}\|(X - I)q_1 - (X - I)q_2\|_a &= \|(X - I)q_1 - (X - I)q_2\| \\ &\leq \|\overline{q}_1(a) - \overline{q}_2(a)\| + \|q_1 - q_2\|_0 \\ &= \|q_1(a) - q_2(a)\| + \|q_1 - q_2\|_0 \\ &= \|q_1 - q_2\|_a.\end{aligned}$$

Since $\|\Gamma_0 N\|_a < 1$, the operator $\Gamma_0 N(X - I)$ is a contraction on C_h. Hence (5.7.5) has a unique solution for every $\varphi \in C_h$. This implies $\big(I + \Gamma_0 N(\Phi - I)\big)^{-1}$ exists. The continuity follows and thus

$$(M+NX)^{-1} = \big(I + \Gamma_0 N(X - I)\big)^{-1}\Gamma_0$$

is continuous. An application of Theorem 5.6.1 concludes the proof.

EXAMPLE 5.7.1. We show the existence of a solution of the BVP

$$y'(t) = f(t,y_t), \qquad My_a + Ny_b = 0,$$

where $Mq(\theta) = q(\theta) - \frac{1}{4}\,\overline{q}(a)$, $Nq(\theta) = \frac{1}{4}\,\overline{q}(a) + \frac{1}{4}\int_\theta^a q(\tau)\,d\tau$, $\theta \in [a-h,a]$ and $h < b-a$. We see $M + N = I + \frac{1}{4}A$, where $Aq(\theta) = \int_\theta^a q(\tau)\,d\tau$. Let us assume that $b-a > 1 = h$. Then $\|A\|_a < 1$. Thus $(M+N)^{-1}$ exists and satisfies $\|(M+N)^{-1}\|_a \le \frac{4}{3}$. Moreover, $\|N\|_a < \frac{3}{4}$. Hence $\|(M+N)^{-1}N\|_a \le \|(M+N)^{-1}\|_a \|N\|_a < 1$ and we may apply Theorem 5.7.1 to conclude the existence of a solution of the BVP for any $f(t,y_t)$ satisfying (H_2) and (H_3).

We now consider (5.7.1) when $A(t) \not\equiv 0$. For the case in which $A(t) \equiv 0$, the assumption on Γ_0 rules out the possibility of having periodic solutions. This is reasonable since the equation $x' \equiv 0$ has any constant as a solution with periodic boundary conditions. Now we will be interested in establishing the existence results of periodic solutions. Since we are now looking at the reduced equation

$$x'(t) = A(t)x(t), \qquad t \in [a,b], \tag{5.7.6}$$

the mapping X will assign the same value to the different functions $q_1, q_2 \in C_h$ as long as $q_1(a) = q_2(a)$. Hence it is unreasonable to expect the operator $M + NX$ to be invertible. In our previous result we imposed the condition $\|\Gamma_0 N\| < 1$ in order to insure the invertibility of $M + NX$. Here we present a result which will allow for periodic boundary conditions. We now assume that M and N are $n \times n$ matrices and let $U(t)$ be the fundamental matrix of (5.7.6) with $U(a) = I$, the identity matrix. Define the operator $X_U\colon C_h \to C_h$ by

$$X_U q(\theta) = U(b-a+\theta)q(\theta), \qquad \theta \in [a-h,a].$$

THEOREM 5.7.2. Let M and N be $n \times n$ matrices. Suppose that for all $\theta \in [a-h,a]$ the matrix $(M + NU(b-a+\theta))^{-1}$ is nonsingular and that $\Gamma(\theta) \equiv (M+NU(b-a+\theta))^{-1}$ satisfies

$$\|\Gamma N X_U\|_0 = \sup_{a-h \le \theta \le a} \|\Gamma(\theta) N U(b-a+\theta)\| < 1. \tag{5.7.7}$$

Then the BVP (5.7.1), (5.5.2) has at least one solution.

<u>Proof</u>: The solution $y_t(s)$ of (5.7.1) can be represented as

$$y_t(\theta) = U(t-a+\theta)[q(a) + \int_a^{t-a+\theta} U^{-1}(s) f(s, y_s(q))\, ds \tag{5.7.8}$$

for $\theta \in [a-h, a]$. Let T_0 be a mapping defined on C_h by

$$T_0 q(\theta) = q(a) + \int_a^{b-a+\theta} U^{-1}(s) f(s, y_s(q))\, ds. \tag{5.7.9}$$

From (5.7.8), (5.7.9), we have $y_b(q) \equiv X_U T_0 q$. Thus a solution of the BVP (5.7.1), (5.5.2) can be found by solving the functional equation

$$(M + N X_U T_0) q = \psi, \qquad \psi \in C. \tag{5.7.10}$$

Let Γ be an operator defined by

$$(\Gamma q)(\theta) = \Gamma(\theta) q(\theta), \qquad \theta \in [a-h, a].$$

Clearly, Γ is invertible. Therefore (5.7.10) is equivalent to

$$\big(I + \Gamma N X_U (T_0 - I)\big) q = \Gamma \psi, \qquad \psi \in C_h. \tag{5.7.11}$$

Let $T_0 - I \equiv T_1 + T_2$ defined by

$$T_1 q(\theta) = q(a) - q(\theta), \tag{5.7.12}$$

and

$$T_2 q(\theta) = \int_a^{b-a+\theta} U^{-1}(s) f(s, y_s(q))\, ds. \tag{5.7.13}$$

To obtain a solution of (5.7.11), we will use Lemma 5.5.4 and show that $\Gamma N \chi_U T_1$ is a contraction and $\Gamma N \chi_U T_2$ is completely continuous and satisfies (5.5.12). We again work with the norm $\|\cdot\|_a$ introduced in (5.7.3). Let $q_1, q_2 \in C_h$. Then from (5.7.12), one obtains, for any $\theta \in [a-h, a]$

$$\begin{aligned}\|(T_1 q_1 - T_1 q_2)(\theta)\| &\leq \|q_1(a) - q_2(a)\| + \|q_1(\theta) - q_2(\theta)\| \\ &\leq \|q_1(a) - q_2(a)\| + \|q_1 - q_2\|_0 \\ &= \|q_1 - q_2\|_a,\end{aligned}$$

and thus it follows that

$$\|T_1 q_1 - T_1 q_2\|_a \leq \|q_1 - q_2\|_a.$$

This together with (5.7.7) implies $\Gamma N \chi_U T_1$ is a contraction.

Observe, from (5.7.13) and (H_2), that the operator T_2 is completely continuous. Since $\Gamma N \chi_U$ is continuous, we immediately have that $\Gamma N \chi_U T_2$ is completely continuous.

From (5.7.9), we obtain

$$\|y_t(q)\|_0 \leq K\left(\|q(a)\| + \int_a^t \|U^{-1}(\theta) f(\theta, y_\theta(q))\| \, d\theta\right), \tag{5.7.14}$$

where $K = \sup_{a-h \leq t \leq b} \|U(t)\|$. Using (5.5.4) and the Gronwall inequality we deduce the existence of positive constants K_1 and K_2 such that $\|y_t(q)\|_0 \leq K_1 \|q\|_0 + K_2$ for all $t \in [a, b]$. It then follows that

$$\lim_{\|q\|_a \to \infty} \|T_2 q\|_a / \|q\|_a = 0$$

and hence with (5.7.7), we conclude that

$$\lim_{\|q\|_a \to \infty} \|\Gamma N \chi_U T_2 q\|_a / \|q\|_a = 0.$$

An application of Lemma 5.5.5 completes the proof.

EXERCISE 5.7.1. Show that the invertibility of the matrix $M + NU(b - a + \theta)$ for each $\theta \in [a - h, a]$ can be accomplished by the following requirement: Eq. (5.7.6) subject to the boundary condition $MX(a) + NX(\tau) = 0$ has only the trivial solution for each $\tau \in [b - h, b]$.

EXAMPLE 5.7.2. Consider the one-dimensional equation

$$y'(t) = -y(t) + f(t, y_t), \qquad t \geq 0, \tag{5.7.15}$$

where $f(t + \pi, y_t) = f(t, y_t)$ for all $t \geq 0$ and satisfies (H_2), (H_3), together with the periodic boundary conditions

$$y_0 = y_\pi, \qquad y_0, y_\pi \in C[-1, 0]. \tag{5.7.16}$$

Here $M = 1$, $N = -1$, $U(t) = e^{-t}$ and thus $\left(M + NU(b - a + \theta)\right)^{-1} = (1 - e^{-(\pi+\theta)})^{-1} = \Gamma(\theta)$. Moreover,

$$\|\Gamma N X_U\|_0 = \sup_{-1 \leq \theta \leq 0} |e^{-(\pi+\theta)}/(1 - e^{-(\pi+\theta)})| \leq 1/(e^{\pi-1} - 1) < 1.$$

Hence (5.7.7) holds and an immediate application of Theorem 5.7.2 yields the existence of a solution of the BVP (5.7.15), (5.7.16).

5.8 NOTES AND COMMENTS

The material contained in Sections 5.1 and 5.2 is taken from Grimm and Schmitt [1], while the work of Section 5.3 is due to De Nevers and Schmitt [1]. Section 5.4 consists of the results of Schmitt [9]. The contents of Sections 5.5 - 5.7 are based on the work of Waltman and Wong [4]. For the techniques of operator theory employed in these sections, see Granas [1].

For related results concerning functional differential equations, refer to Gustafson and Schmitt [1], Fennel [2],

Fennel and Waltman [1], Schmitt [4], Norkin [1], Mawhin [3], and Hale [1].

Chapter 6
SELECTED TOPICS

6.0 INTRODUCTION

This chapter is concerned with some selected topics of current interest. First of all, we present in a unified setting Newton's method which is of particular value in applications. We discuss as illustrations two numerical techniques of solving boundary value problems. These are the Goodman-Lance method and the method of quasilinearization, which have attracted considerable attention in recent years. We then consider nonlinear eigenvalue problems as an application of the angular function technique. The n-point boundary value problem is investigated in detail by studying n-parameter families of solutions. Under the assumption of uniqueness, some results on existence of solutions are derived.

6.1 NEWTON'S METHOD

Here we wish to present an abstract formulation of Newton's method in a unified setting and then in subsequent sections, deduce, from the general results for Newton's method, as illustrations, two numerical techniques of solving boundary value problems. These are the Goodman-Lance method and the method of quasilinearization, which have attracted considerable attention in recent years.

Let X and Y denote real Banach spaces and suppose that the basic equation to be solved,

$$f(x) = 0, \tag{6.1.1}$$

is given by a mapping of an open subset U of X into Y, which is continuously differentiable. This means that f has a derivative $f'(x)$ for every $x \in U$ such that $f'(\cdot)$ is a continuous mapping of U into the Banach space $L(X,Y)$ of continuous linear mappings of X into Y. As usual, $f'(x)$ is defined as the unique element in $L(X,Y)$ for which

$$\lim_{\|h\|\to 0} (1/\|h\|)\ \|f(x+h) - f(x) - f'(x)h\| = 0$$

so that, in particular,

$$\|f'(x)\| = \sup[\|f'(x)h\|:\ \|h\| \leq 1].$$

It follows then from the mean value theorem that there exists, for every $x_0 \in U$ and every $\varepsilon > 0$, an open ball $B(x_0,r) \subset U$ with center x_0 and radius r such that

$$\|f(x_2) - f(x_1) - f'(x_0)(x_2 - x_1)\| \leq \varepsilon\|x_2 - x_1\| \tag{6.1.2}$$

for any two points $x_1,x_2 \in B(x_0,r)$.

If f is twice differentiable in U, the second derivative $f''(x)$ of f at any $x \in U$ may be defined in a natural way with a continuous bilinear mapping of $X \times X$ into Y, and consequently

$$\|f''(x)\| = \sup[\|f''(x)(h_1,h_2)\|:\ \|h_1\| \leq 1,\ \|h_2\| \leq 1].$$

Furthermore, if in addition $\|f''(x)\| \leq k$, $x \in U$, we deduce, as a consequence of the mean value theorem, that

$$\|f'(x_2) - f'(x_1)\| \leq k\|x_2 - x_1\| \tag{6.1.3}$$

for every open ball $B(x_0,r) \subset U$ and $x_1,x_2 \in B(x_0,r)$. Also, as a consequence of Taylor's theorem, there results the inequality

$$\|f(x_2) - f(x_1) - f'(x_1)(x_2 - x_1)\| \leq \tfrac{1}{2}\, k\|x_2 - x_1\|^2. \tag{6.1.4}$$

When $X = R^n$, $Y = R^m$, our assumption implies that each partial derivative D_jf_i, $1 \leq i \leq m$, $1 \leq j \leq n$ exists and is continuous in U, and $f'(x)$ at $x = (x_1,x_2,\ldots,x_n)$ is the linear mapping of R^n into R^m given by the Jacobian matrix

$$J(x_1,x_2,\ldots,x_n) = D_j f_i(x_1,\ldots,x_n).$$

In particular, when $X = Y = R^m$, observe that $f'(x)$ is a linear homeomorphism of R^m onto itself if and only if the Jacobian, $\det J(x_1,x_2,\ldots,x_n)$ is different from zero.

Let us now state a result concerning the modified Newton's method.

THEOREM 6.1.1. Suppose that f is continuously differentiable in U and that there is an $x_0 \in U$ for which $f'(x_0)$ is a linear homeomorphism of X onto Y. Then, for any $\alpha \in (0,1)$, there is an open ball $B(x_0,r) \subset U$ such that, if

$$\|f'^{-1}(x_0)f(x_0)\| \leq r(1-\alpha), \tag{6.1.5}$$

there exists a unique $\tilde{x} \in B(x_0,r)$ for which $f(\tilde{x}) = 0$. Moreover, for any sequence $\{T_n\}$ of linear homeomorphisms of X onto Y satisfying

$$\|f'^{-1}(x_0)\| \; \|T_n - f'(x_0)\| \leq \frac{1}{3}\alpha \tag{6.1.6}$$

for every $n \geq 0$, the successive iterations

$$x_{n+1} = x_n - T_n^{-1}f(x_n), \qquad n = 0,1,2,\ldots, \tag{6.1.7}$$

define a sequence $\{x_n\}$ of points in $B(x_0,r)$ which converges to $\tilde{x}$ such that

$$\|\tilde{x} - x_{n+1}\| \leq \big(2\alpha/(3-\alpha)\big) \; \|\tilde{x} - x_n\| \tag{6.1.8}$$

and

$$\|\tilde{x} - x_n\| \leq \big(1/(1-\alpha)\big) \; \big(2\alpha/(3-\alpha)\big)^n \; \|f'^{-1}(x_0)f(x_0)\| \tag{6.1.9}$$

for every $n \geq 0$.

Proof: Since f is continuously differentiable in U, there is an open ball $B(x_0,r) \subset U$ for which $x \in B(x_0,r)$ implies

$$\|f'^{-1}(x_0)\| \; \|f'(x) - f'(x_0)\| \leq \frac{1}{3}\alpha. \tag{6.1.10}$$

We prove by induction that the iterations (6.1.7) generate a sequence $\{x_n\}$ in $B(x_0,r)$. Indeed, from (6.1.6) we deduce that

$$\|T_n^{-1}\| \leq (3/(3-\alpha)) \; \|f'^{-1}(x_0)\| \tag{6.1.11}$$

for every $n \geq 0$. Hence, if x_1 is determined from (6.1.7) with $n = 0$, then

$$\begin{aligned} \|x_1 - x_0\| &\leq \|I - T_0^{-1}(T_0 - f'(x_0))\| \; \|f'^{-1}(x_0)f(x_0)\| \\ &< 3r\,(1-\alpha)/(3-\alpha) \end{aligned} \tag{6.1.12}$$

by (6.1.5), (6.1.6) and (6.1.11), so that $x_1 \in B(x_0,r)$. Thus, suppose $x_1,\ldots,x_p$ are points in $B(x_0,r)$ satisfying (6.1.7) for $n = 0,1,\ldots,p-1$. Then (6.1.2) and (6.1.6) show that

$$\begin{aligned} \|f(x_{n+1})\| &= \|f(x_{n+1}) - f(x_n) - T_n(x_{n+1} - x_n)\| \\ &\leq (2\alpha/3\|f'^{-1}(x_0)\|) \; \|x_{n+1} - x_n\| \end{aligned} \tag{6.1.13}$$

holds for $n = 0,1,\ldots,p-1$.

Therefore, if x_{p+1} is determined from (6.1.7) with $n = p$, then (6.1.11) and (6.1.13) imply that

$$\|x_{n+2} - x_{n+1}\| \leq (2\alpha/(3-\alpha)) \; \|x_{n+1} - x_n\| \tag{6.1.14}$$

for $n = 0,1,\ldots,p-1$. Hence, in particular,

$$\|x_{p+1} - x_0\| \leq \sum_{i=0}^{p} \left(\frac{2\alpha}{3-\alpha}\right)^i \|x_1 - x_0\| \tag{6.1.15}$$

and so, by (6.1.12), $x_{p+1} \in B(x_0, r)$. This proves our assertion.

It follows that the inequalities (6.1.13), (6.1.14) hold for every $n \geq 0$, and therefore

$$\text{(6.1.16)} \qquad \|x_{n+p} - x_n\| \leq \big((3-\alpha)/3(1-\alpha)\big)\big(2\alpha/(3-\alpha)\big)^n \, \|x_1 - x_0\|$$

for all $n \geq 0$ and $p \geq 1$. This implies that $\{x_n\}$ converges to a point $\tilde{x} \in B(x_0, r)$ for which $f(\tilde{x}) = 0$, as a consequence of (6.1.13) and the continuity of f. If there were another point $\bar{x} \in B(x_0, r)$ for which $f(\bar{x}) = 0$, we could deduce from (6.1.2) and (6.1.10) that

$$\text{(6.1.17)} \qquad \begin{aligned} \|\tilde{x} - \bar{x}\| &\leq \|f'^{-1}(x_0)\| \, \|f(\tilde{x}) - f(\bar{x}) - f'(x_0)(\tilde{x} - \bar{x})\| \\ &\leq \frac{1}{3}\alpha \, \|\tilde{x} - \bar{x}\|, \end{aligned}$$

which is absurd unless $\tilde{x} = \bar{x}$. Clearly, (6.1.12) and (6.1.16) imply that (6.1.9) is satisfied for every $n \geq 0$. Since, by construction,

$$\text{(6.1.18)} \qquad \|\tilde{x} - x_{n+1}\| < \|T_n^{-1}\| \, \|f(x_n) - f(\tilde{x}) - T_n(x_n - \tilde{x})\|,$$

(6.1.2), (6.1.6), and (6.1.11) show that (6.1.8) holds for every $n \geq 0$. The proof is complete.

REMARK 6.1.1. If the open ball $B(x_0, r) \subset U$ is chosen as in the proof of Theorem 6.1.1, then (6.1.10) implies that $f'(x)$ is, for every $x \in B(x_0, r)$, a linear homeomorphism of X onto Y. Thus, we may select in Theorem 6.1.1 the sequence $\{T_n\}$ of linear homeomorphisms of X and Y such that, for every $n \geq 0$, $T_n = f'(z_n)$ for some $z_n \in B(x_0, r)$. Since (6.1.6) is then automatically satisfied, the assertions of Theorem 6.1.1 remain valid without change. In particular, the proof of Theorem 6.1.1 shows that we may always take $z_n = x_n$ for every $n \geq 0$, in which case the iterations (6.1.7) assume the form

$$x_{n+1} = x_n - f'^{-1}(x_n)f(x_n), \qquad n = 0,1,\ldots . \tag{6.1.19}$$

These are the iterations of Newton's method.

REMARK 6.1.2. In Theorem 6.1.1, a particularly simple choice for the sequence $\{T_n\}$ of linear homeomorphisms of X and Y is to let $T_n = T_0$ for every $n \geq 0$. In that case the condition (6.1.6), which implies

$$\|T_0^{-1} \cdot f'(x_0) - I\| \leq \alpha/(3-\alpha) < \tfrac{1}{2},$$

may be replaced by a weaker requirement that T_0 be a linear homeomorphism of X onto Y for which

$$\|T_0^{-1} \cdot f'(x_0) - I\| < 1,$$

and the conclusions of Theorem 6.1.1 may be strengthened as follows: for any α_0 satisfying $\|T_0^{-1} \cdot f'(x_0) - I\| < \alpha_0 < 1$, there is an open ball $B_0(x_0,r) \subset U$ such that, if

$$\|T_0^{-1}f(x_0)\| < r(1-\alpha_0),$$

there exists a unique point $\tilde{x} \in B_0(x_0,r)$ for which $f(\tilde{x}) = 0$, and the iterations (6.1.7) generate a sequence $\{x_n\}$ of points in $B_0(x_0,r)$ which converge to $\tilde{x}$ such that

$$\|\tilde{x} - x_{n+1}\| \leq \alpha_0 \|\tilde{x} - x_n\|,$$

$$\|\tilde{x} - x_n\| \leq \frac{\alpha_0^n}{1-\alpha_0} \|T_0^{-1}f(x_0)\|,$$

for every $n \geq 0$. Observe that, in particular, we may take $T_0 = f'(x_0)$ in which case these assertions remain valid for any $\alpha_0 \in (0,1)$. This is the so-called modified Newton's method.

COROLLARY 6.1.1. If f is continuously differentiable in U and, in addition, there is a constant $k > 0$ satisfying (6.1.3), then the assertions of Theorem 6.1.1 hold relative to any open ball $B(x_0, r)$ contained in U for which

$$k\|f'^{-1}(x_0)\| < \alpha/3r.$$

Moreover, under this additional assumption, the convergence estimates (6.1.8), (6.1.9) can be considerably improved for the iterations (6.1.19) of Newton's method.

EXERCISE 6.1.1. Suppose that f is twice differentiable in U and that there is a point $x_0 \in U$ for which $f'(x_0)$ is a linear homeomorphism of X onto Y. If there is an $\alpha \in (0, \frac{1}{2}]$ and a closed ball $\overline{B}(x_0, r) \subset U$ such that

$$\|f'^{-1}(x_0)f(x_0)\| \leq r(1 - \alpha)$$

and

$$\|f'^{-1}(x_0)\| \; \|f''(x)\| \leq (2/r)\alpha$$

for every $x \in \overline{B}(x_0, r)$, then show that there exists at least one point $\tilde{x} \in \overline{B}(x_0, r)$ for which $f(\tilde{x}) = 0$. Also show that the iterations (6.1.19) are defined for every $n \geq 0$ and generate a sequence $\{x_n\}$ of points in $\overline{B}(x_0, r)$ which converges to $\tilde{x}$ such that, for every $n \geq 0$,

$$\|\tilde{x} - x_{n+1}\| \leq (\alpha_n/r_n) \; \|\tilde{x} - x_n\|^2 \leq 2^n \; (\alpha/r) \; \|\tilde{x} - x_n\|^2$$

and

$$\|\tilde{x} - x_n\| \leq r_n \leq (r/2^n)(2\alpha)^{2^n - 1},$$

where the sequences $\{\alpha_n\}$, $\{r_n\}$ are obtained by setting

$\alpha_0 = \alpha$, $r_0 = r$ and $\alpha_{n+1} = \alpha_n^2/(1 - 2\alpha_n + 2\alpha_n^2)$, $r_{n+1} = \alpha_n r_n$, for every $n \geq 0$.

Hint: First show that if the point x_p, for some $p \geq 0$, is such that the assertions $\overline{B}(x_n, r_n) \subset U$, $f'(x_n)$ is a linear homeomorphism of X onto Y, $\|f'^{-1}(x_n)f(x_n)\| \leq r_n(1 - \alpha_n)$, $\|f'^{-1}(x_n)\| \; \|f''(x_n)\| \leq (2/r_n)\, \alpha_n$ for every $x \in \overline{B}(x_n, r_n)$ are true with $n = p$, then they remain true with $n = p + 1$ for the point x_{p+1} determined from (6.1.19) with $n = p$. Then, since these assertions are true for the point x_0, use the induction argument to complete the proof.

6.2 THE GOODMAN-LANCE METHOD

Newton's method provides a convenient framework for deriving convergence criteria for a variety of techniques for the numerical solution of two point boundary value problems of ordinary differential equations. As illustration, we discuss briefly, in this section, the Goodman-Lance method and in the next section, the method of quasilinearization.

These two methods are typical examples for the two groups into which all such numerical techniques may be divided. Though they are basically iterative in the sense that the solutions of the given problem appears as the limit of a sequence of solutions of auxiliary problems, they differ in the way these auxiliary problems are chosen. In the Goodman-Lance method, these latter problems are iteratively generated initial value problems for the same given differential equation, whereas in the method of quasilinearization, they are boundary value problems involving the same boundary conditions, for iteratively generated linear differential equations.

In order to avoid umimportant details, let us restrict ourselves to the problem of determining, in a given compact interval $J = [0,T]$, a solution $\tilde{u}$ of the differential equation

$$x'' = f(t,x) \tag{6.2.1}$$

which satisfies the given boundary conditions

$$\tilde{u}(0) = a, \quad \tilde{u}(T) = b. \tag{6.2.2}$$

We will assume throughout that $f \in C[J \times R^m, R^m]$, $f_x(t,x)$ exists and is continuous on $J \times R^m$ and for (t,x_1), $(t,x_2) \in J \times R^m$,

$$\|J(t,x_1) - J(t,x_2)\| \leq \gamma \|x_1 - x_2\|, \tag{6.2.3}$$

where the Jacobian matrix at (t,x) of f with respect to x is identified with $J(t,x)$. The norm $\|J(t,x)\|$ is, as usual, the matrix norm induced by the given norm in R^m.

The Goodman-Lance method is based on the following construction. Suppose that u_n, for some integer $n \geq 0$, is a solution of (6.2.1) in J for which $u_n(0) = a$, $u_n'(0) = x_n$ and $u_n(T) \neq b$. Then find, for every integer $i \in [1,m]$, the value $w_i(0)$ of the solution w_i in J of the adjoint variational equation

$$x'' = J^*(t,u_n(t))x \tag{6.2.4}$$

for which $w_i(T) = 0$, $w_i'(T) = e_i$, and form the matrix $W(u_n) = (w_i(0) \cdot e_j)$, where e_i is an orthogonal basis of R^m. If there is a point $x_{n+1} \in R^m$ satisfying

$$W(u_n)(x_{n+1} - x_n) = u_n(T) - b \tag{6.2.5}$$

such that the solution u of (6.2.1) with $u(0) = a$, $u'(0) = x_{n+1}$ is defined in J, let that solution be u_{n+1}. Obviously, our aim is to generate in this way, from a given solution u_0 of (6.2.1) in J with $u_0(0) = a, u_0'(0) = x_0$, a sequence $\{u_n\}$ of solutions u_n of (6.2.1) in J with

$u_n(0) = a$, $u_n'(0) = x_n$, such that $\{u_n\}$ converges uniformly in J to a solution $\tilde{u}$ of (6.2.1) for which (6.2.2) is satisfied. We will show that this is indeed possible, under suitable assumptions.

Let us first observe that, if (6.2.1) is a linear differential equation, say

$$x'' = A(t)x, \tag{6.2.6}$$

then any solution v of (6.2.6) in J is related, for every integer $i \in [1,m]$, to the solution w_i of the adjoint linear equation

$$x'' = A^*(t)\,x \tag{6.2.7}$$

for which $w_i(T) = 0$, $w_i'(T) = e_i$ by the equation

$$w_i(0)\cdot v'(0) = w_i'(0)\cdot v(0) - e_i\cdot v(T), \tag{6.2.8}$$

as a consequence of the Green's formula. Hence, if (6.2.6) has a solution $\tilde{u}$ in J such that $\tilde{u}(0) = a$, $\tilde{u}'(0) = \tilde{x}$ and $\tilde{u}(T) = b$, and if u is any other solution of (6.2.6) for which $u(0) = a$, $u'(0) = x$, then in particular

$$w_i(0)\cdot(\tilde{x} - x) = e_i\cdot\big(u(T) - b\big) \tag{6.2.9}$$

for every integer $i \in [1,m]$; that is, $\tilde{x} - x$ satisfies the equation

$$W_0(\tilde{x} - x) = u(T) - b, \tag{6.2.10}$$

where W_0 is the matrix $\big(w_i(0)\cdot e_j\big)$. Thus, in this case, the Goodman-Lance method does yield, in one step, the missing initial value $\tilde{x} = \tilde{u}'(0)$ for the solution $\tilde{u}$ of (6.2.6) that satisfies the boundary conditions (6.2.2).

In the general case, our assumptions imply, among others, the following facts. If (6.2.1) admits a solution u_0 with $u_0(0) = a$, $u_0'(0) = x_0$ which is defined in J, then there exists an open neighborhood $U \subset R^m$ of x_0 such that for every $x \in U$, the unique solution u of (6.2.1) with $u(0) = a$, $u'(0) = x$ is also defined in J. Moreover, the relation which associates with every $x \in U$ the value $u(T)$ of the solution of (6.2.1) satisfying $u(0) = a$, $u'(0) = x$ is a continuously differentiable mapping h of U into R^m.

We assert that the Goodman-Lance iteration is precisely the iteration of Newton's method for determining a point $\tilde{x} \in U$ for which

$$h(\tilde{x}) - b = 0. \tag{6.2.11}$$

Clearly, any such point $\tilde{x}$ gives rise to a solution $\tilde{u}$ of (6.2.1) with $\tilde{u}(0) = a$, $\tilde{u}'(0) = \tilde{x}$ which is defined in J and satisfies $\tilde{u}(T) = b$.

Indeed, if $x_n \in U$ for some integer $n \geq 0$ and if u_n is the corresponding solution of (6.2.1) in J with $u_n(0) = a$, $u_n'(0) = x_n$, then the partial derivative $D_j h(x_n)$ of h at x_n, for each $j \in [1,m]$, is given by

$$D_j h(x_n) = v_j(T) \tag{6.2.12}$$

where v_j is the solution in J of the variational equation

$$x'' = J\big(t,u_n(t)\big)x \tag{6.2.13}$$

for which $v_j(0) = 0$, $v_j'(0) = e_j$. Since by the Green's formula, each $v_j(T)$ is related, for every integer $i \in [1,m]$, to the solution w_i of (6.2.4) by the equation

$$e_i \cdot v_j(T) = -w_i(0) \cdot e_j, \tag{6.2.14}$$

it follows that the derivative $h'(x_n)$ of h at x_n is given by

$$h'(x_n) = -W(u_n) \tag{6.2.15}$$

where, as before, $W(u_n)$ is the matrix $\big(w_i(0) \cdot e_j\big)$. Thus, if $h'(x_n)$ is nonsingular, the iteration of Newton's method (6.1.19) is

$$x_{n+1} = x_n - h'^{-1}(x_n)\big(h(x_n) - b\big),$$

and this is equivalent to the iteration (6.2.5) because $h(x_n) = u_n(T)$, by definition.

Therefore, any of the results for Newton's method, such as Corollary 6.1.1 or Exercise 6.1.1 will yield automatically sufficient conditions for the convergence of the iterations of the Goodman-Lance method. We shall formulate below only a direction consequence of Corollary 6.1.1.

THEOREM 6.2.1. Suppose that (6.2.1) has a solution u_0 with $u_0(0) = a$, $u_0'(0) = x_0$ which is defined in J, and let $U \subset R^m$ be an open neighborhood of x_0 such that for every $x \in U$ the unique solution u of (6.2.1) with $u(0) = a$, $u'(0) = x$ exists in J. Define h as the mapping $x \to u(T)$ of U into R^m and let $\beta > 0$ be any constant for which

$$\|h'(x_1) - h'(x_2)\| \le \beta\|x_1 - x_2\|$$

holds for any $x_1, x_2 \in U$. If the Jacobian matrix $h'(x_0)$ is nonsingular and there are constants $\alpha \in (0,1)$ and $r > 0$ such that $B(x_0, r) \subset U$, $\|h'^{-1}(x_0)h(x_0)\| < r(1 - \alpha)$, and $\beta\|h'^{-1}(x_0)\| \le \alpha/r$, then there exists a unique point $\tilde{x} \in B(x_0, r)$ for which the solution $\tilde{u}$ of (6.2.1) in J with $\tilde{u}(0) = a$, $\tilde{u}'(0) = \tilde{x}$ satisfies (6.2.2). Moreover, the iterations of Goodman-Lance method are defined for every $n \ge 0$ and generate a sequence $\{x_n\}$ of points in $B(x_0, r)$ which converges to $\tilde{x}$ such that the corresponding sequence $\{u_n\}$ of solutions u_n of (6.2.1) in J with $u_n(0) = a$, $u_n'(0) = x_n$ converge to $\tilde{u}$ uniformly in J, and that

$$\|\tilde{x} - x_{n+1}\| \le \big(\beta/(1 - \alpha)\big)\ \|h'^{-1}(x_0)\|\ \|\tilde{x} - x_n\|^2$$

and

$$\|\tilde{x} - x_n\| \leq \frac{1}{\alpha} \frac{\alpha^{2^n}}{1 - \alpha^{2^n}} \|h'^{-1}(x_0)h(x_0)\| \quad \text{for every } n \geq 0.$$

6.3 THE METHOD OF QUASILINEARIZATION

Let us now discuss the method of quasilinearization as it applies to the problem of constructing a solution $\tilde{u}$ of (6.2.1) which is defined in $J = [0,T]$ and satisfies (6.2.2). As before, we shall assume that $f \in C[J \times R^m, R^m]$, $f_x(t,x)$ exists and is continuous on $J \times R^m$ and (6.2.3) holds.

Our aim, again, is to generate, from a given continuous mapping u_0 of J into R^m, a sequence $\{u_n\}$ of continuous mappings of J into R^m such that $\{u_n\}$ converges uniformly in J to a mapping $\tilde{u}$ which is a solution of (6.2.1) in J satisfying (6.2.2). However, in contrast to the Goodman-Lance method, the mapping u_{n+1}, for every $n \geq 0$, is not chosen as a solution in J of the nonlinear equation (6.2.1) but, instead, is to be a solution in J of the linear equation

$$(6.3.1) \qquad x'' = J\big(t,u_n(t)\big)x + f\big(t,u_n(t)\big) - J\big(t,u_n(t)\big)u_n(t)$$

for which

$$(6.3.2) \qquad u_{n+1}(0) = a, \qquad u_{n+1}(T) = b.$$

This requires, of course, that a continuous mapping u_0 of J into R^m can be found such that not only the equation (6.3.1) with $n = 0$ but, moreover, each iteratively generated equation (6.3.1) with $n \geq 1$ admits a solution u_{n+1} which satisfies the boundary conditions (6.3.2). We will now show that this demand can indeed be met, under appropriate assumptions.

We note first that a continuous mapping $\tilde{u}$ of J into R^m is a solution of (6.2.1), (6.2.2) if and only if

$$(6.3.3)\quad \tilde{u}(t) = a + \frac{t}{T}(b-a) - \int_0^T G(t,s)f\big(s,\tilde{u}(s)\big)\,ds, \qquad t \in J,$$

where G is the Green's function in $J \times J$ such that

$$G(t,s) = \begin{cases} (1/T)(T-t)s, & 0 \leq s \leq t \leq T, \\ (1/T)\,t(T-s), & 0 \leq t \leq s \leq T. \end{cases}$$

This may be formulated in the following manner. Consider the Banach space $C = C[J,R^m]$ with the usual norm $\|u\| = \sup_{t\in J} \|u(t)\|$. Define φ as the mapping of C into itself such that

$$(6.3.4)\quad \varphi(u)\colon t \to a + \frac{t}{T}(b-a) - \int_0^T G(t,s)f\big(s,u(s)\big)\,ds$$

for every $u \in C$, and set

$$(6.3.5)\qquad \Phi(u) = u - \varphi(u).$$

then any point $\tilde{u} \in C$ is a solution of (6.2.1) which satisfies (6.2.2) if and only if

$$(6.3.6)\qquad \Phi(\tilde{u}) = 0.$$

Our assumptions on f imply that the mapping φ is continuously differentiable in C. Indeed, from (6.1.2) we infer immediately that the derivative $\varphi'(u)$ of φ at $u \in C$ is the continuous linear mapping of C into itself such that, for every $v \in C$,

$$(6.3.7)\quad \varphi'(u)v\colon t \to -\int_0^T G(t,s)J\big(s,u(s)\big)v(s)\,ds.$$

Hence the mapping Φ of C into itself, as defined by (6.3.5), is also continuously differentiable in C. In fact, since evidently

(6.3.8) $$\Phi'(u) = e - \varphi'(u)$$

for every $u \in C$, where e is the identity mapping in C, we deduce from (6.2.3) and (6.3.7) that

(6.3.9) $$\|\Phi'(u_1) - \Phi'(u_2)\| \leq \frac{1}{8} \gamma T^2 \|u_1 - u_2\|$$

for any $u_1, u_2 \in C$. Observe that, as a consequence of (6.3.7) and (6.3.8), $\Phi'(u)$ is, for some $u \in C$, a linear homeomorphism of C onto itself if and only if $\Phi'(u)v = 0$ implies $v = 0$ or, equivalently, if and only if the relation

(6.3.10) $$v(t) = -\int_0^T G(t,s)J\big(s,u(s)\big)v(s)\, ds$$

for every $t \in J$ implies $v(t) = 0$ in J. Thus, $\Phi'(u)$ is a linear homeomorphism of C onto itself precisely for those $u \in C$ for which the homogeneous linear differential equation

(6.3.11) $$x'' = J\big(t,u(t)\big)x$$

has no nontrivial solution v in J such that $v(0) = 0$, $v(T) = 0$ or equivalently, for which the nonhomogeneous linear differential equation

(6.3.12) $$x'' = J\big(t,u(t)\big)x + w(t)$$

has, for every $w \in C$, a unique solution v such that $v(0) = a$, $v(T) = b$. In particular, $\Phi'(u)$ is a linear homeomorphism of C onto itself for any point $u \in C$ for which $\|\varphi'(u)\| < 1$, and in that case

$$\|\Phi'^{-1}(u)\| \leq \big(1 - \|\varphi'(u)\|\big)^{-1}.$$

We claim that, if the iterations of Newton's method for determining a point $\tilde{u} \in C$ for which (6.3.6) holds, namely the iterations

(6.3.13) $$u_{n+1} = u_n - \Phi'^{-1}(u_n)\Phi(u_n), \qquad n = 0,1,2,\ldots,$$

are defined for every $n \geq 0$, then they are exactly the iterations of the method of quasilinearization. More precisely, the point $u_{n+1} \in C$ obtained from (6.3.13) is then, for every $n \geq 0$, the unique continuous mapping of J into R^m which is a solution of (6.3.1) satisfying (6.3.2).

Clearly, if the iterations (6.3.13) are defined for every $n \geq 0$, then $\Phi'^{-1}(u_n)$ exists for every $n \geq 0$. Hence (6.3.13) implies by (6.3.5) that

$$(6.3.14) \qquad u_{n+1} = \varphi'(u_n)(u_{n+1} - u_n) + \varphi(u_n)$$

which, in turn, implies by (6.3.4) and (6.3.7) that

$$(6.3.15) \quad u_{n+1}(t) = a + (t/T)(b - a) - \int_0^T G(t,s)\Big[J\big(s,u_n(s)\big)\big(u_{n+1}(s) - u_n(s)\big) + f\big(s,u_n(s)\big)\Big]\, ds$$

for every $t \in J$. This shows that u_{n+1} is a solution of the linear differential equation (6.3.1) for which (6.3.2) holds. In fact, u_{n+1} is the only such solution, because the existence of $\Phi'^{-1}(u_n)$ assures, by the preceding remarks, that the homogeneous linear differential equation (6.3.11) corresponding to (6.3.1) has no nontrivial solution v in J such that $v(0) = 0$, $v(T) = 0$. This proves our claim.

Thus, sufficient conditions for the convergence of the iterations of the method of quasilinearization can be deduced directly from any of the general results for Newton's method, such as Corollary 6.1.1 or Exercise 6.1.1, by simply applying the latter to the mapping Φ as defined by (6.3.5).

We state here the following result which is a direct consequence of Corollary 6.1.1.

THEOREM 6.3.1. Let u_0 be the mapping of J into R^m

for which

$$u_0(t) = a + (t/T)(b - a)$$

for every $t \in J$, let α, β be positive constants such that

$$\|f(t, u_0(t))\| \leq \alpha, \qquad \|J(t, u_0(t))\| \leq \beta, \quad t \in J,$$

and suppose that $(1/8)\beta T^2 < 1$. If there exist positive constants $\alpha_0 < 1$ and r such that

$$\frac{1}{8} T^2 \frac{\alpha}{1 - \frac{1}{8}\beta T^2} < r(1 - \alpha_0), \qquad \frac{1}{8} T^2 \frac{\gamma}{1 - \frac{1}{8}\beta T^2} < \frac{\alpha_0}{r},$$

where $\gamma > 0$ is a constant for which (6.2.3) holds, then there exists a unique continuous mapping $\tilde{u}$ of J into R^m with $\|\tilde{u} - u_0\| < r$ which is a solution of (6.2.1) and (6.2.2) in J. Moreover, the iterations of the method of quasilinearization generate an infinite sequence $\{u_n\}$ of continuous mappings of J into R^m which converges to $\tilde{u}$ uniformly in J such that

$$\|\tilde{u} - u_{n+1}\| \leq \frac{1}{1 - \alpha_0} \frac{\frac{1}{8}\gamma T^2}{1 - \frac{1}{8}\beta T^2} \|\tilde{u} - u_n\|^2$$

and

$$\|\tilde{u} - u_n\| \leq \frac{1}{\alpha_0} \frac{\alpha_0^{2n}}{1 - \alpha_0^{2n}} \frac{\frac{1}{8}\alpha T^2}{1 - \frac{1}{8}\beta T^2}$$

for every $n \geq 0$.

6.4 NONLINEAR EIGENVALUE PROBLEMS

Employing the angular function technique, discussed in Chapter 2, we shall consider the boundary value problem P_λ

$$x' = f(t, x, y, \lambda), \tag{6.4.1}$$

$$(6.4.2) \qquad y' = g(t,x,y,\lambda),$$

$$(6.4.3) \qquad A_1x(a) + A_2y(a) = 0,$$

$$(6.4.4) \qquad B_1x(b) + B_2y(b) = 0,$$

where $f,g \in C[J \times R^3, R]$, $J = [a,b]$, $A_1^2 + A_2^2 > 0$, and $B_1^2 + B_2^2 > 0$. We seek values of the parameter λ for which P_λ possesses a nontrivial solution. Let us list the following assumptions.

(H_1) $f,g \in C[J \times R^3, R]$, the solutions of (6.4.1), (6.4.2) can be continued over J, and $x(t) \equiv 0$, $y(t) \equiv 0$ is the only solution of (6.4.1), (6.4.2) with $x(t_0) = 0$, $y(t_0) = 0$ for any $t_0 \in J$.

(H_2) There exists a continuous function $H(t,u,v,\lambda)$ defined on $J \times D \times R$, where $D = [(u,v)\colon u^2 + v^2 = 1]$, such that

$$(6.4.5) \qquad xg(t,x,y,\lambda) - yf(t,x,y,\lambda) \le r^2 H(t,x/r,\ y/r,\lambda)$$

on $J \times R^3$, where $r^2 = x^2 + y^2$.

We shall first prove an elementary result.

THEOREM 6.4.1. Let (H_1) and (H_2) hold. Suppose that

$$(6.4.6) \qquad \liminf_{\lambda \to +\infty} \int_a^b G(t,\lambda)\, dt = -\infty,$$

where

$$G(t,\lambda) = [\sup H(t,u,v,\lambda)\colon (u,v) \in D].$$

Then there exists a sequence $\{\lambda_k\}$ of eigenvalues of P_λ such that

$$\lambda_m < \lambda_{m+1} < \cdots \qquad \text{with } \lim_{k\to\infty} \lambda_k = +\infty,$$

where m is nonnegative integer. If $x(t,\lambda_k)$, $y(t,\lambda_k)$ is an eigenfunction pair for λ_k, $k \geq m$, then $x(t,\lambda_k)$ has at least k zeros, while $y(t,\lambda_k)$ has at least $k-1$ zeros on J.

Proof: Let $(x(t),y(t))$ be a solution of (6.4.1) - (6.4.3). We set $x = r\cos\theta$, $y = r\sin\theta$, with θ normalized so that it satisfies $-\pi/2 < \theta(a,\lambda) \leq \pi/2$. A simple calculation shows, in view of (H_2) and the definition of $G(t,\lambda)$, that

$$\theta'(t,\lambda) \leq H(t, \cos\theta, \sin\theta,\lambda) \leq G(t,\lambda),$$

which, when integrated, yields

$$\theta(b,\lambda) \leq \theta(a,\lambda) + \int_a^b G(t,\lambda)\,dt.$$

Recalling that $\theta(a,\lambda) = \text{arc tan}(-A_1/A_2)$, the condition (6.4.6) gives $\liminf_{\lambda\to\infty} \theta(b,\lambda) = -\infty$.

For each $k = 0,1,2,\ldots,$ define

$$A_k = \left[\lambda \geq 0 \colon \theta(a,\lambda) = \text{arc tan}\left(-\frac{A_1}{A_2}\right),\ \theta(b,\lambda) = \text{arc tan}\left(-\frac{B_1}{B_2}\right) - k\pi\right],$$

where $\theta(t,\lambda)$ is arc tan $y(t)/x(t)$ for the corresponding solution of (6.4.1) - (6.4.3). Since $\theta(b,\lambda)$ is continuous in λ and $\lim_{\lambda\to\infty} \theta(b,\lambda) = -\infty$, the range of $\theta(b,\lambda)$ contains the interval $(-\infty,\theta(b,0))$. Let $m \geq 0$ be the smallest integer such that A_m is nonempty. Then for $k \geq m$, we may define

$$\lambda_m = \inf A_m, \qquad \lambda_k = \inf[\lambda\colon \lambda \in A_k,\ \lambda > \lambda_{k-1},\ k > m].$$

The sets defining λ_k, $k > m$, are also nonempty, since $\theta(b,\lambda_{k-1}) > \text{arc tan}(-B_1/B_2) - k\pi$ and $\theta(b,\lambda)$ is continuous in λ with $\liminf_{\lambda\to\infty} \theta(b,\lambda) = -\infty$. Finally, notice that $\theta(t,\lambda_k)$

crosses each of the lines $\theta = -(2j+1)\pi/2$, $j = 0,1,\ldots,k-1$, at least once to give $x(t,\lambda_k)$ at least k zeros. Also, $\theta(t,\lambda_k)$ crosses each of the lines $\theta = -j\pi$, $j = 1,2,\ldots,k-1$, so that $y(t,\lambda_k)$ has at least $k-1$ zeros. The proof is therefore complete.

EXERCISE 6.4.1. Prove other possibilities of Theorem 6.4.1, namely (i) replace (6.4.6) by $\liminf_{\lambda\to-\infty} \int_a^b G(t,\lambda)dt = -\infty$, (ii) reverse the inequality in (H_2) and assume that $\limsup_{\lambda\to\pm\infty} \int_a^b G(t,\lambda)\,dt = +\infty$.

A drawback of Theorem 6.4.1 lies in the function $G(t,\lambda)$. For very simple systems, $G(t,\lambda)$ may become constant for large λ and consequently the condition (6.4.6) cannot be satisfied. For example in the case $x' = y$, $y' = -(\lambda+1)x$, we have $H(t,x,y,\lambda) = -(\lambda+1)x^2 - y^2$ so that $G(t,\lambda) = -1$ for large positive λ. Clearly one needs to eliminate a neighborhood of $x = 0$ from the consideration to get better results. This is exactly the point of the exceptional set in the following hypotheses.

On the unit circle D, let $E = [\alpha_1,\ldots,\alpha_p]$ denote a finite set of points, where D is parametrized by the central angle $-\pi < \alpha \le \pi$. For each $\delta > 0$, let

$$E_\delta = \bigcup_{i=1}^{p} (a_i - \delta, \alpha_i + \delta).$$

(H_3) There exists a finite set E of points on D such that

(i) there is a $K > 0$ for which

$$H(t, \cos\alpha, \sin\alpha, \lambda) < -K$$

for $t \in J$, $\lambda \ge 0$, $\alpha \in E_\delta$, for all sufficiently small $\delta > 0$,

(ii) if $G_\delta(\lambda) = \sup[H(t, \cos\alpha, \sin\alpha,\lambda)\colon t \in J,\ \alpha \in D - E_\delta]$, then $\liminf_{\lambda\to\infty} G_\delta(\lambda) = -\infty$, for each sufficiently small $\delta > 0$.

In the example, (H_3) is satisfied, if we take $E = [\pi/2, -\pi/2]$, $\delta \leq \frac{1}{2}$ and $K = 1$.

THEOREM 6.4.2. If (H_1) - (H_3) hold, then the assertions of Theorem 6.4.1 remain valid.

Proof: We need only to show that $\liminf_{\lambda\to\infty} \theta(b,\lambda) = -\infty$, since the remainder of the proof of Theorem 6.4.1 applies. Suppose the contrary, that is, for some M, $\theta(b,\lambda) \geq -2M\pi$, for $\lambda \geq 0$. Let $E = [\alpha_1,\ldots,\alpha_p]$ be the exceptional set and define

$$F = [\varphi_1,\ldots,\varphi_p] = [\alpha_i - 2\pi j\colon 1 \leq i < p,\ j = 0,1,\ldots,M].$$

Choose $\delta < K(b-a)/4p(M+1)$. For each $\lambda \geq 0$, let t_i, $i = 1,\ldots,p$, denote the unique time such that $\theta(t_i^*,\lambda) = \varphi_i - \delta$. Notice that t_i, t_i^* depend on λ and are well defined because $\theta' \leq -K$ when $\theta = \varphi_i \pm \delta$. In fact,

$$\theta'(t,\lambda) \leq -K, \qquad t \in [t_i, t_i^*],$$

so that $\theta(t,\lambda)$ is strictly decreasing in t on $[t_i, t_i^*]$. We have also

$$-2\delta = \theta(t_i^*,\lambda) - \theta(t_i,\lambda) \leq -K(t_i^* - t_i)$$

and, as a result, we conclude that $t_i^* - t_i \leq 2\delta/K < (b-a)/2(M+1)p$. Since there are at most $p(M+1)$ such intervals in J, the measure of the set

$$T_1(\lambda) = [t\colon t \in J, \theta(t,\lambda) \in E_\delta]$$

satisfies $m\big(T_1(\lambda)\big) < (b-a)/2$, so the measure of

$$T_2(\lambda) = [t\colon t \in J,\ \theta(t,\lambda) \notin E_\delta]$$

is greater than $(b - a)/2$. However,

$$\theta(b,\lambda) = \theta(a,\lambda) + \int_{T_1(\lambda)} \theta'(t,\lambda)\,dt + \int_{T_2(\lambda)} \theta'(t,\lambda)\,dt$$

so that by (H_3), it follows that

$$\theta(b,\lambda) \leq \theta(a,\lambda) - Km(T_1(\lambda)) + G_\delta(\lambda)m(T_2(\lambda)).$$

Since $m(T_2(\lambda)) > (b - a)/2$ for every λ and $\liminf_{\lambda\to\infty} G_\delta(\lambda) = -\infty$, there exists a $\lambda > 0$ for which $\theta(b,\lambda) < -2M\pi$. This contradiction proves the theorem.

6.5 n-PARAMETER FAMILIES AND INTERPOLATION PROBLEMS

In Section 2.1, it was shown for second-order scalar equations that the uniqueness of the two-point BVP implies its existence. An analogous theorem exists for the third-order equation; however, the question of uniqueness implying existence remains open for the n-point BVP, $n \geq 4$. In this section, we prove under the additional assumption of "local solvability, that the uniqueness does in fact imply the existence for the n-point BVP. Our main technique is the theory of a n-parameter family of functions.

We state our result for the first order n-dimensional system; a special case of this, of course, is the nth order scalar equation.

THEOREM 6.5.1. Consider the boundary value problem

(6.5.1) $$x' = f(t,x),$$

(6.5.2) $$x_0(t_j) = c_j, \qquad j = 1,\ldots,n,$$

$t_1 < t_2 < \cdots < t_n$, where $x = (x_0,x_1,\ldots,x_{n-1})$. Assume

(A) $f(t,x) \in C[(a,b)\times R^n,R^n]$;

(B) all solutions of (6.5.1) exist on (a,b);

(C) there exists at most one solution of (6.5.1) and (6.5.2) for all $c_j \in R$ and $t_j \in (a,b)$ where $t_1 < t_2 < \cdots < t_n$;

(D) the BVP (6.5.1), (6.5.2) are locally solvable at every point $t_0 \in (a,b)$; this is, for every $t_0 \in (a,b)$ there exists an open interval $I(t_0), t_0 \in I(t_0) \in (a,b)$ such that (6.5.1) and (6.5.2) has a solution for all distinct $\{t_i\}_{i=1}^n \in I(t_0)$ and all $\{c_i\}_{i=1}^n \in R$.

Then every BVP (6.5.1), (6.5.2) has a unique solution.

We shall need some definitions and lemmas first.

DEFINITION 6.5.1. A set F of functions $f \in C[(a,b),R]$ is said to be an <u>n-parameter family</u> on (a,b) if for every set of n distinct points $t_1 < t_2 < \cdots < t_n$ of (a,b) and every set of n numbers $c_1,\ldots,c_n$ there is one and only one element $f_0 \in F$ satisfying

$$(6.5.3) \qquad f_0(t_j) = c_j, \qquad j = 1,\ldots,n.$$

DEFINITION 6.5.2. A family F of functions $f \in C[(a,b),R]$ is said to be a <u>local n-parameter family</u> if for every $t_0 \in (a,b)$ there is an open interval, $I(t_0), t_0 \in I(t_0) \subset (a,b)$ such that the set of restrictions $f|I(t_0)$ of the elements of $f \in F$ is an n-parameter family on $I(t_0)$.

Observe that the conditions (B) and (D) imply that the solutions of the BVP (6.5.1), (6.5.2) form a local n-parameter family F. In order to obtain the result of Theorem 6.5.1 it is sufficient to show that F is indeed an n-parameter family on (a,b). Our next result shows this is indeed so. Thus Theorem 6.5.1 is incorporated in the following more general setting.

THEOREM 6.5.2. Let $F = \{f\}$ be a family of continuous functions on (a,b) satisfying

(i) F is a local n-parameter family on (a,b) and

(ii) if f,g are distinct elements of F, then $f-g$ has at most $n-1$ zeros on (a,b).

Then F is an n-parameter family on (a,b).

To prove Theorem 6.5.2 we shall need some lemmas.

LEMMA 6.5.1. Let F be an n-parameter family on (a,b).

(i) If f,g are distinct elements of F such that $f-g$ has $n-1$ zeros on (a,b), then $f-g$ changes signs at each of its zeros.

(ii) if $f_0(t) = f_0(t,t_1,\ldots,t_n,c_1,\ldots,c_n)$ is the unique element of F determined by (6.5.3), then f_0 is a continuous function of its $2n+1$ variables for $t \in (a,b)$, $a < t_1 < \cdots < t_n < b$, $(c_1,\ldots,c_n) \in R^n$.

We shall provide part of the proof in the following exercise.

EXERCISE 6.5.1. Prove Lemma 6.5.1.

Hints: For part (i) assume $f-g$ does not change sign at some x_1. Let (a',b') be an interval having x_1 in its interior and containing none of the abcissas of the other $n-2$ points of intersection. Choose $x' \not\subset (a',b')$ such that $f(x') \neq g(x')$. Let F' be the one parameter family of functions restricted to (a',b') of the members of F passing through the other $n-2$ points of intersection and through $(x',g(x'))$. Show that if f intersects any element $g \in F'$ in (a',b') exactly one time, then $f-g$ changes sign. This can be done by taking two points x_2,x_3 such that $a' < x_2 < x_1 < x_3 < b'$. Then choose f_1 and $f_2 \in F'$ such that $f_1(x_2) = g(x_2)$ and

$f_2(x_3) = g(x_3)$. Show that if $f - g$ does not change signs, then $f_1 = f_2$ at some point on (x_2, x_3). But this is impossible since f_1 and f_2 belong to F'.

For (ii) show that for any sequence $\{\tau_{1_k}, \tau_{2_k}, \ldots, \tau_{n_k}, s_{1_k}, \ldots, s_{n_k})$ converging to $(\tau_1, \ldots, \tau_n, s_1, \ldots, s_n)$ then $f(t, \tau_{1_k}, \ldots, \tau_{n_k}, s_{1_k}, \ldots, s_{n_k})$ converges to $f(t, \tau_1, \ldots, \tau_n, s_1, \ldots, s_n)$ uniformly for $t \in (a,b)\}$. This can be shown by using the results of the first part of the lemma.

LEMMA 6.5.2. Assume the conditions of Theorem 6.5.2 and let

$$\Delta = \{(t_1, \ldots, t_n) \in R^n,\ a < t_1 < t_2 < \cdots < t_n < b\},$$

$$\Omega = \Big\{\big(t_1, \ldots, t_n, f(t_1), \ldots, f(t_n)\big) \in \Delta \times R^n \subset R^{2n},\ f \in F\Big\}.$$

Then Ω is an open subset of R^{2n}.

<u>Proof</u>: Let $t_0 \in (a,b)$ and $s_1 < s_2 < \cdots < s_n$ be points of the open interval $I(t_0)$. Recall $f/I(t_0)$ forms an n-parameter family and let

$$f(t) = f(t, s_1, \ldots, s_n, c_1, \ldots, c_n)$$

be the unique element of F satisfying $f(s_j) = c_j$, $j = 1, \ldots, n$. For fixed t_0 and $s_1, \ldots, s_n$ consider the mapping $\varphi\colon \Delta \times R^n \to \Omega$ defined by

$$\varphi(t_1, \ldots, t_n, c_1, \ldots, c_n) = \big(t_1, \ldots, t_n, f(t_1), \ldots, f(t_n)\big).$$

We first show that φ is a continuous function. It is sufficient to show that $f(t, s_1, \ldots, s_n, c_1, \ldots, c_n)$ is a continuous function of $(t, c_1, \ldots, c_n)$. Let J be any closed interval contained in (a,b) such that t_i is contained in the interior of J and $t_0 \in J$. For each point $x \in J$, define an open interval $I(x)$ such that F restricted to $I(x)$ form an n-parameter family. There exists a finite number of x_i, $i = 1,\dots,k$ such that $\bigcup_{i=1}^{k} I(x_i)$ covers J. Let $S = \bigcup_{i=1}^{k} I(x_i) \cup I(t_0)$. Observe that $f(t, s_1,\dots,x_n, c_1,\dots,c_n)$ is a continuous function of $(t, c_1,\dots,c_n)$ on $I(t_0) \times R^n$ because of Lemma 6.5.1. Now $I(t_0)$ intersects $I(x_k)$ for some k. On $I(x_k)$ apply Lemma 6.5.1 by considering any points $r_1,\dots,r_n \in I(t_0) \cap I(x_k)$ having the values ℓ_j, that is, $\ell_j = f(r_j, s_1,\dots,s_n, c_1,\dots,c_n)$, $j = 1,\dots,n$. Apply now Lemma 6.5.1 to $f(t, r_1,\dots,r_n, \ell_1,\dots,\ell_n) = f(t, s_1,\dots,s_n, c_1,\dots,c_n)$ on $I(x_k)$. Hence $f(t, s_1,\dots,s_n, c_1,\dots,c_n)$ is continuous for $(t, c_1,\dots,c_n)$ on $I(t_0) \cup I(x_k)$. We may continue this process a finite number of times and obtain the continuity of $f(t, s_1,\dots,s_n, c_1,\dots,c_n)$ on $S \times R^n$. Hence we see that φ is continuous on $\Delta \times R^n$.

We immediately have that φ is one-to-one on $\Delta \times R^n$ because of condition (ii) in Theorem 6.5.2. Since $\Delta \times R^n$ is open in R^{2n} then $\varphi(\Delta \times R^n) = \Omega$ is open in R^{2n} by the Brouwer invariance of domain theorem.

We now generalize Lemma 6.5.1 as follows.

LEMMA 6.5.3. Let F be a family of continuous functions $f(t)$ on an open interval (a,b) such that

(i) the set $\Omega \subset R^{2n}$ is open and

(ii) if f,g are distinct elements of F, then $f - g$ has at most $n - 1$ zeros on (a,b).

Then

(a) If $f,g \in F$ and $f - g$ has exactly $n - 1$ zeros on (a,b), then $f - g$ changes signs at each of its zeros.

(b) If $f(t) = f(t, t_1,\dots,t_n, c_1,\dots,c_n)$ is the unique element of F satisfying $f(t_i) = c_i$, then f is a continuous function of its $2n+1$ variables for $t \in (a,b)$ and $(t_1, \ldots, t_n, c_1, \ldots, c_n) \in \Omega$.

Proof: Let $f, g \in F$ be distinct elements of F with $f-g$ vanishing at $t = t_1, \ldots, t_{n-1}$ where $t_i < t_{i+1}$. Suppose there exists $k > 0$ such that $f-g > 0$ in some deleted neighborhood about t_k. Choose any $s \in (t_{k-1}, t_k)$ if $k > 1$ or contained in (a, t_k) if $k = 1$. Because of the openness of Ω we may pick an $\varepsilon > 0$ so small such that there exists an $h \in F$ having the property that

$$h(t_j) = f(t_j) = g(t_j) \qquad \text{for } j \neq k$$

$$h(s) = f(s) \qquad \text{and} \qquad h(t_k + \varepsilon) = g(t_k + \varepsilon).$$

Observe that $f-h$ and $g-h$ each vanish at $n-1$ points. However, the continuity of $f, g,$ and h imply that either $f-h$ or $g-h$ vanishes on $(s, t_k+\varepsilon)$ since $f-g > 0$ in some deleted neighborhood of t_k (which we may assume contains $t_k + \varepsilon$). This contradicts (ii).

The proof of the second part is the same as in Exercise 6.5.1.

LEMMA 6.5.4. Let F be a set of continuous functions $f(t)$ on an open interval (a,b). Then F is an n-parameter family on (a,b) if it has properties (i) and (ii) of Lemma 6.5.3, the property

(iii) if $f_1, f_2, \ldots$ is any sequence of F satisfying

$$(6.5.4) \qquad \begin{aligned} f_i(t) &\leq f_{i+1}(t) \qquad \text{for } i = 1,2,\ldots \text{ or} \\ f_i(t) &\geq f_{i+1}(t) \qquad \text{for } i = 1,2,\ldots, \end{aligned}$$

on a compact set $[\alpha, \beta] \subset (a,b)$, then either

(6.5.5) $\quad f(t) = \lim f_i(t) \quad$ exists on (a,b) and $f \in F$,

or

(6.5.6) $\quad \lim_{i\to\infty} |f_i(t)| = \infty \quad$ on a dense set of (a,b);

and the property

(iv) for all $t_0 \in (a,b)$ the set $S(t_0) = \{f(t_0): f \in F\}$ is not bounded above or below.

<u>Proof</u>: To show F is an n-parameter family it is sufficient to show that $\Omega = \Delta \times R^n$. Let k, where $1 \leq k \leq n$, $(t_1,\ldots,t_n) \in \Delta$, and $(c_1,\ldots,c_{k-1}, c_{k+1},\ldots,c_n)$ be fixed and let $S = \big(c_k: (t_1,\ldots,t_n, c_1,\ldots,c_n) \in \Omega\big)$. We immediately have that S is open from (i) of Lemma 6.5.3.

We now show that S is also closed. If S is not empty consider a sequence $\{c_{k_j}\}$ satisfying $c_{k_j} \in S$ for $j = 1,2,\ldots$ and $c_{k_j} \to c_k$ as $j \to \infty$. We may assume $\{c_{k_j}\}$ is a monotone sequence and without loss of generality $c_{k_j} \leq c_{k_{j+1}}$ for all j. By Lemma 6.5.3, if $f_j \in F$ and satisfies $f_j(t_m) = c_m$ for $m \neq k$, $f_j(t_k) = c_{k_j}$, then

$$(-1)^{r+k-1} f_j(t) \leq (-1)^{r+k-1} f_{j+1}(t) \quad \text{for} \quad t \in (t_r, t_{r+1}), \quad r = 0,\ldots,k-2,$$

(6.5.7) $$f_j(t) \leq f_{j+1}(t) \quad \text{for} \quad t \in (t_{k-1}, t_{k+1}),$$

$$(-1)^{r+k} f_j(t) \leq (-1)^{r+k} f_{j+1}(t) \quad \text{for} \quad t \in (t_r, t_{r+1}), \quad r = k+1,\ldots,n,$$

where $a = t_0$ and $b = t_{n+1}$. Thus by condition (iii) either (6.5.5) holds, and $c_k \in S$ or (6.5.6) holds.

Assume (6.5.6) holds. By (iv) we can find some $f \in F$ satisfying $f(t_k) > c_k > f_j(t_k)$ for all j. For sufficiently large j, we have in view of (6.5.6) and (6.5.7) that $f - f_j$ vanishes on either side of t_k arbitrarily near t_k and also arbitrarily near t_m for $m \neq k$. Thus, for large j, $f - f_j$ has n zeros on (a,b), a contradiction to (ii) of Lemma 6.5.3. Thus S is closed. Hence $S = (-\infty,\infty)$ implying $\Omega(a,b) = \Delta(a,b) \times R^n$. The proof of Theorem 6.5.2 thus follows from Lemmas 6.5.1 - 6.5.4.

For our further discussion we shall consider the BVP

$$x^{(n)} = f(t,x,x',\ldots,x^{(n-1)}), \tag{6.5.8}$$

$$x(t_i) = c_i, \qquad i = 1,\ldots,n. \tag{6.5.9}$$

The results of Theorem 6.5.1, applied to the BVP (6.5.8) and (6.5.9), remain true without the local solvability condition (iii) for $n = 3$ (as well as $n = 2$). Thus, we have the following theorem.

THEOREM 6.5.3. Consider the BVP

$$\begin{aligned} &x''' = f(t,x,x',x'') \\ &x(t_i) = c_i, \qquad i = 1,2,3, \\ &a < t_1 < t_2 < t_3 < b, \qquad c_i \in R. \end{aligned} \tag{6.5.10}$$

Assume that

(i) $f(t,x_1,x_2,x_3) \in C[(a,b) \times R^3, R]$

(ii) solutions of initial values of $x''' = f(t,x,x',x'')$ extend on (a,b)

(iii) there exists at most one solution of the BVP (6.5.10) for each t_i, $i = 1,2,3$, where $t_1 < t_2 < t_3$ and each $c_i \in R$.

Then there exists exactly one solution of the BVP (6.5.10).

As we have pointed out, we have from Theorem 6.5.1 that (A) - (D) imply the existence of solutions of BVP (6.5.8) and (6.5.9). Under assumptions (A) - (C) we have seen that the following condition is equivalent to (D):

(D_1) If $\{x_k(t)\}_{k=1}^{\infty}$ is a sequence of solutions of (6.5.8) which is uniformly bounded and montone on a compact interval $[c,d] \subset (a,b)$, then limit $x_k(t) = x(t)$ is a solution of (6.5.8) on $[c,d]$.

Another condition under assumptions (A) - (C) equivalent to (D) and (D_1) is

(D_2) If $\{x_k(t)\}_{k=1}^{\infty}$ is a sequence of solutions of (6.5.9) which is uniformly bounded on a compact interval $[c,d] \subset (a,b)$, then there is a subsequence $\{x_{k_j}(t)\}$ such that $\{x_{k_j}^{(i)}(t)\}$ converges uniformly for each $0 \leq i \leq n-1$.

For $n = 2,3$ it can be proved that if the BVP (6.5.8), (6.5.9) satisfies (A) - (C), then it also satisfies (D_2) (from which the existence of solutions follows).

For completeness of discussion we now state a result for the k-point BVP, (6.5.8) and

$$x^i(t_j) = c_{ij}, \qquad 0 \leq i \leq n_j - 1, \quad 1 \leq j \leq k, \tag{6.5.11}$$

$$a < t_1 < t < \cdots < t_k < b, \qquad \sum_{j=1}^{k} n_j = n, \qquad 2 \leq k \leq n-1.$$

THEOREM 6.5.4. If solutions of initial value problems for (6.5.8) are unique and if the n-point BVP (6.5.8), (6.5.9) have unique solutions which extend throughout (a,b), then the k-point BVP (6.5.8), (6.5.11) have unique solutions.

Thus in view of our previous results, conditions (A) - (C), uniqueness of initial value problems for (6.5.9) and either

(D), (D_1), or (D_2) imply the existence and uniqueness of n-point BVP's and k-point BVP's.

6.6 NOTES AND COMMENTS

The contents of Sections 6.1 - 6.3 are taken from Antosiewicz [4]. For a detailed treatment of quasilinearization techniques, see Bellman and Kalaba [1]. Section 6.4 consists of the work of Macki and Waltman [1]. For further results using similar methods, see Macki and Waltman [2] and Hartman [5]. The results of Section 6.5 concerning n-parameter families are adapted from Hartman [4]. In particular, Lemma 6.5.1 is due to Tornheim [1]. See also Klaasen [3] and Jackson [7] for related results. The results relating the k-point and n-point problems may be found in Hartman [1], Jackson and Klaasen [3] and Jackson [8]. The special results on the third-order equation can be found in Jackson and Schrader [4,5]. See also Klaasen [2].

BIBLIOGRAPHY

AKO, K.

[1] Subfunctions for ordinary differential equations V, Funkcial. Ekvac. 12(1969), 239-249.

ANTOSIEWICZ, H.

[1] Boundary value problems for nonlinear ordinary differential equations, Pacific J. Math. 17(1966), 191-197.

[2] Un analogue du principe du point fixe de Banach, Ann. Mat. Pura Appl. 74(1966), 61-64.

[3] Linear problems for nonlinear ordinary differential equations, Proc. U.S.-Japan Seminar on Diff. Func. Eqs., W. A. Benjamin, New York (1967), 1-10.

[4] Newton's method and boundary value problems, J. Comput. System Sci. 2(1968), 177-202.

BABKIN, B.

[1] Solutions of a boundary value problem for an ordinary differential equation of second order by Chaplygin's method, Prikl. Mat. Meh. 18(1954), 239-242.

BAILEY, P.

[1] with L. Shampine and P. Waltman, Existence and uniqueness of solutions of the second order boundary value problem, Bull. Amer. Math. Soc. 72(1966), 96-98.

[2] with L. Shampine and P. Waltman, The first and second boundary value problems for nonlinear second order differential equations, J. Differential Equations 2(1966), 399-411.

[3] with L. Shampine and P. Waltman, "Nonlinear two point boundary value problems," Academic Press, New York, 1968.

[4] with L. Shampine, Existence from uniqueness for two point boundary value problems, J. Math. Anal. Appl. 25(1969), 569-574.

BIBLIOGRAPHY

BARBĂLAT, I.

[1] with A. Halanay, Solutions périodiques des systèmes d'equations différentielles non lineaires, Rev. Math. Pures Appl. 3(1958), 395-411.

BEBERNES, J.

[1] A subfunction approach to a boundary value problem for ordinary differential equations, Pacific J. Math. 13(1963), 1053-1066.

[2] with L. Jackson, Infinite interval boundary value problems for $y'' = f(x,y)$, Duke Math. J. 34(1967), 39-48.

[3] with R. Gaines, Dependence on boundary data and a generalized boundary-value problem, J. Differential Equations 4(1968), 359-368.

[4] with R. Gaines, A generalized two-point boundary value problem, Proc. Amer. Math. Soc. (1968), 749-754.

[5] with R. Wilhelmsen, A technique for solving two dimensional boundary value problems, SIAM J. Appl. Math. 17(1969), 1060-1064.

[6] with R. Wilhelmsen, A general boundary value problem technique, J. Differential Equations 8(1970), 404-415.

[7] with J. Schuur, The Wazewski topological method for contingent equations, Ann. Mat. Pura Appl. 87(1970), 271-280.

[8] with R. Wilhelmsen, A remark concerning a boundary value problem, J. Differential Equations 10(1971), 389-391.

[9] with R. Fraker, A priori bounds for boundary sets, Proc. Amer. Math. Soc. 29(1971), 313-318.

[10] with W. Kelley, Some boundary value problems for general differential equations, SIAM J. Appl. Math. 25(1973), 16-23.

[11] with K. Schmitt, Periodic boundary value problems for systems of second order differential equations, J. Differential Equations 13(1973), 32-47.

BELLMAN, R.

[1] with Kalaba, R., "Quasilinearization and nonlinear boundary value problems," Amer. Elsevier, New York, 1965.

BERNFELD, S.

[1] and V. Lakshmikantham and Leela, S., Nonlinear boundary value problems and several Lyapunov functions, J. Math. Anal. Appl. 42(1973), 545-553.

[2] and V. Lakshmikantham, Existence of solutions of ordinary differential equations with generalized boundary conditions, Trans. Amer. Math. Soc., to appear.

[3] with G. Ladde and V. Lakshmikantham, Existence of solutions of two point boundary value problems for nonlinear systems, to appear.

[4] with G. Ladde and V. Lakshmikantham, Differential inequalities and boundary value problems for nonlinear differential systems, to appear.

[5] with G. Ladde and V. Lakshmikantham, Lyapunov functions and boundary value problems for functional differential equations, to appear.

CASTAING, C.

[1] Sur les equations differentielles multivoques, C. R. Acad. Sci. Paris Ser. A 263(1966), 63-66.

CHANDRA, J.

[1] with B. A. Fleishman, Existence of a solution of a boundary value problem for a class of discontinuous nonlinear systems, J. Math. Anal. Appl. 33(1971), 562-573.

CHOW, S. N.

[1] with A. Lasota, An implicit function theorem for non-differentiable mappings, Proc. Amer. Math. Soc. 34(1972), 141-147.

COLES, W.

[1] with T. Sherman, Convergence of successive approximations for non-linear two-point boundary value problems, SIAM J. Appl. Math. 15(1967), 426-433.

CONTI, R.

[1] Problèmes lineaires pour les equations differentielles ordinaires, Math. Nachr. 23(1961), 161-178.

[2] Recent trends in the theory of boundary value problems for ordinary differential equations, Boll. Un. Mat. Ital. 22(1967), 135-178.

CORDUNEANU, C.

[1] Problèmes aux limites lineaires, Ann. Mat. Pura Appl. 124(1966), 65-73.

DE LA VALLÉE POUSSIN, C.

[1] Sur l'équation differentielle lineaire du second ordre, J. Math. Pures Appl. 8(1929), 125-144.

deNEVERS, K.

[1] with K. Schmitt, An application of the shooting method to boundary value problems for second order delay equations, J. Math. Anal. Appl. 36(1971), 588-597.

ERBE, L.

[1] Nonlinear boundary value problems for second order differential equations, J. Differential Equations 7(1970), 459-472.

[2] Boundary value problems for ordinary differential equations, Rocky Mountain Math. J. 1(1971), 709-729.

FENNEL, R.

[1] with P. Waltman, A boundary value problem for a system of nonlinear functional differential equations, J. Math. Anal. Appl. 26(1969), 447-453.

[2] Periodic solutions of functional differential equations, J. Math. Anal. Appl. 39(1972), 198-201.

FOUNTAIN, L.

[1] with L. Jackson, A generalized solution of the boundary value problem $y'' = f(x,y,y')$, Pacific J. Math. 12(1962), 1251-1272.

GAINES, R.

[1] Continuous dependence for two-point boundary value problems, Pacific J. Math. 28(1969), 327-336.

[2] Differentiability with respect to boundary value problems for nonlinear ordinary differential equations, SIAM J. Appl. Math. 20(1971), 754-762.

[3] A priori bounds and upper and lower solutions for nonlinear second order boundary value problems, J. Differential Equations 12(1972), 291-312.

[4] A priori bounds for solutions to nonlinear two-point boundary value problems, J. Applicable Anal., to appear.

GEORGE, J.

[1] with W. Sutton, Applications of Lyapunov theory to boundary value problems, Proc. Amer. Math. Soc. 25(1970), 666-671.

GRANAS, A.

[1] The theory of compact vector fields and some of its applications to topology of function space, Rozprawy Matematyczne Polish Acad. Sci., Warsaw, 30(1962).

GRANDOLFI, M.

[1] Problems ai Limiti per le equazioni differenziali multivoche, Atti Accad. Naz. Lincei Rend. C. Sci. Fis. Mat. Natur. 42(1967), 355-360.

GRIMM, L.

[1] with K. Schmitt, Boundary value problems for differential equations with deviating arguments, Aequationes Math. 4(1970), 176-190.

GUDKOV, B.

[1] with A. Ja. Lepin, Necessary and sufficient conditions for existence of the solutions of some boundary value problems for ordinary differential equations of the second order with discontinuous right hand side, Latviski Math. 9(1971), 47-72.

[2] On a boundary value problem for ordinary differential equations of second order, Latviski Math. 10(1972), 15-31.

GUSTAFSON, G.

[1] with K. Schmitt, Nonzero solutions of boundary value problems for second order ordinary and delay differential equations, J. Differential Equations 12(1972), 129-147.

HALE, J.

[1] "Functional differential equations," Springer-Verlag, Berlin and New York, 1971.

HALIKOV, E.

[1] Certain boundary value problems with conditions at infinity, Differencial'nye Uravnenija 6(1970), 435-442.

HARTMAN, P.

[1] Unrestricted n-parameter families, Rend. Circ. Mat. Palermo 7(1958), 123-142.

[2] On boundary value problems for systems of ordinary nonlinear, second order differential equations, Trans. Amer. Math. Soc. 96(1960), 493-509.

[3] "Ordinary differential equations," Wiley (Interscience), New York, 1964.

[4] On N-parameter families and interpolation problems for nonlinear ordinary differential equations, Trans. Amer. Math. Soc. 154(1971), 201-226.

[5] Boundary value problems for second order ordinary differential equations involving a parameter, J. Differential Equations 12(1972), 194-212.

[6] On two point boundary value problems for nonlinear second order systems, to appear.

HEIDEL, J.

[1] A counterexample in nonlinear boundary value problems, Proc. Amer. Math. Soc. 34(1972), 133-137.

HEIMES, K.

[1] Boundary value problems for ordinary nonlinear second order systems, J. Differential Equations 2(1966), 449-463.

HETHCOTE, H.

[1] Geometric existence proofs for nonlinear boundary value problems, SIAM Rev. 14(1972), 121-129.

INGRAM, S.

[1] Continuous dependence on parameters and boundary data for nonlinear two point boundary value problems, Pacific J. Math. 41(1972), 395-408.

JACKSON, L.

[1] with K. Schrader, Comparison theorems for nonlinear differential equations, J. Differential Equations 3(1967), 248-255.

[2] Subfunctions and second order ordinary differential inequalities, Advances in Math. 2(1968), 307-363.

[3] with G. Klaasen, Uniqueness of solutions of boundary value problems for ordinary differential equations, SIAM J. Appl. Math. 19(1970), 542-546.

[4] with K. Schrader, Subfunctions and third order differential inequalities, J. Differential Equations 8(1970), 180-194.

[5] with K. Schrader, Existence and iniqueness of solutions of boundary value problems for third order ordinary differential equations, J. Differential Equations 9(1971), 46-54.

[6] with G. Klaasen, A variation of the topological method of Wazewski, SIAM J. Appl. Math. 20(1971), 124-130.

[7] Uniqueness and existence of solutions of boundary value problems for ordinary differential equations, Proc. N.R.L-M.R.C. Conference, Academic Press, New York, 1971.

[8] Uniqueness of solutions of boundary value problems ordinary differential equations, SIAM J. Appl. Math. 24(1973), 535-538.

[9] Existence and uniqueness of solutions of boundary problems for third order differential equations, J. Differential Equations 13(1973), 432-437.

KAPLAN, J.

[1] with A. Lasota and J. Yorke, An application of the Wazewski retract method to boundary value problems, Zeszyty Nauk. Univ. Jagiello Prace Mat., to appear.

KELLER, H.

[1] Existence theory for two point boundary value problems, Bull. Amer. Math. Soc. 72(1966), 728-731.

KLAASEN, G.

[1] Dependence of solutions on boundary conditions for second order ordinary differential equations, J. Differential Equations 7(1970), 24-33.

[2] Differential inequalities and existence theorems for second and third order boundary value problems, J. Differential Equations 10(1971), 529-537.

[3] Existence theorems for boundary value problems for nth order ordinary differential equations, Rocky Mountain J. Math. 3(1973), 457-472.

KNOBLOCH, H.

[1] Comparison theorems of nonlinear second order differential equations, J. Differential Equations 1(1965), 1-26.

[2] Second order differential inequalities and a nonlinear boundary value problem, J. Differential Equations 4(1969), 55-71.

[3] On the existence of periodic solutions for second order vector differential equations, J. Differential Equations 9(1971), 67-85.

KRASNOSEĹSKII, M.

[1] "Topological methods in the theory of nonlinear integral equations," Macmillan, New York, 1964.

[2] with A. Perov, A. Provodockii, and P. Zabreiko, "Plane vector fields," Academic Press, New York, 1966.

KY FAN

[1] Fixed-point and minimax theorems in locally convex topological linear spaces, Proc. Nat. Acad. Sci. U.S.A. 38(1952), 121-126.

LAKSHMIKANTHAM, V.

[1] with S. Leela, "Differential and integral inequalities," Vol. I, Academic Press, New York, 1969.

LASOTA, A.

[1] Sur les problèmes linéaires aux limites pour un système d'équations différentielles ordinaires, Bull. Acad. Polon. Sci. Ser. Sci. Math. Astronom. Phys. 10(1962), 565-570.

[2] with Z. Opial, L'existence et l'unicité des solutions du problème d'interpolation pour l'équation différentielle ordinaire d'ordre n, Ann. Polon. Math. 15(1964), 253-271.

[3] with Z. Opial, An application of the Kakutani-Ky Fan theorem in the theory of ordinary differential equations, Bull. Acad. Polon. Sci. Ser. Sci. Math. Astronom. Phys. 13(1965), 781-786.

[4] with Z. Opial, On the existence of solutions of linear problems for ordinary differential equations, Bull. Acad. Polon. Sci. Ser. Sci. Math. Astronom. Phys. 14(1966), 371-376.

[5] Une generalisation du premier théorème de Fredholm et ses applications à la théorie des equations différentielles ordinaires, Ann. Polon. Math. 18(1966), 65-77.

[6] with C. Olech, An optimal solution of the Nicolletti boundary value problem, Ann. Polon. Math. 18(1966), 131-139.

[7] with Z. Opial, On the existence and uniqueness of solutions of nonlinear functional equations, Bull. Acad. Polon. Sci. Ser. Sci. Math. Astronom. Phys. 15(1967), 797-800.

[8] with Z. Opial, On the existence and uniqueness of solutions of a boundary value problem for an ordinary second-order differential equation, Colloq. Math. 18(1967), 1-5.

[9] Contingent equations and boundary value problems, Centro. Inter. Matematico Estivo (1967), 257-266.

[10] with M. Lucyzyński, A note on the uniqueness of two point boundary value problems I, Zeszyty Nauk. Uniw. Jagiello Prace Mat. 12(1968), 27-29.

[11] with M. Lucyzyński, A note on the uniqueness of a two point boundary value problem II, Zeszyty Nauk. Uniw. Jagiello Prace Mat. 13(1969), 44-48.

[12] Boundary value problems for second order differential equations, Seminar on Differential Equations and Dynamic Systems II, Lecture Notes in Math. 144, Springer-Verlag, Berlin and New York, 1970.

[13] with J. Yorke, Existence of solutions of two point boundary value problems for nonlinear systems, J. Differential Equations 11(1972), 509-518.

LEES, M.

[1] A boundary value problem for nonlinear ordinary differential equations, J. Math. Mech. 10(1961), 423-430.

LEPIN, A. Ja.

[1] with A. D. Myshkis, A method of treating nonlinear boundary value problems for ordinary differential equations, Differencial'nye Uravenija 3(1967), 1882-1888.

[2] Necessary and sufficient conditions for the existence of a solution of the two point boundary value problem for a second order nonlinear ordinary differential equation, Differencial'nye Uravenija 6(1970), 1384-1388.

[3] The application of topological methods for nonlinear boundary value problems for ordinary differential equations, Differencial'nye Uravnenija 8(1969), 1390-1397.

LEVIN, A. Ju.

[1] On linear second order differential equations, Dokl. Akad. Nauk. SSSR 153(1963), 1257-1260.

MACKI, J.

[1] with P. Waltman, A nonlinear boundary value problem of Sturm-Liouville type for a two dimensional system of ordinary differential equations, SIAM J. Appl. Math. 21(1971), 225-231.

[2] with P. Waltman, A nonlinear Sturm-Liouville problem, Indiana Univ. Math. J. 22(1972), 217-225.

MAMEDOV, Ja. D.

[1] One sided estimates in the conditions for investigating the solutions of differential equations in Banach spaces, Akad. Nauk. Azerbaidzan. SSR. "ZIM," Baky, 1971.

MARKUS, L.

[1] with N. Amundsen, Nonlinear boundary value problems arising in chemical reactor theory, J. Differential Equations 4(1968), 102-113.

MAWHIN, J.

[1] Degré topologique et solutions periodiques des systems differentiels non lineaires, Bull. Soc. Roy. Sci. Liege 38(1969), 308-398.

[2] Existence of periodic solutions for higher order differential systems that are not of class D, J. Differential Equations 8(1970), 523-530.

[3] Periodic solutions of nonlinear functional differential equations, J. Differential Equations 10(1971), 240-261.

MC CAUDLESS, W.

[1] Nonlinear boundary value problems for ordinary differential equations, Dissertation, University of Waterloo, Canada, 1972.

MOYER, R.

[1] Second order differential equations of monotonic type, J. Differential Equations 2(1966), 281-292.

NAGUMO, M.

[1] Eine Art der Randwertaufgabe von systemen gewöhulicher differentialgleichungen, Proc. Phys. Math. Soc. Japan 25(1943), 221-226.

NICOLSON, L.

[1] Boundary value problems for systems of ordinary differential equations, J. Differential Equations 6(1969), 397-407.

NORKIN, S.

[1] Differential equations of the second order with retarded argument, Amer. Math. Soc. Transl. 31(1972).

OPIAL, Z.

[1] Sur une inégalité de Ch. de la Vallée Poussin dans la théorie de l'équation différentielle linéaire du second ordre, Ann. Polon. Math. 6(1959), 87-91.

[2] Linear problems for systems of nonlinear differential equations, J. Differential Equations 3(1967), 580-594.

PEROV, A.

[1] On a boundary value problem for a system of two differential equations, Dokl. Akad. Nauk SSSR 144(1962), 493-496.

[2] with A. Kibenko, On a certain general method for investigation of boundary value problems, Izv. Akad. Nauk SSSR 30(1966), 249-264.

PICARD, E.

[1] Sur L'application des methodes d'approximation successives a l'etude de certaines equations differentielles ordinaires, J. Math. 9(1893), 217-271.

PLÍS, A.

[1] Measurable orientor fields, Bull. Acad. Polon. Sci. Ser. Sci. Math. Astronom. Phys. 13(1965), 565-569.

SCHMITT, K.

[1] Periodic solutions of nonlinear second order differential equations, Math. Z. 98(1967), 200-207.

[2] Boundary value problems for nonlinear second order differential equations, Monatsh. Math. 72(1968), 347-354.

[3] Bounded solutions of nonlinear second order differential equations, Duke Math. J. 36(1969), 237-244.

[4] On solutions of nonlinear differential equations with deviating arguments, SIAM J. Appl. Math. 17(1969), 1171-1176.

[5] A nonlinear boundary value problem, J. Differential Equations 9(1970), 527-537.

[6] A note on periodic solutions of second order ordinary differential equations, SIAM J. Appl. Math. 21(1971), 491-494.

[7] Periodic solutions of systems of second order differential equations, J. Differential Equations 11(1972), 180-192.

[8] Periodic solutions of small period of a system of nth order differential equations, Proc. Amer. Math. Soc. 36(1972), 459-463.

[9] Boundary value problems for functional differential equations, to appear.

SCHRADER, K.

[1] Boundary value problems for second order ordinary differential equations, J. Differential Equations 3(1967), 403-413.

[2] Solutions of second order ordinary differential equations, J. Differential Equations 4(1968), 510-518.

[3] A note on second order differential inequalities, Proc. Amer. Math. Soc. 19(1968), 1007-1012.

[4] with P. Waltman, An existence theorem for nonlinear boundary value problems, Proc. Amer. Math. Soc. 21(1969), 653-656.

[5] Existence theorems for second order boundary value problems, J. Differential Equations 5(1969), 572-584.

[6] Second and third order boundary value problems, Proc. Amer. Math. Soc. 32(1972), 247-252.

SCHUUR, J.

[1] The existence of proper solutions of a second order ordinary differential equation, Proc. Amer. Math. Soc. 17(1966), 595-597.

SEDZIWY, S.

[1] Dependence of solutions on boundary data for a system of two ordinary differential equations, J. Differential Equations 9(1971), 381-389.

[2] Uniqueness and existence of solutions of a boundary value problem for a system of differential equations, Zeszyty Nauk. Uniw. Jagiello. Prace Mat. 15(1971), 147-149.

[3] On the existence of a solution of a certain non-linear boundary value problem, to appear.

SHAMPINE, L.

[1] Existence and uniqueness for nonlinear boundary value problems, J. Differential Equations 5(1969), 346-351.

SHERMAN, T.

[1] Uniqueness for second order two point boundary value problems, J. Differential Equations 6(1969), 197-208.

TORNHEIM, L.

[1] On n-parameter families of functions and associated convex functions, Trans. Amer. Math. Soc. 69(1950), 457-467.

TUTAJ, E.

[1] On boundary value problems for ordinary differential equations of second order, Zeszyty Nauk. Uniw. Jagiello Prace Mat., to appear.

WALTMAN, P.

[1] A nonlinear boundary value problem, J. Differential Equations 4(1968), 597-603.

[2] Existence and uniqueness of solutions to a nonlinear boundary value problem, J. Math. Mech. 18(1968), 585-586.

[3] Existence and uniqueness of solutions of boundary value problems for two dimensional systems of nonlinear differential equations, Trans. Amer. Math. Soc. 153(1971), 223-234.

[4] with J. S. W. Wong, Two point boundary value problems for nonlinear functional differential equations, Trans. Amer. Math. Soc. 164(1972), 39-64.

WAZEWSKI, T.

[1] Une méthode topologique de L'examen du phénomène asymptotique relativement aux equations differentielles ordinaires, Rend. Accad. Lince 3(1947), 210-215.

WHYBURN, W.

[1] Differential equations with general boundary conditions, Bull. Amer. Math. Soc. 48(1942), 692-704.

YORKE, J.

[1] Invariance of contingent equations, "Lecture notes in operations research and mathematical economics 12: mathematical systems theory and economics II," Springer-Verlag, Berlin and New York, 1969.

YOSHIZAWA, T.

[1] "Stability theory by Lyapunov's second method," Math. Soc. Japan, Tokyo, 9, 1966.

ADDITIONAL BIBLIOGRAPHY

BEBERNES, J.

[12] A simple alternative problem for finding periodic solutions of second order ordinary differential systems, Proc. Amer. Math. Soc., Dec. (1973), to appear.

BERNFELD, S.

[6] with V. Lakshmikantham, Estimation of number of solutions of boundary value problems of ordinary differential equations with nonlinear boundary conditions, to appear.

DENKOWSKI, Z.

[1] On a criterion of uniqueness for periodic solutions of linear second order difference equations, Ann. Polon. Math. 27(1972), 41-49.

[2] On the existence and uniqueness of periodic solutions for difference equations of second order, Ann. Polon. Math., to appear.

GAINES, R.

[5] Difference equations associated with boundary value problems for second order nonlinear ordinary differential equations, SIAM J. Num. Anal., to appear.

GUSTAFSON, G.

[2] with K. Schmitt, A note on periodic solutions for delay differential systems, to appear.

HEIDEL, J.

[2] A second order nonlinear boundary value problem. J. Math. Anal. Appl. (to appear).

HEIMES, K.

[2] Two point boundary value problems in Banach space, J. Differential Equations 5(1969), 215-225.

JACQUEZ, J.

[1] with E. Daniels, Differential equations in physiology and how to find them, Conf. on Functional Differential Equations, Univ. of Utah, Salt Lake City, pp. 161-183, Academic Press, New York, 1973.

ADDITIONAL BIBLIOGRAPHY

KARTSATOS, A.

[1] Boundary value problem for ordinary differential systems on infinite intervals, to appear.

KELLER, H.

[2] Shooting and embedding for two-point boundary value problems, J. Math. Anal. Appl. 36(1971), 598-610.

MAWHIN, J.

[4] Boundary value problems for nonlinear second order vector differential equations, to appear.

MULDOWNEY, J.

[1] with D. Willett, An intermediate value property for operators with applications to integral and differential equations, to appear.

SCHMITT, K.

[10] Boundary value problems and comparison theorems for ordinary differential equations, SIAM J. Appl. Math., to appear.

[11] Applications of variational equations to ordinary and partial differential equations-multiple solutions of boundary value problems, to appear.

[12] with R. Thompson, Boundary value problems for infinite systems of second order differential equations, to appear.

STERNBERG, R.

[1] with J. Shipman and S. Zohn, Multiple Fourier analysis in rectifier problems, Quart. Appl. Math. 26(1959), 335-360.

THOMPSON, R.

[1] Boundary value problems for differential in Banach spaces, Ph.D. Thesis, Univ. of Utah, Salt Lake City, 1973.

UMAMHESHWARAM, S.

[1] Boundary value problems for n-th order ordinary differential equations, Ph.D. Thesis, Univ. of Missouri, Columbia, 1973.

INDEX

K

L

M

N

O

P

Q

S

T

U

W